K.-H. Henning · M. Störr

Electron micrographs (TEM, SEM) of clays and clay minerals

Schriftenreihe für Geologische Wissenschaften

Series in Geological Sciences

Herausgegeben
von der Gesellschaft
für Geologische Wissenschaften
der DDR

Heft 25

K.-H. Henning
M. Störr

Electron micrographs (TEM, SEM) of clays and clay minerals

with
602 figures and 1 table

Akademie-Verlag Berlin 1986

Verfasser:

Doz. Dr. sc. nat. Karl-Heinz Henning
Prof. Dr. sc. nat. Manfred Störr

Ernst-Moritz-Arndt-Universität Greifswald
Sektion Geologische Wissenschaften
Friedrich-Ludwig-Jahn-Straße 17a
Greifswald
DDR - 2200

ISBN 3-05-500168-0
ISSN 0323-8946

Erschienen im Akademie-Verlag Berlin, DDR - 1086 Berlin, Leipziger Straße 3—4

Lizenznummer: 202 · 100/449/86
Printed in the German Democratic Republic
Geamtherstellung: VEB Druckhaus „Maxim Gorki", 7400 Altenburg
Lektor: Dipl.-Phys. Helga Müller
Umschlagentwurf: Peter Fischer-Sternaux
LSV 1415
Bestellnummer: 763 605 7 (2170/25)
06800

Contents

1. Introduction

During the past two decades electron microscopy has become an indispensable and widely used tool for investigating rocks, minerals and mineral raw materials. This method is of special importance to the investigation of clays and clay minerals, because the small particles of these minerals cannot usually be resolved by light microscopy. For this reason electron microscopy has a specific role to play in the determination of clay minerals and their admixtures in clays, bentonites, kaolins etc.. Frequently, the identification of certain minerals is only possible by means of electron microscopy, e.g. the identification of halloysite beside kaolinite or of very small amounts of admixtures, e.g. of montmorillonite in kaolin. In addition to using electron microscopy in mineral analysis, however, specific electron optical methods have been developed for the detailed characterization of certain minerals, including morphometry and surface decoration of kaolinite. Using such methods, it is possible to determine the technological properties of clays and to elicit the relationships between morphology and technological properties.

Modern electron optical investigations take advantage of both transmission electron microscopy (TEM) and scanning electron microscopy (SEM). In the last few years the latter has been developed to such technical perfection and ease of operation that it is often preferred to TEM because of simpler specimen preparation and larger permissible specimen size (up to several mm^2). Thus, SEM provides useful information on the texture of clays and rocks and the distribution of minerals in them, whereas the higher resolution of TEM permits a more detailed characterization of mineral morphology. Selection of proper electron optical imaging techniques and preparation method depends on the material itself and on the kind of information that is required. In this atlas an attempt is made to use the customary electron microscopic methods in combination, the main aim being to show the great morphological variety that is displayed by the clay minerals. With its about 600 micrographs the atlas illustrates a wide range of clay minerals, kaolins, bentonites, clays and their admixtures, especially from well known localities. Special emphasis is given to admixtures in the form of iron, titanium, and salt minerals, which play an increasing role in the selection of the proper beneficiation technology and in the interpretation of genetic processes. However, the Al-hydroxides, which sometimes occur as admixtures of clayey rocks, are not discussed in greater detail.

There are only few monographs on electron optical studies of clay minerals, of which those by Gritsaenko et al. (1961), Rekshinskaya (1966), Beutelspacher & van der Marel (1968), Gard (1971) and Sudo et al. (1981) are mentioned here. However, all of them primarily contain transmission pictures. In addition, a great number of specific SEM and TEM studies have been published, some of which are referred to in the text. In recent years, high resolution electron microscopes have become available, and these are able to resolve the base planes of the clay mineral lattices (e.g. KÖSTER 1983). Unfortunately, such high-resolution pictures cannot be presented in this atlas. Furthermore, microprobe analysis is not included so as to keep the scope of this atlas within reasonable limits. The main chapter, which deals with the individual clay minerals, is preceded by short introductory sections on electron microscopes and the preparation techniques that were used in this study. Detailed information on equipment and methods is provided in the textbooks by KAY (1965), PICHT & HEYDENREICH (1966), REIMER (1967), SCHIMMEL & VOGELL (1975) and GLAUERT (1976).

Concluding these introductory remarks, we present a short list of some frequently used terms and their abbreviations:

SEM — Scanning electron microscope
— Scanning electron microscopy
— Scanning electron micrograph

TEM — Transmission electron microscope
— Transmission electron microscopy
— Transmission electron micrograph, especially transmission picture of the specimen

Replica — Replica of the specimen
— Transmission picture of the replica

Acknowledgements

The authors would like to express their sincere appreciation to Mrs. ERIKA GROTHE of the laboratory for electron microscopic preparation, Mr. FRANZ RASSAU, head of the laboratory for electron microscopy of the Department of Geological Sciences and Mr. ERNST FISCHER, head of the laboratory for electron microscopy of the Department of Biological Sciences of the Ernst-Moritz-Arndt-University for preparing the specimens and making the micrographs.

Thanks are also due to Dr. BERND ULLRICH and Mr. HARRY BECK who helped investigate the secondary minerals (efflorescences) and to Dr. PETER ADOLPHI and Mr. JÖRN KASBOHM for the preparation of the coals (low-temperature ashing).

Special acknowledgements are extended to Mrs. MARGARETHE JANTZEN, Mrs. IRMGARD LANGBEIN and Mrs. ALMA SPRINDT of the photographic laboratory of the Department of Geological Sciences, who made the photoprints and Mrs. EVA LASS for preparing the typescript.

Finally, thanks are due to the Akademie-Verlag Berlin for complying with the special wishes of the authors.

2. Instrumentation and methods used in electron optical investigations of clay minerals and their admixtures

2.1. Equipment necessary for producing electron micrographs

2.1.1. Electron microscopes and accessories

A wide range of high-resolution transmission electron microscopes, produced by various firms, is available on the world market. While accessories and ease of operation may vary, the main criterion remains the resolving power of the instrument. At acceleration voltages of 20—100 kV point resolution ranges from 0.3 to 1.0 nm and line resolution from 0.2 to 0.8 nm. Most of the micrographs included in this atlas were made by means of the transmission electron microscope BS 500 manufactured by Tesla (Brno, Czechoslovakia). The remaining micrographs were made using the electron microscope EF of VEB Carl Zeiß Jena (Jena, GDR).

The main parameters of TEM BS 500 are as follows:

- Point resolution: 0.9 nm guaranteed
- Accelerating voltages: 60 kV, 90 kV
- Magnification: at a focal distance of 3 mm 1.000—100.000 × (in 12 steps)
  at a focal distance of 6 mm 50—50.000 × (in 12 steps)

For the SEM studies a scanning electron microscope, type BS 300, manufactured by Tesla (Brno, Czechoslovakia) was used. This instrument has a resolving power of 15 nm and an accelerating voltage of 15—25 kV.

For specimen preparation additional equipment is required including:

- High-vacuum evaporation chamber for shadowcasting and coating, type HVB 30.1 of VEB Hochvakuum Dresden
- Ultrasonic generator
- Hoods for hydrofluoric acid

The evaporation device is suited for all evaporation modes necessary for preparation, i.e. for vertical coating of the rotating specimen for making replicas, for shadowcasting to improve contrast, for heavy-metal decoration and for making supporting films by means of glow-discharge.

2.1.2. Supporting films

Among the indispensable accessories of an electron microscope are the specimen holders, such as copper and bronze grids or platinum specimen holders with one or more minute holes (30—70 μm in diameter). For fixing the fine mineral particles to the holders supporting films are placed on these holders. Due to their adhesive properties, these films become firmly adherent to the holders on drying. The films should resist electron beaming, have some mechanical stability, be electron transparent, be structureless, and have a smooth surface. In common use are supporting films that consist of Formvar (Polyvinylformaldehyde), collodion, SiO or carbon. Most of the supporting films for the specimens shown in the atlas are of the Formvar type, the rest being collodion. For producing Formvar films a clean glass slide is dipped in a 0.3% solution of Formvar in chloroform. After the glass slide has been picked up, the surplus solution is allowed to flow off. The dried film is then scratched with a needle and by cautiously immersing the slide in water the film is floated. Next, the support grids are placed onto the floating Formvar film and by means of filter paper the Formvar film, together with the support grids, is lifted from the water surface. On drying, the Formvar film adheres firmly to the support grids.

A convenient way to make collodion films is to place one drop of a 2—3% solution of collodion in amylacetate onto a water surface, where it spreads evenly. After evaporation of the solvent a thin collodion film is floating on the surface. By lowering the water level the collodion film is brought into contact with the support grids placed on the bottom of the dish or funnel. On drying, the film tightly adheres to the grid. Carbon and SiO films are produced in the evaporation device using the carbon evaporator and glow discharge.

2.1.3. Preparation techniques for specimen transmission

The natural particle size of the clay minerals and of several of their admixtures permits preparation for specimen transmission to be made from suspension. The suspension is made by soaking the clay particles in distilled water to which NH_4OH is added as a dispersing agent until a 0.01 n concentration is reached. If necessary, the particles are cautiously comminuted. After dilution to a solid content of 2—4% the clay suspension is dispersed using an electric stirrer and then an ultrasonic vibrator. The particles and aggregates of 3—4 μm are separated from the suspension by sedimentation or centrifugation. Thereafter, the suspension is again diluted and, if necessary, once more placed in the ultrasonic vibrator. Of this highly diluted suspension a single droplet is placed on a grid-mounted supporting film and dried. If the specimen shows poor contrast, Pt-shadowcasting is carried out. By means of a carbon backing it is possible to strengthen supporting films which have lost stability due to specimen preparation. From specimens that have strong electron scattering (minerals which contain heavy elements or are relatively thick) thin-cut sections can be made for direct transmission, using an ultramicrotome with diamond knife.

2.1.4. *Preparation of replicas*

For a closer investigation of surface textures replicas are made, using metal, carbon, resin and oxide coats. There are one-stage and multi-stage replicas. For the investigation of clay minerals one-stage replicas with Pt-shadowing and carbon backing are most commonly used. These one-stage replicas can be made from suspensions as well as from natural specimen surfaces or cleavage faces. In the first case some droplets of a clay suspension, prepared as described in Section 2.1.3., are placed onto a glass slide and dried. These glass slides and the untreated or etched specimens are put in the bell-jar of the vacuum device and are carbon coated. For better contrast the carbon coat is then shadowcast with platinum or another heavy metal at an angle of about 30°. After this, the carbon backing on the specimen is cut into little squares using a needle, the squares being floated in hydrofluoric acid (1%) in a platinum dish. When clay minerals are prepared, the mineral particles adhering to the carbon are washed off or dissolved in higher concentrations of hydrofluoric acid (20—30%) at 80°C. Finally, the replicas are washed in several changes of distilled water, picked up from the water on grids, and dried.

2.1.5. *Preparation for scanning electron microscopy*

In scanning electron microscopy (SEM) the image is produced directly from the surface of fragments or sections of specimens, which may have a size of several square millimeters. Thus, compared to the dimensions of the clay particles, the specimens are 10^3—10^5 times as large so that the findings made with regard to arrangement and distribution of clay minerals in rocks or manufactured clay products can be considered as representative. Another advantage of SEM is that at sufficient resolution great field depth is achieved so that even rather rough surfaces are depicted in all details.

When preparing clay mineral specimens for SEM, attention should be paid to:

— operational aspects of SEM, such as image formation, resolution, field depth
— specific properties of the material to be investigated, such as small grain size, high porosity, low conductivity, and low hardness
— aims of the investigation and selection of samples suited for these aims.

For clay minerals the following preparation techniques, which have been used for all SEMs, have proved appropriate.

Specimen selection

In general, fragments are used as specimens. Usually, they are pinched from the rock using pliers. The specimen should have such a shape that the area to be depicted can be positioned parallel to the upper face of the specimen holder. Moreover, the specimens should be sufficiently thin.

In bedded specimens, such as sedimentary rocks, it is easier to pinch off the specimen parallel to the layers rather than perpendicular to them. If the texture is to be investigated, the specimens should be prepared parallel as well as vertical to the layers.

Fixing the specimen to the specimen holder

The specimen is fixed to the specimen holder using a vacuumproof adhesive so that specimen and holder are firmly connected. Small specimens are fixed by means of conductive paint. After the specimens have been fixed, any particles that may adhere are removed using purified compressed air.

Improving conductivity

Since clay or rock fragments have a very low electrical conductivity, the specimens should be provided with a carbon or metal coat using one of the vacuum evaporation techniques. In our investigations the vertical coating technique with gold or copper was employed. The main purpose of these coats is to increase the yield of secondary electrons and to enhance resolution, because owing to increased conductivtiy, electron energies of 10—30 keV can be used, and local charge build-up on the specimen surface is avoided. Well covered surfaces with good conductivity are an indispensable prerequisite for scanning electron microscopy.

For the clay mineral specimens vertical sputtercoating of the rotating specimen with heavy metals has proved effective. If the samples are thick, the coat is additionally connected to the specimen holder using conductive paint. Care should be taken that samples containing hydration water are allowed sufficient time for drying in the vacuum before sputtering so that the coat will not break. Moreover, any loose particles should be removed from the specimen surface, lest their high beam-induced charges interfere with image formation.

2.2. Treatment of sample material and preparation of specimens

2.2.1. Mineral separation and enrichment

For electron optical studies, as for other techniques of mineral analysis, the mineral mixtures need to be separated according to particle size and type of mineral. Sometimes, mineral enrichment is necessary as well. The methods used are largely those employed in mineralogy, and include:

— manual selection by macroscopic or microscopic means
— grain-size selection by various methods of sedimentation or separation according to density, as customary in heavy-mineral analysis
— magnetic separation by permanent and electromagnets as well as by high-degree magnetic separation
— chemical selection by selective solution techniques (see HARMAN 1969)
— electrostatic selection
— flotation.

2.2.2. Surface treatment of the specimen

For an electron microscope study of sample structure and mineral surface textures special treatment of the specimen surface is necessary. The techniques used include:

- etching (chemical, ion-beam)
- etching by low-temperature ashing (coals etc.)
- electrolysis
- decoration.

In this study, besides etching with solvents and acids (hydrofluoric, hydrochloric, and acetic) in various concentrations, the various techniques of Au-decoration were used. These techniques are based on research by BASSET (1960), BETHGE (1960) and BETHGE et al. (1962) concerning methods of making visible the dislocations and slip steps on alkali halide cleavage faces and other materials. GRITSAENKO & SAMOTOIN (1966), ČEKIN et al. (1974), SUNAGAWA & KOSHINA (1975) and RUSKO (1976) have applied these methods to kaolinites and other phyllosilicates. Using the decoration techniques lattice defects such as screw dislocations can be recognized in the form of growth spirals. Moreover, atomic steps and other inhomogeneities or lattice defects as small as the lattice plane distance can be made visible. The decoration effect is due to the increased deposition of gold particles on slip steps and other inhomogeneities, which then become visible under the electron microscope. For Au-decoration the specimen is heated to about 350 °C in the evaporation device for 10 minutes at a pressure of 1.3×10^{-3} Pa. Then, replicas are made of the decorated specimens, as outlined in Section 2.2.4. After Au-decoration the surfaces of three-layer silicates and feldspars show a random distribution of the gold particles, whereas minerals of the kaolinite group show the decoration effect.

2.3. Evaluation of electron micrographs

The evaluation of electron micrographs provides qualitative and quantitative data on morphology, crystal structure, surface texture and chemical composition of the samples. For obtaining these data the appropriate procedures of preparation and the appropriate image systems must be chosen. However, direct data on chemical composition can be obtained only by means of microprobe analysis.

2.3.1. Counting and measuring methods

Morphology

For the quantitative evaluation of morphological differences, e.g. of the kaolinite crystals, the morphometric analysis of electron micrographs, as described by EITEL & SCHUSTERIUS (1940) for determining the grain-size distribution of clays and kaolins was modified. First, one-stage replicas, shadowcast with platinum at an angle of 30°, were made. Then, the length, width and thickness of about 1000 pseudohexagonal kaolinite particles were measured, the length being defined as the greatest distance between two parallel edges and the width as the shortest distance between two parallel edges. Thickness can be calculated from the length of the shadow at known shadowing angle. These data were evaluated by means of a special computing programme (HEN-

NING & REINHARDT 1974). Using this programme, the following morphometric parameters can be calculated:

- grain-size distribution of particles smaller than 2 μm in grain-% and weight-% at 0.2 μm intervals
- diameter $\left(\frac{\text{length} + \text{width}}{2}\right)$
- form factor (diameter/thickness)
- degree of distortion (length/width)
- geometrical surface ($m^2 \cdot g^{-1}$ or cm^{-1})
- number of kaolinite crystals per unit volume (number/μm^3).

In comparison to the granulometric analysis, which uses the equivalent spherical diameter, the method determines the particle parameters in a way that reflects the real grain dimensions far more exactly. The differences between granulometric and morphometric values of grain-size distribution in the < 2 μm grain fractions are given in Table 1.

Table 1. Comparison of grain-size distribution according to centrifugation analysis (granulometry — equivalent spherical diameter) and morphometry (morphometry — approximate real diameter) in kaolins No. 1 and No. 8 in weight-% (after HENNING, REINHARDT & STÖRR 1981)

Grain fraction μm	Kaolin No. 1 Granulometry	Morphometry	Kaolin No. 8 Granulometry	Morphometry
0.4	22	1	21	1
0.4—0.8	24	8	27	11
0.8—1.2	12	15	12	27
1.2—1.6	13	22	13	27
1.6—2.0	4	20	4	12
2.0—6.3	18	—	21	—
6.3—63	7	—	2	—

In the same manner, a more exacting statistical evaluation of the dimensions of other clay minerals is also possible, e.g. with regard to the length:width ratio of elongate crystal forms.

Photogrammetry

Another measuring technique especially suited for rough surfaces is the photogrammetric measurement of electron microscopic stereopictures (HENNING 1979). First, a replica of the specimen is made using the technique described above. The stereo-pair is made by tilting the specimen holder on the goniometer stage through an angle of 5 degrees between two successive photographs. Magnification varies with roughness of surface and size of selected area. As customary in photogrammetry, the stereo-pairs are evaluated using a precision stereo-comparator. The stereoscopic measurement of the coordinates x', y' and the parallaxes nx' is done by setting a manually operated optical marker onto discrete points (change of slope) of the photographed sample surface. The values are automatically recorded on perforated tape. Depending on magnification the measured distance can be varied by measuring several profiles in one stereo-pair or by examining several stereo-pairs. The coordinates x (distance between points in the profile) and z (height) and the profile characteristics are calculated from the measured values using a special computing programme. The characteristics of a profile include the maximum height and depth, the mean roughness, and the standard deviation from a compensation line representing the mean surface.

The computer provides the control tapes for the drawing of profiles by means of an automated mapping device. Thus, this measuring method can provide the following information about the mineral surface:

— graphs of the surface profiles
— table of maximum heights and depths
— mean roughness (mean deviation from the calculated mean surface in μm)
— standard deviation from the mean surface in μm.

Fig. 1. Stereo-pair: Replica of a kaolin coated paper surface (offset printing paper)

An example of photogrammetric measurement of stereo-pair micrographs of a kaolin-coated paper surface is presented in Figure 1, which shows a stereo-pair of the surface replica of offset printing paper with clearly recognizable pseudohexagonal kaolinites. Figure 2 gives the line profiles of this surface and their characteristics.

The photogrammetric measurement or mapping of stereo-pairs also permits image reconstruction (holography) in the form of maps with isophyses.

2.3.2. *Mineral analysis*

Electron microscopy is an indispensable tool in mineral analysis, especially of clay minerals and their admixtures. Some examples of this analytic technique are outlined below:

The discrimination of halloysite that occurs alongside kaolinite is possible on the basis of morphologic differences between halloysite (tubular) and kaolinite (pseudo-hexagonal). As illustrated in Figure 515 (Figs. 7—602 see plates, pp. 45—342), this method permits the detection even of minute amounts of halloysite, such as cannot be accomplished by other techniques. Furthermore, X-ray data are not always suitable for clear-cut differentiation between disordered kaolinite and halloysite, not even in mono-mineralic samples. In the trioctahedral two-layer silicates, the identification of chrysotile alongside antigorite is also difficult, especially if one of these minerals

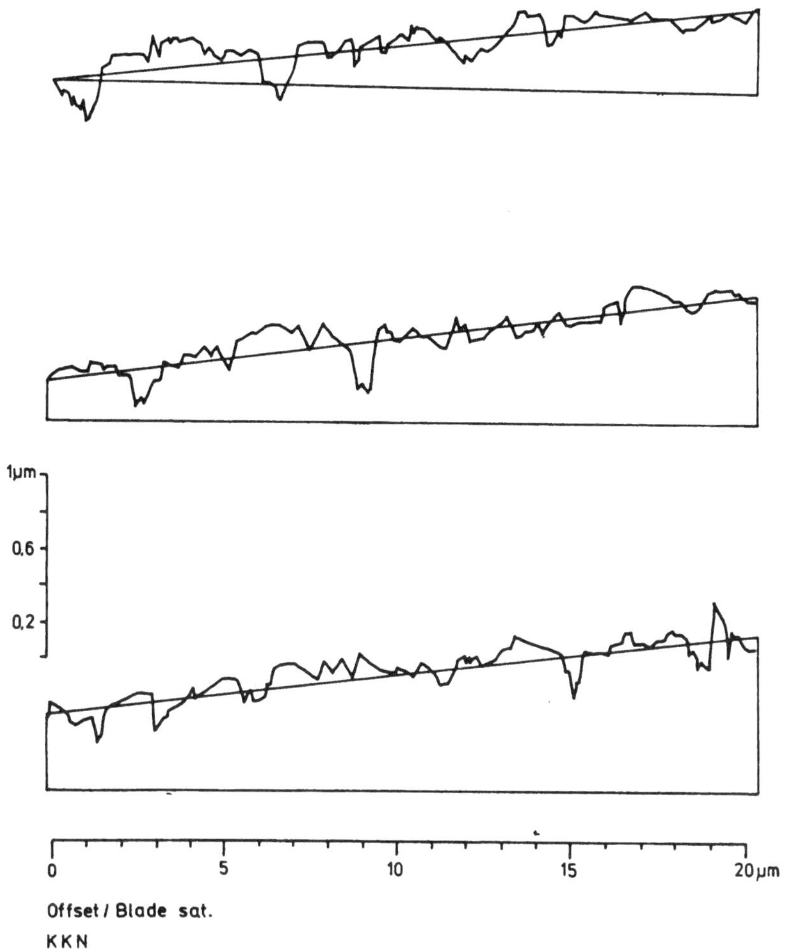

Fig. 2. Photogrammetric surface profiles of the paper surface shown in Fig. 1

occurs only in small amounts. In Figure 155 the lath- to slat-shaped chrysotile can be clearly differentiated from the platy antigorite. Moreover, electron microscopy helps one to identify minute amounts of impurities, which is particularly important for beneficiation. Figure 397 shows whiteness-diminishing anatase crystals in a kaolin.

2.3.3. Genesis

Electron micrographs can be interpreted with regard to the genesis of minerals and rocks. An example of the first is the detection of the rolling-up mechanism for the formation of halloysite tubules (RUSKO & PETROV 1976). With the aid of the Au-decoration technique these workers were able to elucidate the rolling-up process and to develop a model thereof. With regard to lithogenesis the example given here is the morphometric investigation of various kaolins by HENNING (1976). Three genetic types of kaolinite-bearing sediments were investigated:

— kaolins formed by areal weathering of various parent rocks (primary kaolins)

- kaolinite clays formed by redeposition of kaolins formed by weathering (secondary kaolins)
- joint fillings and kaolinite-containing sandstones in which the kaolinite is of diagenetic origin.

The morphometric analysis of the kaolinites of these sediments revealed significant differences in the distribution of the grain numbers, as shown by the histograms and cumulative curves. With regard to primary kaolins the main finding has been that the grain percent maxima, which range from 30 to 40 grain-%, are in the 0.4—0.8 μm size fractions, the remainder being distributed through the other grain fractions. From these data conclusions can be drawn concerning the nature of parent rocks.

Secondary kaolin has a maximum of 70 grain-% in the finest size fraction (< 0.4 μm), the remaining size fractions having only small grain percentages.

In contrast to this, diagentic kaolinites have the largest grain sizes. The maximum grain size, comprising 15—20 grain-% of the material, occurs in the 1.2—1.6 μm size fraction. The other size fractions, especially the fractions > 2 μm, have relatively high grain percentages as well.

2.3.4. *Beneficiation and technological properties*

Electron optical techniques were employed e.g. by ZWAHR and STÖRR (1981) for studies on Fe- and Ti-impurities in GDR kaolins and on technologies for their removal by means of high-degree magnetic separation. Using a Sala-HGMS-Magnetic Separator, type 10—15—20, manufactured by Sala Magnetic Inc. of Cambridge, Mass., USA, they enriched these impurities and examined them under an electron microscope. As a result the following morphologic features could be established: Anatase mostly occurs in separate crystals, especially in tetragonal bipyramids and intergrowths of these crystals. Occasionally barrel-shaped crystals and also globular aggregates are seen. Rutile usually occurs as needles, to which small anatase crystals may be attached. Siderite forms globular aggregates or idiomorphous steep scalenohedrons, usually 20 μm or more in length. Pyrite forms cubes and globular aggregates. Goethite occurs in needles which may form star-shaped triplets. Of special interest has been the detection of epitactic intergrowths of iron (?) minerals and kaolinite, which also reduce whiteness. On the basis of these findings and of whiteness measurements it could be shown that high degree magnetic separation is not only a technique for enriching minerals occurring in minute amounts but also an effective commercial technology for removing impurities. In the filler and coating kaolins investigated a reduction of Fe_2O_3 by 15% and of TiO_2 by 60—70% led to a 4% increase in whiteness. High-degree magnetic separation has also proved effective in the beneficiation of ceramic kaolins.

Concerning the technological properties of clay minerals reference is made to the results obtained by HENNING, REINHARDT & STÖRR (1981). They found that the rheology of kaolin suspensions containing no measurable content of swelling three-layer silicates is governed by the morphometric parameters of the kaolinite. High-viscosity concentrations of kaolin suspensions are related to the following morphometric features of the kaolinite:

- isometric kaolinite particles, i.e. a form factor that is as small as possible
- small geometric surfaces in the fine fractions (e.g. < 2.5 m^2/g in the grain fraction of 0.2—0.4 μm)
- small grain numbers per unit volume in the fine fractions.

Thus, a relatively large geometric or specific surface of the sample does not necessarily lead to reduced dispersity. Reliable indicators of dispersity are the specific values derived from the form factor, grain number and mass distribution, especially in the fine fractions (< 1 μm). The rheological properties of kaolins are of great importance to the behaviour of coating slurries for paper and of casting slips for ceramics.

3. Interpretation of the electron micrographs

3.1. Morphology of clay minerals and their admixtures in electron microscopy

Like all minerals, the clay minerals and their admixtures are also governed by the correspondence principle, i.e. the structure, which depends on the crystal chemistry and crystal physics of the mineral, corresponds to mineral morphology and properties. The clay minerals as such (i.e. in the more precise sense) are pseudohexagonal phyllosilicates with great variations in crystal structure and chemistry. Thus, under the electron microscope, clay minerals show great variations in morphology as well. Pertinent data are given in the text sections dealing with the individual clay minerals.

In this introductory section Figures 3—6 present the basic morphology of the clay minerals. The left sides of Figure 3, 4, and 5 are schematic illustrations of the micrographs of the minerals, whereas the right sides show replicas of the crystals or particles. Figure 6 shows some typical parallel intergrowths and aggregates that are composed of the basic forms given in Figures 3—5.

In Figure 3 the following seven basic forms are presented:

a) Pseudohexagonal idiomorphic crystals
b) Pseudohexagonal hypidiomorphic crystals
c) Lath-shaped crystals with idiomorphic ends
d) Lath-shaped crystals with irregular, partly ragged ends
e) Slat-shaped crystals with idiomorphic ends
f) Slat-shaped crystals with irregular, partly ragged ends
g) Fibrous crystals

The basic forms given in lines a and b are typical of minerals of the dioctahedral kaolinite group, but are also found in the di- and trioctahedral minerals of the mica group.

The specimen transmission pictures show that the transparency of the crystals varies in relation to thickness and ions contained. This is indicated by various shades of grey and sometimes by moiré structures. The surfaces of the replicas are smooth, and with standard preparation they show only few inhomogeneities. The elongates of lines c—g differ primarily in their length:width ratios. Moreover, they show differences in the shapes of their ends. These elongates may have different thicknesses, which causes differences in transparency. Both in specimen and replica transmission longitudinal cleavage cracks are observable. Elongate forms are found in some minerals of the serpentine group, of the montmorillonite-saponite group, of the mica group, and in some minerals with mixed-layer structure, especially in the regular and irregular montmorillonite-muscovite mixed-layer minerals. The fibrous crystals are typical of the palygorskite-sepiolite group

In Figure 4 the following basic forms are presented:

a) Tubular crystals
b) Rod-shaped crystals
c) Platy xenomorphic crystals
d) Platy xenomorphic crystals with cleavage cracks
e) Platy crystals with folds
f) Nepheloid crystals

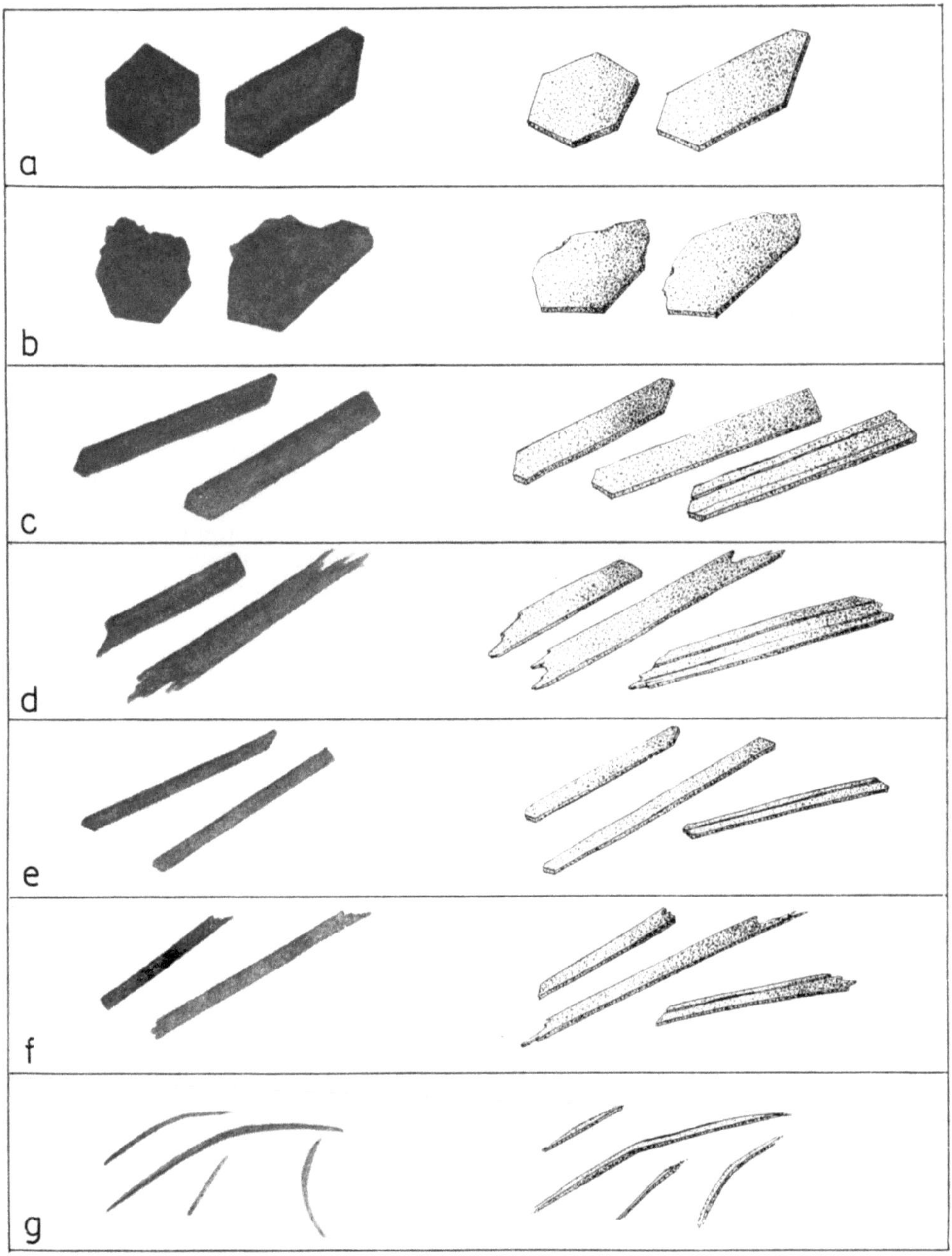

Fig. 3. Schematic illustration of the basic forms of the clay minerals and their admixtures
For explanation see text

In specimen transmission the tubular crystals are depicted as having opaque longitudinal contours that are caused by stronger scattering in the two peripheral regions where the electrons have to pass through a greater thickness due to the curvature. In contrast, the replicas clearly show that the tubular crystals are hollow bodies. The

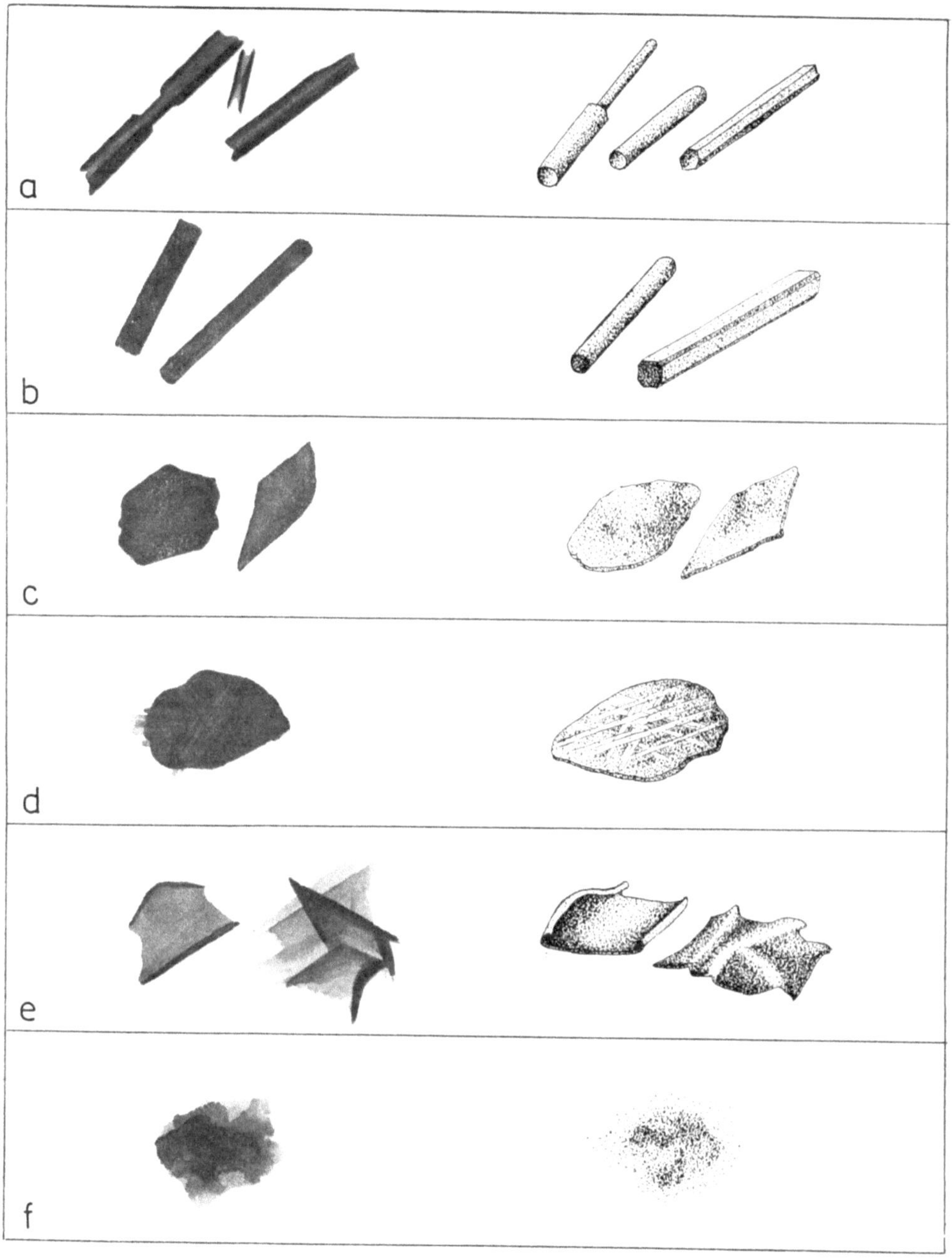

Fig. 4. Schematic illustration of the basic forms of the clay minerals and their admixtures
For explanation see text

external wall is commonly shown to be rounded, but may have streaks that indicate faces. Occasionally, tubules of smaller diameter are stuck into those of larger diameter. The length/width ratio may also vary widely. Some minerals of the serpentine group and in particular of the kaolinite group (halloysite) occur as tubules.

The channels that occur in the minerals of the palygorskite-sepiolite group are oriented in a longitudinal direction because of crystal structure. They have diameters of only 40–110 nm and are hardly recognizable at standard TEM magnification. Usually these minerals appear as fibres.

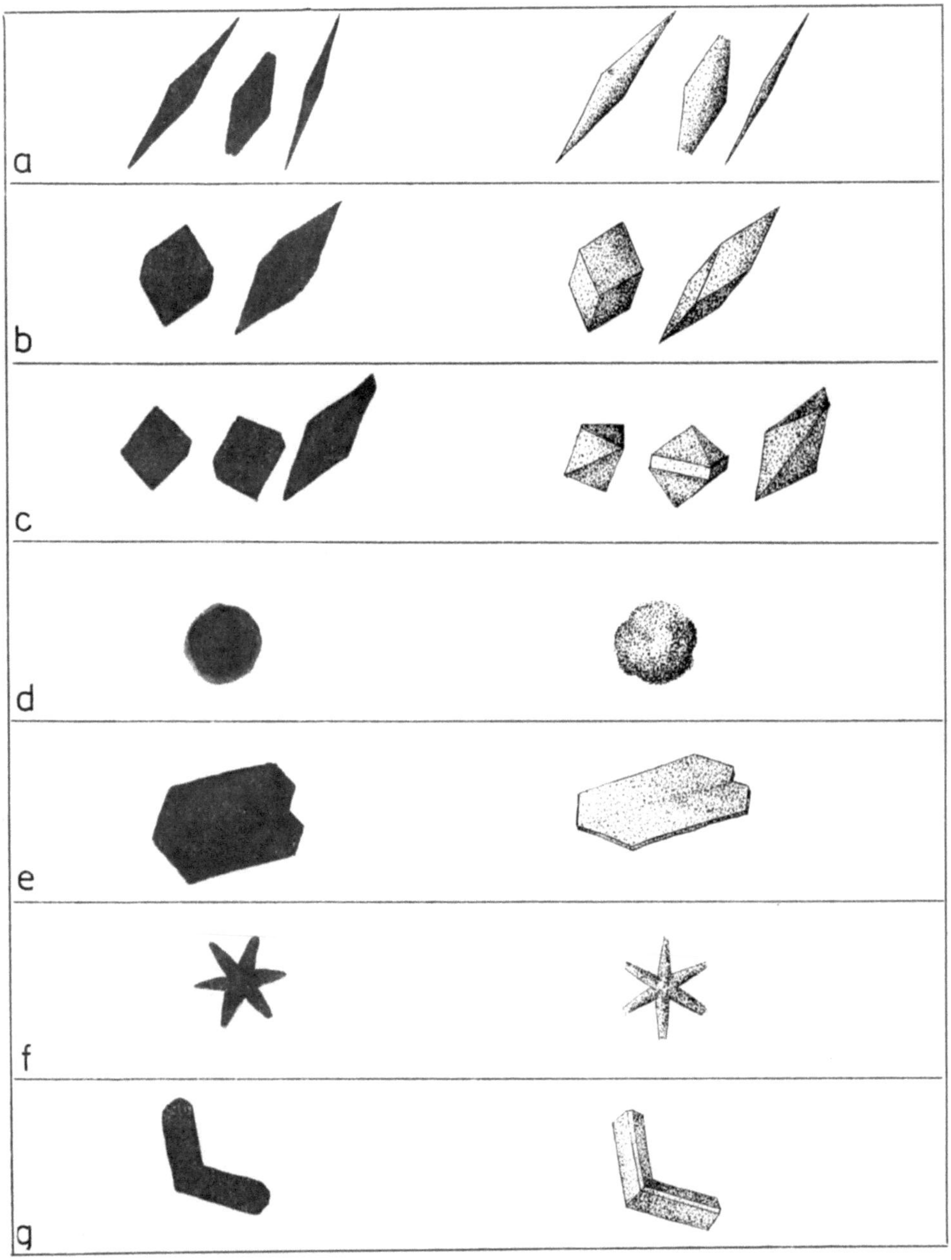

Fig. 5. Schematic illustration of the basic forms of the clay minerals and their admixtures
For explanation see text

In specimen transmission the rod-shaped crystals are usually opaque due to strong electron scattering, but the replicas show rounded forms and forms with longitudinal faces. Such forms occur in certain minerals of the trioctahedral three-layer silicates and in some admixtures of the clay minerals, e.g. in rutile.

The platy crystals that have irregular contours and occasional cleavage cracks show great variations in transparency. The replicas show smooth surfaces with few cleavage cracks. Such forms are typical of many minerals of the talc-pyrophyllite group, the mica and the chlorite group, but are far less common in the montmorillonite-saponite and kaolinite-serpentine groups.

The individual crystals of the plates with folds and of the nepheloid particles are usually very thin and thus always transparent. However, there are often no observable crystals due to film formation. This morphology is confirmed by the replicas, which show that film formation after optimal dispersion occurs in several minerals. This behaviour is typical of some minerals of the montmorillonite-saponite and vermiculite groups.

In Figure 5 the following basic forms are presented:

a) Needles
b) Rhombohedrons
c) Tetragonal bipyramids
d) Spherical particles
e) Pseudohexagonal twins (contact twins)
f) Triplets of needles (penetration triplets)
g) Knee-shaped twins (contact twins)

The needles do not always have sharp points, nor do they usually show clearly recognizable crystal faces. Occurring in needle form are the iron minerals hematite and goethite, which are common admixtures of clays and weathering products. In specimen transmission they appear as opaque bodies.

The elongate to isometric rhombohedrons and the isometric tetragonal bipyramids can be assigned to defined crystal systems, in these cases to the trigonal and tetragonal system, respectively. In specimen transmission these crystals are usually opaque and cannot be identified by their contours. The replicas, however, make an identification possible.

The forms shown in lines b and c, together with other similar forms, are due to admixtures of clay minerals, e.g. to carbonate minerals (line b) and to anatase, a titanium mineral (line c).

Spherical particles frequently occur in amorphous or X-ray amorphous admixtures of clay minerals. In specimen transmission they are usually opaque, their spherical shape being shown only in the replicas.

In the three following lines schematic illustrations are given of the occasional twin and triplet crystals. Contact twins occur sparingly in the phyllosilicates, especially in the hexagonal basic forms (kaolin and some mica minerals), whereas the star-shaped triplets are found in the admixtures, e.g. in goethite or hematite. The contact twins occur, according to the rutile law, in the titanium mineral rutile. The twins can easily be recognized in the replicas, because here the reentrant angles are clearly observable. Similar contours, however, also occur in parallel intergrowths and aggregates. If the admixtures of the clay minerals show distinguishable crystal forms, twins may be observed as well.

Figure 6 depicts some typical examples of parallel intergrowths and aggregates of clay minerals. In the parallel intergrowths all delimiting morphological elements, and thus the crystal lattices, are exactly parallel to one another, whereas in the aggregates the individual crystals or particles show an irregular arrangement relative to one another.

In Figure 6 the following forms are shown:

a) Book-shaped parallel intergrowth
b) Vermicular aggregates
c) Platy parallel intergrowth

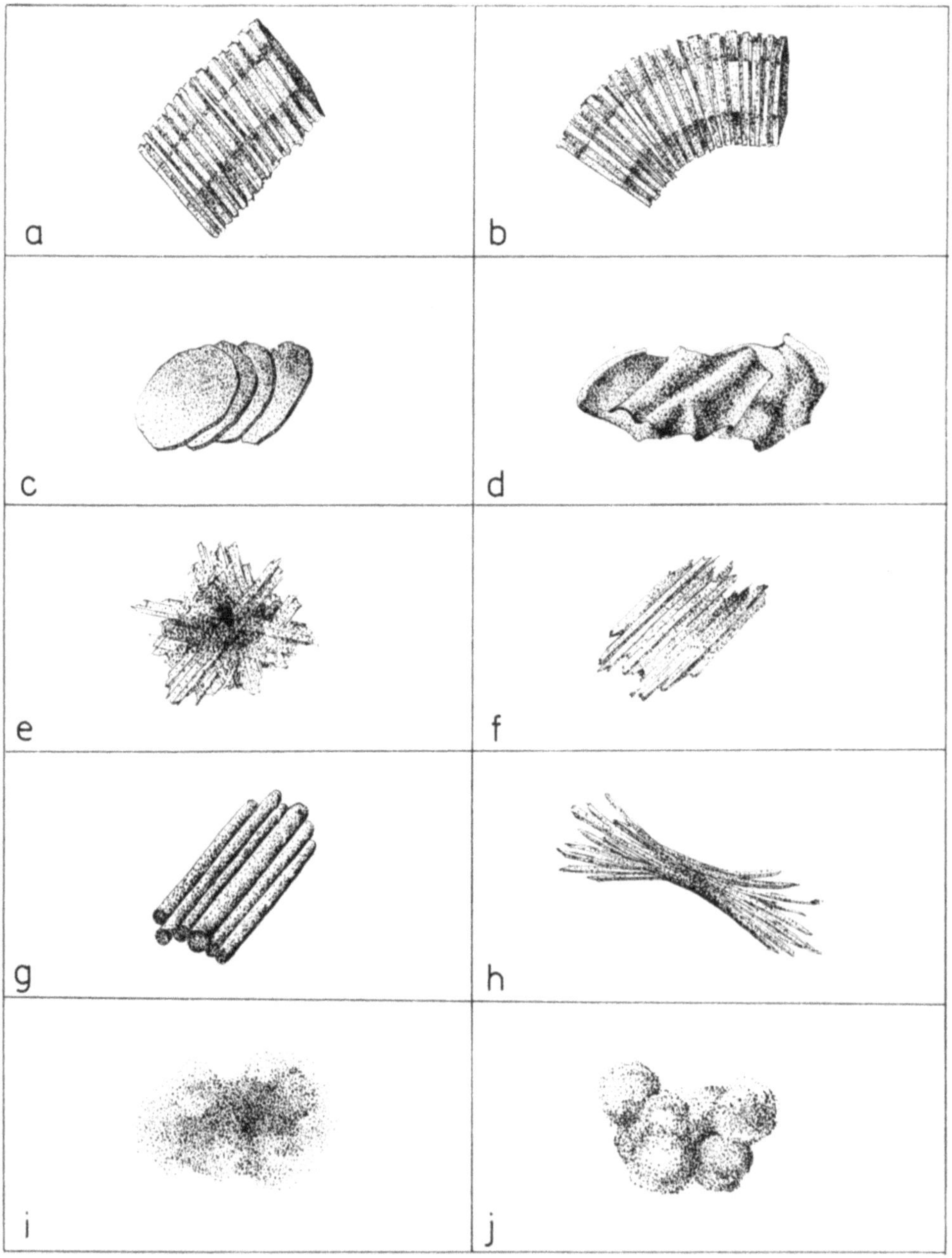

Fig. 6. Schematic illustration of the basic forms of the clay minerals and their admixtures
For explanation see text

d) Platy aggregates with folds
e) Aggregates of slat-shaped crystals
f) Parallel intergrowths of slat-shaped crystals
g) Parallel intergrowths of rod-shaped and tubular crystals
h) Fibrous aggregates (tufts)
i) Nepheloid aggregates (partly with folds)
j) Globular aggregates

All these forms are shown as replicas, because this technique gives a more detailed picture of the characteristic features of parallel intergrowths and aggregates.

The books commonly have pseudohexagonal form, especially in the kaolin minerals. The same is true of vermicular aggregates, which are assumed to have developed from books.

The platy parallel intergrowths and also the platy aggregates and crystals typically occur in many minerals of the talc-pyrophyllite, mica, and chlorite groups. The platy aggregates with folds, however, are usually found in minerals of the montmorillonite-saponite and vermiculite groups. The slat-shaped aggregates and slat-shaped parallel intergrowths are assumed to have developed from platy crystals and parallel intergrowths by segregation. They occur in various clay minerals that have lath-shaped or slat-shaped morphology (Fig. 3). The globular aggregate shown in Figure 6 is only one variety among many that may occur in clays. Several clay minerals and their admixtures can combine, via aggregation of such basic forms, as plates, slats, fibres and needles, to form various kinds of globular aggregates.

3.2. Two-layer silicates (1:1)

3.2.1. Kaolinite, dickite, halloysite — dioctahedral

Amongst the minerals of the kaolinite group kaolinite, halloysite, and dickite are dealt with in greater detail (Figs. 7—146).

Kaolinite — Figs. 7—108
$Al_6[(OH)_8|Si_4O_{10}]$

Kaolinite is a common mineral, occurring in weathering crusts of acid or intermediate rocks (kaolins), sedimentary rocks (kaolinite clays), in products of hydrothermal alteration or hydrothermal genesis of minerals and may also be a product of sediment diagenesis. Commonly, the kaolinite forms rocks or is the only clay mineral in rocks.

Kaolinite occurs in a wide range of morphological varieties (Figs. 7—108). Usually, the triclinic kaolinite forms platy crystals with pseudohexagonal borders that are diagnostic of this mineral. Some of the main morphological varieties are:

— Idiomorphous, pseudohexagonal plates of varying size. In sedimentary clays the kaolinites, which are usually poorly ordered, occur as fine grains < 0.5 µm in diameter (Figs. 106, 563—565). Diagenetic kaolinites show a high degree of order and form crystals several microns across (Figs. 7—24, 35, 36). In residual kaolins we find idiomorphous pseudohexagonal kaolinite plates, but these are outnumbered by the intergrowths of such platelets.

— Idiomorphous crystals or intergrowths that are elongate on [010] or [110] and always display pseudohexagonal contours, have been found in several samples, especially those of residual kaolins. A number of kaolins contain elongates (slats) of mica minerals, which also have pseudohexagonal contours. Compared to kaolinite, they show reduced thickness and width and random Au-distribution, as can be seen from Figures 41 and 67.

- Xenomorphous forms commonly occur in fire clays (Figs. 106—108).
- Twins and intergrowths on the base plane are very common and represent the typical kaolinitic morphology, as can be seen in most of the kaolinite micrographs. The reentrant angles (e.g. Figs. 73—75) and the decoration pattern (Figs. 42, 84) are indicative of twins. These intergrowths form kaolinite plates, commonly as much as several tenths of a millimeter across, which are macroscopically observable but may be taken for mica. TEMs, however, reveal the pseudohexagonal contours. Frequently, such scaly intergrowths are built up of kaolinized mica.
- Book-shaped and vermicular intergrowths and aggregates are found in nearly all kaolins, mostly in primary but also in sedimentary kaolins, such as the Georgia kaolins. Such vermiforms and books may be of considerable size and can easily be recognized in the SEMs.

The vermiforms in the primary kaolin of Aue, GDR (Fig. 32) have a length of 30 μm and are 10 μm across. Those in the secondary kaolins of Georgia, USA (Figs. 85—87) are 50 μm long and 25 μm broad. Even larger vermiforms occur locally. Due to very good cleavage along the base plane both the books and the vermiforms disintegrate during the preparation process so that the TEMs show only scaly intergrowths (Figs. 85—87).

In the past few years Au-decoration of kaolinites has produced remarkable mineralogical and genetic findings. Under the conditions described on page 12 the colloidal gold becomes readily attached to lattice steps and lattice defects, whereas on the basal planes of other phyllosilicates it shows random distribution. Thus, kaolinite can easily be differentiated from other phyllosilicates. Among others, RUSKO (1978) and HENNING et al. (1984) found, using the decoration technique, epitaxies of kaolinite and mica. The different decoration patterns of the two minerals (Figs. 52, 54) are diagnostic.

The decoration pattern represents the structure of the base planes. The prism faces show random distribution of the gold particles. Lattice steps of different shapes are recognizable, and can be interpreted in various ways. Commonly one can observe growth spirals (e.g. Figs. 41—44, 55, 83), which indicate crystal growth in aqueous solutions. Particularly well developed growth spirals occur in the synthetic kaolinites that are produced at 575 K in autoclaves from amorphous silica and aluminium oxide. The well ordered diagenetic kaolinites of Keokuk, Brazil and Wyoming, USA do not show typical spirals, but have long, straight and curved lattice steps. If two lattice steps meet, twinning is indicated (Figs. 18, 84). The curved decoration lines are difficult to interpret. There are, however, certain similarities to etched kaolinite surfaces, on which, in contrast to unetched surfaces, there are curved lines as well (Figs. 15, 20, 44).

Textural investigations are the domain of the scanning electron microscope. SEM studies of kaolinites were made among others by W. D. KELLER. He found close relationships between texture and genesis of the kaolins. In residual kaolins the kaolinites show no orientation. There is no grain-size fractionation, i.e. minute crystals are found together with large aggregates (vermiforms) (Figs. 32, 517, 538). Moreover, the SEMs (Figs. 64, 517, 522) show that due to mass losses by weathering many pores have developed, thus making the texture rather loose. In sedimentary kaolins the texture is governed by the effects of grain-size fractionation and sedimentation. Frequently, the particles are arranged parallel to each other (Fig. 88) and the kaolinite particles, which were deposited together with the quartz grains, cling tightly to the latter (Fig. 524). In contrast, quartz grains and kaolinite in residual kaolins have no such intimate contact with one another (Fig. 523).

The occurrence of large kaolinite vermiforms in residual kaolins can be explained by in-situ growth. For sedimentary kaolins, however, it is hardly imaginable that these readily cleavable vermiforms, which may be of considerable length, are transported and deposited particles. Thus, a diagenetic neoformation seems more likely (KELLER, 1976). However, a sedimentation of the more compact books may well have taken place (MURRAY, 1976). As outlined above, diagenetic kaolinites can be recognized by their distinct pseudohexagonal, platy morphology (Figs. 7, 8).

Dickite — Figs. 109—120
$Al_6[(OH)_8|Si_4O_{10}]$

Dickite is a rather rare mineral. It may occur in hydrothermally altered rocks and hydrothermal veins, e.g. in druses, pores or clefts of diagenetically altered sediments.

Usually dickite forms pseudohexagonal, idiomorphous, platy crystals which, like other phyllosilicates, show moiré patterns in the TEMs. Because the crystals are rather thick, they usually are opaque in the TEMs. In some SEMs the (110)- and (010)-faces can be recognized easily. Some crystals are elongate along the *c*-axis (Figs. 113, 114). The dickite plates are as much as 20—30 μm across and several microns thick. Thus, they are easy to recognize even under a light microscope.

Halloysite — Figs. 121—146
$\left\{\begin{matrix} Al_4[(OH)_8|Si_4O_{10}] \\ (H_2O)_4 \end{matrix}\right\}$

Halloysite is a widespread, well described mineral that has been found at many localities, but occurs in minable quantities and in mono-mineralic form only in a few deposits (e.g. at Djebel Djebar, Algeria). Its genetic occurrence is in weathering crusts, soils, sedimentary rocks and in low-thermal or exogenously altered volcanic rocks. As a secondary mineral halloysite occurs along joints in the oxidized zone of various ore deposits.

The electron optical habit has repeatedly been described. Usually, halloysite is identified by its tubular shape; with other methods this mineral is hardly detectable alongside kaolinite. In the vacuum of the electron microscope halloysite certainly occurs as metahalloysite, which, however, cannot be differentiated morphologically from hydrohalloysite (BEUTELSPACHER & VAN DER MAREL 1968). In the TEMs the tubules are so conspicuous in appearance that the halloysite content is usually grossly overestimated. The tubules show great morphological variations. Only in rare cases do they exceed 2 μm in length whereas their diameter usually is < 0.1 μm. In some samples their ends are unrolled (Fig. 138), or more slender tubules emerge from the larger ones (Figs. 107, 123, 124). Occasionally, pseudohexagonal plates come loose from the tubules (Fig. 139). Some replicas show hexahedral tubules. The halloysite particles of one and the same sample may differ in size by more than one order of magnitude. Tubules with a length greater than 1 μm occur together with very short and slender tubules. From the SEMs it becomes evident that the larger crystals occur on joints and in pores, whereas the smaller halloysite tubules occur in the more compact parts of the sample (Figs. 143—146).

The spherical forms in the SEMs suggest primary gel-like segregations that crystallized to halloysite. The habit of these halloysites is comparable to that of whiskers. The different environmental conditions may produce micro-nodes of halloysite close to, or amidst, associated kaolinite, but the origin of this association cannot be related to the primary mineral. Such micronodes have been observed by W. D. KELLER (1978) in the granitic kaolin of Sparta.

3.2.2. *Serpentine minerals (trioctahedral): Chrysotile, antigorite, garnierite* (7 Å)
$Mg\ [(OH)_8|Si_4O_{10}]$

The serpentine minerals are primarily formed by serpentinization of basic and ultrabasic rocks, i.e. by hydrothermal (metasomatic) transformation of magnesium silicates, especially olivine and orthopyroxene, into serpentine. In the weathered zone of serpentines these minerals are relatively stable. However, they take up metal ions, partly by absorption and partly by partial substitution of the Mg-ions in the octahedral sheet. Thus, nickel deposits formed by weathering contain, besides nickel chlorites and vermiculites, also nickel serpentines, e.g. garnierite (7 Å).

In this atlas the following serpentine minerals are considered:

- clinochrysotile (Figs. 147–150)
- orthochrysotile (Figs. 151–154)
- antigorite (Figs. 155–158)
- garnierite (7 Å) (Figs. 159–162).

Due to their rolled lattices the chrysotiles (clino-, ortho-, parachrysotile) are depicted in the TEMs as tubular particles or fibres. The diameter of the tubules is only fractions of a micron. Their length varies widely; in unweathered rocks they may be as long as several centimeters, whereas in the weathered zone they are commonly broken and their length ranges from less than one micron to several microns.

As evident from the micrographs, the tubules tend to form parallel intergrowths. The SEMs of chrysotile show parallel intergrowths of fibres, which are, however, without recognizable tubular structure.

Antigorite has a corrugated layer-lattice. The electron micrographs show platy particles as well as platy parallel intergrowths and aggregates (Figs. 155, 156). In addition, lath-shaped crystals with irregular, or even ragged ends can be seen (Fig. 158).

Lizardite, with its plane layer-lattice, forms platy particles. The garnierite (7 Å) forms aggregates of minute tubules, as depicted in Figure 161. In addition, the micrographs show platy particles (possibly unrolled tubules) and aggregates of plates. SEMs of the fracture pattern reveal the irregular arrangement of the coarser aggregates (Fig. 162).

3.3. Three-layer silicates (2:1)

3.3.1. *Pyrophyllite (dioctahedral)* — Figs. 163–172
$Al_2[(OH)_2|Si_4O_{10}]$

Pyrophyllite is more widespread in the earth's crust than supposed earlier. It forms in various genetic conditions, including:

- hydrothermal veins in paragenesis with quartz, carbonates, hematite and other minerals
- argillaceous metamorphic schists, where it commonly occurs in lenticular forms, e.g. in the green schist and the albite-epidote-hornfels facies
- a hydrothermal alteration of acidic volcanic rocks.

The basic structural unit (unit layer) of pyrophyllite consists of two SiO_4 tetrahedral sheets linked by Al surrounded by 4 O and 2 OH in octahedral arrangement (octahedral sheet). The valences of the unit layer are saturated so that the layer is linked to the neighbouring unit layers only by van der Waals' forces. Therefore, the platy crystals show good cleavage along (001).

With some variation in ion composition the unit layer of pyrophyllite occurs in all dioctahedral three-layer silicates.

Due to the good cleavage of pyrophyllite, micrographs of this mineral usually reveal thin xenomorphic plates, which are characterized by a high degree of flexibility. This is shown in detail in the TEMs (Figs. 163–166, 169) and in the SEMs (Figs. 168, 171, 172). In the TEMs there are commonly moiré patterns of the rotational type (e.g. Fig. 165). These are electron diffraction effects which are caused by slight twisting of successive lattice planes, and can be used for indirect imaging of the lattice planes.

3.3.2. *Talc, garnierite* (10 Å) — *trioctahedral*

Talc — Figs. 173 and 174
$Mg_3[(OH)_2|Si_4O_{10}]$

Talc occurs
- as talc schist in the metamorphic series
- very commonly as a product of hydrothermal alteration of Mg-rich ultrabasic rocks
- as a product of contact-metasomatic transformation in the contact zone between dolomite and intrusive rocks
- as a secondary product of the weathering of ultrabasic rocks (frequently as garnierite (10 Å))
- with other clay minerals in Permian salt rocks.

As in pyrophyllite, the unit layer of talc consists of two SiO_4 tetrahedral sheets linked by Mg surrounded by 4 O and 2 OH in octahedral arrangement. The valences of the unit layer are saturated and thus the unit layers are held together only by van der Waals' forces. Talc, too, is therefore characterized by very good cleavage along (001). With some variation in ion composition the unit layer of talc is found in all trioctahedral three-layer silicates.

Because of its very good cleavage talc appears in the micrographs as thin xenomorphic platelets and as platy parallel intergrowths (Figs. 173, 174). The TEMs of talc commonly show moiré patterns of the rotational type. In electron micrographs talc may be mistaken for pyrophyllite, or vice versa.

Garnierite (10 Å) — Figs. 175 and 176
$(Ni, Mg)_3[(OH)_2|Si_4O_{10}]$

The garnierite (10 Å), which was formed by weathering, has a markedly smaller grain size. The xenomorphic platelets, which may have diffuse borders, tend to form aggregates (Figs. 175, 176). Moreover, they have a tendency towards flaking.

3.3.3. *Smectite (dioctahedral): Montmorillonite, beidellite, nontronite*

Montmorillonite — Figs. 177—210

$$\left\{\begin{array}{l}(Al, Mg)_2[(OH)_2|Si_4O_{10}]^{-0.33} \\ Na_{0.33}(H_2O)_4\end{array}\right\}$$

Montmorillonite is a typical weathering and decomposition product of volcanic ashes and intermediate to basic volcanic rocks. The bentonites formed by these processes contain montmorillonite as dominant mineral, together with minor mica, quartz and feldspar. Moreover, montmorillonite is an essential constituent of subtropical to tropical, recent and fossilized weathering profiles and soils. Montmorillonite can also be formed in the late hydrothermal periods of volcanism, e.g. from volcanic tuffs by solfatara activity, and in hydrothermal veins.

In contrast to the pyrophyllite the unit layer of the montmorillonite is characterized by an excess negative charge that results from partial substitution of Al for Si in the tetrahedral sheet and of Mg for Al in the octahedral sheet. The excess negative charge is partially but never completely compensated for by surplus Al in the octahedral sites.

These crystallo-chemical conditions give rise to intracrystalline swelling, i.e. water may enter in between two unit layers and is usually coordinated around surplus cations. This property produces the morphological features observable in the electron micrographs.

Because of its world-wide occurence and commercial importance montmorillonite has been the subject of a great number of papers dealing with crystal morphology. Long ago the morphology of this mineral was described by VON ARDENNE, ENDELL & HOFMANN (1940), who observed platelets with folds and curled edges, and the tendency to form thin films. This was probably the first specific electron optical study of a member of the dioctahedral three-layer silicates. GRIM (1953) and HAUSER (1955) suggest a connection between the various cations between the sheets and the shape of the montmorillonite platelets. GRIM & KULBICKI (1961) present a morphological characterization of the "Wyoming" and "Cheto" types of montmorillonite recognized by them. They found that the Wyoming type has a more pronounced capacity to form thin films than the Cheto type. GÜVEN & PEASE (1975) subdivide the particles of the investigated samples into several morphological types (types L, E, A and SH), but a differentiation between Wyoming and Cheto types is not possible on this basis.

Another attempt to solve this problem was made by HENNING & LANDGRAF (1978) who discuss the relationships between morphology and crystal chemistry of dioctahedral three-layer silicates. The text describing the montmorillonite micrographs of this atlas is in its essence taken from their paper.

The electron micrographs of Figures 177—186 show montmorillonites referred to in the literature as Cheto types. Their main characteristics are as follows: the individual particles are platelets with diffuse outlines; the primary particles are difficult to recognize; the platelets have a marked tendency to form aggregates, thus giving rise to nepheloid particles and aggregates; these aggregates show the typical swelling-controlled morphology of montmorillonites, including curled edges and folds.

In the electron micrographs of Figures 187—195 montmorillonites of the Wyoming type are illustrated. These can be characterized in the following way: in general, they occur in platelets, which usually have clearly recognizable contours, so that they look like primary particles; the platelets tend to form thin films; thicker and larger aggregates are less numerous than in the Cheto types; the thin platelets commonly show the typical swelling-related features, especially curled edges but also folds.

The montmorillonites of Figures 198 and 200 may be viewed as transitional types. The montmorillonites of Figures 199, 201 and 202 (Stare Kremnice, Istenmezeje, Krimskiy Kit) are similar to those of the Wyoming type, whereas the Branany montmorillonite (Fig. 197) is similar to those of the Cheto type. Another interesting montmorillonite variety is shown in Figures 204 and 205 (TEMs), Figure 206 (Replica) and Figures 207 to 210 (SEMs). This montmorillonite has formed by hydrothermal segregation and is found at Michalkovo, Bulgaria. As shown in Figure 204, the material resembles the Wyoming type. But, as can be seen in Figures 205 and 206 this montmorillonite also occurs as lath-shaped particles with the customary features indicative of swelling. The SEMs (Figs. 207—210) indicate that the lath-shaped particles may have grown in cavities or pores and the platy particles may have derived from the rock matrix.

Beidellite — Figs. 211—212

$$\left\{\begin{matrix} Al_2[(OH)_2|Al_{0.5}Si_{3.5}O_{10}]^{-0.5} \\ (Ca, Na)_{\sim 0.3}(H_2O)_4 \end{matrix}\right\}$$

Beidellite usually occurs in the weathering profile of basic and ultrabasic rocks. Its morphology resembles that of the montmorillonite and has repeatedly been described in the literature, e.g. by von GÜVEN & PEASE (1975) and by MERING & OBERLIN (1971).

Figures 211 and 212 show the typical crystal habit of beidellite. This mineral occurs as xenomorphic platelets which tend to form small but rather thick aggregates. Compared to the Cheto-type-montmorillonite these aggregates have sharper contours, and unlike the Wyoming-type-montmorillonite the beidellite does not form films. However, aggregates of beidellite do show the typical signs of swelling.

Nontronite — Figs. 213—224

$$\left\{\begin{matrix} Fe_2^{3+}[(OH)_2|Al_{0.33}Si_{3.7}O_{10}]^{-0.33} \\ Na_{0.33}(H_2O)_4 \end{matrix}\right\}$$

Most nontronite occurs in the weathering profile of Fe-rich basic and ultrabasic rocks, but may also originate in low-temperature magmatic bodies. Nontronite has a slat-shaped morphology which is generally viewed as the result of the mismatch between the sheets due to the replacement of Al^{3+} by Fe^{3+} and some Mg^{2+} in the octahedral sheet, (e.g. ROSS & HENDRICKS 1945). Nontronite forms lath-shaped xenomorphic particles which tend to flake because of pronounced longitudinal cleavage (TEMs and replicas). Aggregates and parallel intergrowths of these laths show some degree of bending.

Occasionally, nontronite occurs as small platy xenomorphic particles (Figs. 213, 216, 223). Typical signs of swelling are exhibited especially by the thicker aggregates and intergrowths. The fracture face of nontronite, e.g. of samples from Garfield (Fig. 215) and Dolny Bory (Fig. 218) shows coarse bent leaves, whereas that of the Rajasthan nontronite (Fig. 220) is characterized by pronounced flake formation.

3.3.4. Smectite (trioctahedral): Saponite, hectorite

Saponite — Figs. 225—234

$$\left\{\begin{matrix} (Mg, Fe)_3[(OH)_2|Al_{0.33}Si_{3.67}O_{10}]^{-0.33} \\ (1/2Ca, Na)_{0.33}(H_2O)_4 \end{matrix}\right\}$$

Though saponite is less common in the earth's crust than montmorillonite, it also forms under a wide range of chemical conditions. In particular, it occurs as a weathering product of basic and ultrabasic rocks, especially of serpentinite; as a hydrothermal decomposition product, e.g. on joints in serpentinites; as a product of volcanic-hydrothermal processes; and as a constituent of the oxidized zone e.g. of Zn-, Cu- and Mn-ore deposits. Incorporation of one or another kind of metal in the octahedral sheet gives rise to several varieties. In the TEMs the saponite shows a wide range of morphological features. The saponite of Cornwall (Figs. 225—228) occurs in xenomorphic platelets and aggregates and also forms xenomorphic parallel intergrowths with folds indicating swelling. In addition, moiré patterns can be seen (Fig. 226). The replicas (Figs. 227, 228) show segregation of small slats on the larger platelets. In contrast, the saponite of Szklary (Figs. 229, 230) has nepheloid platelets. Signs of swelling are exhibited especially by the coarser aggregates. Quite another crystal habit is shown by the saponite from Pöhla (Figs. 231 to 233), which forms lath- to slat-shaped idiomorphous crystals that lack signs of longitudinal cleavage. Long parallel intergrowths can be seen especially in the replica (Fig. 233). The SEM of the fracture face (Fig. 234) reveals lamellae and fine elongate crystals.

Hectorite — Figs. 235—237
fluorine-containing Li-*saponite*

Hectorite, named after the type locality in Hector, California, has formed from Li-containing micas. The TEM and replica (Figs. 235, 236) show slat-shaped to fibrous and locally nepheloid particles and tuft-like aggregates, all of which display typical signs of swelling. The SEM (Fig. 237) shows the nepheloid habit, with formation of aggregates.

3.3.5. *Vermiculite (di- and trioctahedral)* — Figs. 238—254

$$\left\{\begin{matrix}(Mg, Fe, Al)_3[(OH)_2|Al_{1.28}Si_{2.72}O_{10}]^{-0.64} \\ Mg_{0.32}(H_2O)_4\end{matrix}\right\}$$

Vermiculites may form by hydrothermal alteration and weathering of muscovite, biotite or phlogopite and other rock-forming minerals. In the vermiculites the typical properties of the montmorillonite-saponite minerals are still more pronounced, e.g. the pyrophyllite and talc unit layers have a higher surplus charge due to substitution of Al for Si in the tetrahedral sheet. This leads to a higher base exchange capacity. The dioctahedral vermiculites, which have formed from muscovite via hydromuscovite, are by far less common than the trioctahedral ones, which derive from biotite via hydrobiotite. The vermiculites occur as rather large platy crystals and in very small particles only a few microns in maximum dimension. The TEM sand replicas reveal that the dominant habit of the vermiculites is xenomorphic platelets and conspicuous parallel intergrowths of these platelets (e.g. Figs. 244, 247 and 248). In addition, lath-shaped particles occur, as can be seen in the TEMs of the Montana vermiculite (Figs. 239 and 240).

The platelets and aggregates commonly have curled edges and tend to flake along the base planes. This can be clearly seen in the TEM (Fig. 243) and in the SEMs (Figs. 241, 245 and 249). The vermiculite from Callenberg, which has been modified by weathering, forms thin platelets that show moiré patterns (Figs. 253 and 254).

3.3.6. *Mica (dioctahedral): Muscovite, paragonite*

Muscovite — Figs. 255—258
$KAl_2[(OH, F)_2|AlSi_3O_{10}]$

Mica minerals make up about 4% of the earth's crust. Among the mica minerals muscovite is second in abundance only to biotite. It is a primary mineral constituent of various igneous and sedimentary rocks. As a secondary constituent, muscovite occurs in rocks that have undergone pneumatolytic, hydrothermal or metamorphic alteration. The muscovite is largely resistant to weathering and becomes enriched in clayey sediments in form of minute platelets. In the tetrahedral sheets of muscovite the Si of the pyrophyllite unit layer is replaced by Al. The resulting negative charge of the unit layers is compensated for by the incorporation of K in the interlayers (about 12% K_2O in muscovite). Muscovite occurs in large platy crystals (in pegmatites) with very good cleavage along (001), and in minute platelets only a few micrometers in maximum dimension.

In the TEMs and replicas the muscovite appears as thin xenomorphic platelets (Figs. 255, 256, 257) and as platy parallel intergrowths; however, there are no significant differences between these polytypes. Figure 258 shows decorated surfaces of muscovite particles with random Au-distribution.

Paragonite — Figs. 259—264
$NaAl_2[(OH, F)_2|AlSi_3O_{10}]$

Paragonite occurs in the greenschists facies of metamorphic rocks. At temperatures above 400 °C, this mineral will form by alteration of other clay minerals, e.g. from muscovite and possibly also from kaolinite, provided that Na is present. In contrast to muscovite the potassium ions in the interlayer are replaced by sodium ions. Macroscopically, paragonite appears to consist of fine flakes; the TEMs, however, show rather thick xenomorphic platelets and parallel intergrowths and aggregates of platelets (Fig. 259). This is particularly evident in some of the replicas (Figs. 260 and 261) and SEMs (Figs. 263 and 264).

3.3.7. *Hydromica (dioctahedral): Hydromuscovite, illite, glauconite, seladonite*

Hydromuscovite — Figs. 265—278 —
Illite
$(K, H_3O^-)Al_2[(OH)_2|AlSi_3O_{10}]$

Hydromuscovite (illite) is common in many weathering profiles and soils, as well as in marine and limnic clays of all ages. It differs from muscovite in that it contains less K_2O. Part of the K-ions in the interlayer is replaced by $(H_3O)^+$-ions. On the whole, "hydromica" and "illite" can be used synonymously; however, the term "illite" is primarily used to denote the hydromuscovite that occurs as very fine flakes.

Xenomorphic platelets and aggregates of platelets are the dominant morphological features of illite. However, some illite has been described as having a slat-shaped morphology.

HAMDY & RADCZEWSKI (1974) related the slat-like shape of some illites to cleavage along (110), the main factors in the development of such structure being the pressure and temperature conditions. The impact of K-removal on crystal morphology has been discussed in the literature. This problem has been studied by HUFF (1972), who, by removing the K, succeeded in producing illite slats on illite platelets. An evaluation of illite morphological data which are presented in the literature is difficult, because the mineralogical characterization is not always adequate, and in many cases no clear-cut differentiation is made between illites and mixed-layer minerals (see 3.5.1.).

In the TEMs of this atlas the illites appear as xenomorphic platelets, parallel intergrowths and aggregates, none of which show signs of swelling (Figs. 265, 266, 269, 270, 275, 276). However, the replicas of some samples reveal cleavage cracks and slat formation on the base planes (Figs. 273, 274, 277, 278). The SEMs also show aggregates of platelets oriented parallel to the bedding (e.g. Figs. 267 and 268).

Glauconite — Figs. 279—283
$(K, Na, Ca, H_3O^-)_{<1}(Al, Fe^{3+}, Fe^{2+}, Mg)_2[(OH)_2|Al_{0.35}Si_{3.65}O_{10}]$

Most glauconite forms authigenically in sandstones, clays, marls and carbonate rocks in off-shore environments. Upon weathering this mineral decomposes to iron hydroxides and silica gel. The chemical composition shows major variations, depending on the locality. The morphological habit of glauconite is usually described as slat-shaped (e.g. MÜLLER & BEHNE 1954; BURST 1958).

Xenomorphic slat-shaped glauconite particles, many of them showing longitudinal cleavage, are depicted in Figures 279 and 281. In addition, glauconite occurs in platy xenomorphic particles with no signs of swelling (Fig. 280). The SEMs (Figs. 282 and 283) show an exfoliating glauconite pellet.

Seladonite — Figs. 284—286

Seladonite is a hydrothermal alteration product of intermediate to basic igneous rocks and commonly occurs in cavities and along joints. Its chemical composition varies somewhat but is largely the same as that of glauconite. The structure of seladonite also resembles that of glauconite. According to JASMUND (1955), seladonite is depicted in the TEM as slat-shaped. HENNING & LANDGRAF (1978) found that the dioctahedral three-layer silicates with a high Al-content (montmorillonite, beidellite, illite) primarily form xenomorphic platelets and those with a high Fe-content (nontronite, glaconite, seladonite) mainly form xenomorphic slats.

In Figures 284–286 xenomorphic slats of seladonite and aggregates of platelets split into slats (easily recognizable in the replica) are shown.

3.4. Four-layer silicates (2:1:1)

3.4.1. Chlorite (dioctahedral with gibbsite layer): Sudoite

Sudoite — Figs. 287–290

$$\left\{\begin{matrix} Al_2[(OH)_2|AlSi_3O_{10}]^{-1} \\ [Al_{2.33}(OH)_6]^{+1} \end{matrix}\right\}$$

Sudoite was detected as late as 1960 in a hydrothermally decomposed rhyolitic tuff in Japan, but may also occur in soils and various sedimentary rocks. The sudoite structure is composed of alternating pyrophyllite and gibbsite layers. The TEMs show thin xenomorphic platelets and parallel intergrowths of these platelets with moiré patterns (Figs. 187 and 288). Especially in the replica (Fig. 289) the parallel intergrowths are easy to recognize. Aggregates of platelets also occur in the SEM of the cleavage face (Fig. 290).

3.4.2. Chlorite (trioctahedral): Chlorite, pennine, clinochlore, prochlorite — Figs. 291–310

The minerals of the trioctahedral chlorite group may form under a great variety of genetic conditions. Prochlorite and clinochlore are rock-forming minerals in metamorphic chlorite schists. Other chlorites occur along joints. Chlorites may also originate from hydrothermal decomposition and weathering of biotite, pyroxene, amphibole and many other minerals.The Fe-rich members are sedimentary in origin but are usually metamorphosed.

The structure of this group is made up of talc and brucite layers in alteration:

$$\left\{\begin{matrix} Mg_3[(OH_2|Si_4O_{10}] \\ Mg_3(OH)_6 \end{matrix}\right\}$$

The octahedral Mg may be replaced by Fe^{2+} and Al^{3+} and the tetrahedral Si by Al^{3+}. The varying degrees of substitution produce a wide range of mixed crystals, such as talc-chlorite, pennine, clinochlore, prochlorite, chamosite and thuringite. Only a few of these chlorite minerals have been investigated in the SEM, since many chlorites occur as crystals of macroscopic size.

In electron micrographs the chlorites usually are depicted as xenomorphic platelets, aggregates, and parallel intergrowths of these platelets (Figs. 291–310). Only in rare cases are they split into laths or slats (e.g. Fig. 310). Such splitting is found in the Callenberg chlorite, which has been modified by weathering. In nearly all SEMs of chlorites there are conspicuous stacks of platelets, which frequently show exfoliation and bending. In addition, well developed idiomorphic crystals occur (e.g. Figs. 301 and 302). Exfoliating stacks are typical of the Vermont prochlorite (Figs. 305 and 306).

3.5. Mixed-layer structures

Clay minerals of simple structural type as well as mixed-layer type have been extensively discussed in the literature, and especially the latter have received increased attention in recent years. There are two types of mixed-layer minerals. One type is characterized by an irregular mixed-layer structure (two different layers occur in variable sequences) which may, for example, be composed of:

— montmorillonite and muscovite (illite) layers (a very common mixed-layer mineral)
— montmorillonite and kaolinite layers (only at a few localities)
— phlogopite and hectorite layers.

The other type has a regular mixed-layer structure (two different layers alternate) which may, for example, be built up of:

— montmorillonite and muscovite layers (1:1) (rectorite, syn. allevardite)
— montmorillonite and chlorite layers (tosudite)
— vermiculite and chlorite layers (corrensite).

The clay minerals with mixed-layer structure have formed by weathering of various rocks and are found in situ or in sedimentary basins, where they occur as moderately altered detritus. Moreover, they have formed as products of hydrothermal alteration of rocks and as a hydrothermal neoformation or as a result of the transformation of phyllosilicates. They may also occur in primary ore deposits, and some have formed by diagenetic and metamorphic processes in sedimentary rocks.

3.5.1. Irregular mixed-layer structures

Montmorillonite-muscovite mixed-layer — Figs. 311—330

The irregular montmorillonite-muscovite mixed-layer has very frequently been described in the literature, e.g. as a constituent mineral of the weathering profile of acid to basic igneous rocks and of many detrital sedimentary rocks. The ratio of montmorillonite to muscovite layers may range widely but is, in general, typical of locality and type of sediment. Also, the swelling capacity varies with the proportion of montmorillonite layers in the mineral structure. The morphology of this mineral has been described as slat-shaped and platy (e.g. HAMDY & RADCZEWSKI 1974; HUFF 1972 and 1974). HENNING & LANDGRAF (1978) have pointed out that the structural stress which is due to the interplay between ion substitution, tetrahedron rotation, K^+-fixation, and layer charge necessarily leads to a slat-like morphology. In certain localities the aggregates or intergrowths show a hexagonal arrangement of the slats (Figs. 327 to 329) suggesting strongly that segregation and splitting have contributed to this morphology. In terms of layer-lattice structure this morphology indicates the cohesion minima on the *a*/*b* plane along [110], [100] and [1$\bar{1}$0], which under structural stress promote cleavage.

The TEMs, replicas and SEMs (Figs. 311—320) depict montmorillonite-muscovite mixed-layers from primary kaolins. These minerals occur usually as irregularly terminated thin slats, some of which form parallel intergrowths and aggregates. The slats show longitudinal cleavage but no signs of swelling. Only compact aggregates exhibit some signs of swelling. One SEM (Fig. 318) shows slats projecting into a cavity.

Figures 323—330 depict montmorillonite-muscovite mixed-layers from clayey sedimentary rocks of various ages. These minerals predominantly occur as thin, in part very fine slats with irregular ends and longitudinal cleavage but without any signs of swelling. Such signs, however, are found on the compact aggregates. In Figures 327, 328 and 329 intergrowths of slats with hexagonal angles (described above) can be seen. The illite of Füzerradvany (sarospatakite), an irregular montmorillonite-muscovite mixed-layer mineral, is a morphological exception in that it occurs in slats with idiomorphous ends and has a much reduced longitudinal cleavage (Figs. 321 and 322).

Kaolinite-montmorillonite mixed-layer — Figs. 331—338

The irregular kaolinite-montmorillonite mixed-layer of Jegłowa is associated with kaolinite in veins occurring in quartzitic schists, and has probably formed hydrothermally. According to WIEWIORA (1971) and SZPILA, WIEWIORA & GADOMSKI (1972) the mineral consists of 86% kaolinite and 14% montmorillonite layers in irregular sequence.

In the TEMs and replicas (Figs. 331—338) this mineral exhibits morphological features of both kaolinite and montmorillonite. The pseudohexagonal edges of the hypidiomorphic crystals (0.02 μm) and the pseudohexagonal edges in aggregates are indicative of kaolinite. The Au-decoration shows, in addition to the pseudo-hexagonal pattern typical of kaolinite, areas of random Au-distribution that are diagnostic of montmorillonite. Folds, curled edges and film formation also indicate montmorillonite. All of this indicates that the particle surfaces have areas with a kaolinite structure as well as such with montmorillonite structure.

3.5.2. *Regular mixed-layer structures*

Rectorite (syn. *allevardite)* — Figs. 339—346

The structure of rectorite comprises a regular alternation of paragonite and montmorillonite layers in a 1:1 ratio.

In the TEM the rectorite is depicted as extremely thin lathshaped particles, which often develop folds during preparation (Figs. 339, 340 and 345). Montmorillonitic features, such as curled edges and folds are especially evident in the replicas (Figs. 341, 346). In the SEM of the fracture fabric bent lamellae are observable (Figs. 342—344).

3.6. Tetrahedral layer-structures with chain-like units

3.6.1. *Palygorskite (syn. attapulgite)* — Figs. 347—368

$(Mg, Al)_2[(OH)|Si_4O_{10}] \cdot 2H_2O + 2H_2O$

Palygorskite occurs as a weathering product of Mg-containing rocks. It may also have formed hydrothermally, in which case the mineral occurs as joint linings or as seamlike layers. It also occurs in certain lime marls, where it formed from solutions that resulted from rock weathering in the salty lagoons of the arid regions.

Palygorskite has a tetrahedral layer structure with pyroxene double chains along [100], which are responsible for its fibrous habit. Rods and parallel intergrowths of rod-like particles have been reported in the literature. Part of the water ($\cdot H_2O$ of the structural formula) is crystal water and the other part ($+ H_2O$) is bound in the channels of the crystals (zeolitically).

The TEMs of the palygorskites from various localities (e.g. Figs. 347, 357, 364 and 365) make evident that the fibrous particles have one or more channels, which permit the zeolitic binding of water. Moreover, parallel intergrowths can be seen, especially in the replicas (Figs. 348, 355, 367). The replica shown in Figure 367 was prepared parallel to, and that of Figure 368 normal to, the fibres. Depending on occurrence and genesis the tubules have widely differing length/width ratios. The SEMs (Figs. 349, 350, 353, 361, 362) show a feltlike fabric composed of fibres that are parallel intergrowths of the much finer tubules, as a comparison with the TEMs reveals. Moreover, it has been established that the length of the intergrown tubules seen in the SEM varies with the length of the individual tubules seen in the TEM.

3.6.2. *Sepiolite (syn. xylotile)* — Fig. 369—376
$Mg_4[(OH)_2|Si_6O_{15}] \cdot 2H_2O + 4H_2O$

Most sepiolite is a weathering product of serpentinites, but this mineral may also form under marine conditions if Mg-rich solutions are present. Thus, sepiolite is much more common in sedimentary rocks than supposed earlier.

Sepiolite is built up of tetrahedral layers with pyroxene triplechains along [100] and contains crystal water as well as water which is bound zeolitically. The ctructural similarity to palygorskite leads to a palygorskite-like morphology in sepiolite. Sepiolite, too, occurs in very fine and usually very long tubules, which tend to form parallel intergrowths (Figs. 369, 375). Examples of extremely long and fine tubules can be seen in Figures 375 and 376. The SEMs (Figs. 371 and 372) show a feltlike fabric composed of partly rather long fibres that represent parallel intergrowths of fine tubules.

3.7. Admixtures

3.7.1. *Iron and titanium minerals* — Figs. 377—430

Iron and titanium minerals are very common admixtures in clayey rocks. It may be rather difficult to detect these minerals, especially if they occur in very small amounts (e.g. in the white kaolins) because it is only by chance that such particles are detectable by electron microscopy. By high-degree magnetic separation or chemical treatment the iron and titanium minerals can be enriched and thus be detected more conveniently. Some iron and titanium minerals are naturally enriched in certain grain fractions of clays and can therefore be detected without difficulty (e.g. pyrite and siderite in the grain fractions $> 20\ \mu m$). The following minerals are discussed in detail:

- goethite α-FeOOH
- lepidocrocite γ-FeOOH
- hematite α-Fe_2O_3
- anatase TiO_2
- rutile TiO_2
- pyrite FeS_2
- siderite $FeCO_3$.

Goethite (Figs. 377—386) occurs as elongate prismatic crystals, which commonly form the stellated intergrowths by which goethite is easy to identity. In the Caminau kaolin there are oriented intergrowths of goethite and kaolinite, the former of which has been identified by means of X-rays (Figs. 407, 408).

Lepidocrocite (Fig. 406) generally forms tabular crystals. In our specimens, however, it occurs as very finc grains that form crusts on other minerals.

Hematite (Figs. 387—392) usually occurs in oblong aggregates of very fine fibres but also in globular aggregates. Due to the intense red of the hematite, even very small amounts produce an appreciable coloration of the clays.

Anatase (Figs. 393—401) is found in nearly all clayey rocks and usually is enriched in the grain fraction $< 2\ \mu m$. Our preparation technique was as follows: after anatase enrichment by high-degree magnetic separation, the remaining silicates were dissolved in hexafluoride titanic acid (after SAYIN & JACKSON 1975). By this method the widely varying morphology of the anatase becomes recognizable. Common forms are tetragonal bipyramids and barrel-shaped or globular aggregates. The aggregates are composed of very fine-grained particles, which may encrust other minerals, such as rutile (Fig. 401).

Rutile (Figs. 401—404) forms the customary idiomorphic needles.

Pyrite (Figs. 409—428) occurs in several forms and is found in clays and kaolins. Common forms are cubes and octahedrons, which make up the aggregates found in pyritized fossils and pyritic framboids. Frequently, the crystal faces show signs of decomposition, and secondary sulfates are present (Figs. 426, 428). Some of the pyrites weather rather rapidly, especially if the ground water level in the clay pits has been lowered, thus exposing the pyrite to exogenous influences. This leads to the formation of Fe-sulfates (see 3.7.2.) in the immediate vicinity of pyrite crystals.

Siderite (Figs. 421—430) is found in many kaolins and clays and occurs there as individual crystals or in concretions, frequently in paragenesis with pyrite. In the Friedland clay the siderite crystals have a coating of fine sulfate needles, which originated from pyrite that was originally attached to these siderite crystals (Figs. 429, 430).

3.7.2. Sulfates — Figs. 431—484

In clay rocks secondary minerals may precipitate if the ion concentrations in the pore solutions exceed the solubility products. Depending on the ions contained, various salts are formed, especially sulfates, carbonates and phosphates. These salt crusts typically occur in deserts and arid regions. But comparable processes take place also in exposures of clays in other climatic zones. These processes can be subdivided as follows:

— precipitation of salts from the ground water on the rock surface, on clefts and in pores. Depending on duration and rate of evaporation, salt crusts of variable thickness are formed. Such secretions may also form in the laboratory when fresh rock samples loose their natural moisture.
— weathering of sulfides, particularly of pyrite in clays which are located in the oxidized zone above the ground water level. Then, various iron sulfates precipitate onto, and in the immediate vicinity of, the pyrite. Frequently, pyrite weathering sets in as a result of lowered ground water level due to mining operations.
— precipitation of sulfates, carbonates and phosphates due to ion mobilization in clay diagenesis.

Well kown examples of precipitations in clayey rocks are gypsum, jarosite and vivianite. Less well known are other salts, such as aluminite (VON FREYBERG 1922, PICARD 1922), alunite (FELDHAUS 1981) and various phosphates (SZPILA 1981). A considerable number of such secondary minerals have been described by STÖRR et al. (1983).

This atlas considers those secondary minerals that the authors have found in clays, kaolins and bentonites and which usually form submicroscopic crystals. Such crystals cannot be seen under the light microscope so that an electron microscope, in particular the SEM, has to be used. With few exceptions the secondary minerals shown here have been identified by X-ray techniques. The following list includes the minerals that have been detected in clays by the authors:

— thenardite $\alpha\text{-}Na_2[SO_4]$

Anhydrous sulfates with foreign anions

– alunite	$KAl_3[(OH)_6 \vert (SO_4)_2]$
– jarosite	$KFe_3^{3+}[(OH)_6 \vert (SO_4)_2]$
– natrojarosite	$NaFe_3^{3+}[(OH)_6 \vert (SO_4)_2]$

Hydrous sulfates without foreign anions

– szomolnokite	$Fe[SO_4] \cdot H_2O$
– rozenite	$Fe[SO_4] \cdot 4H_2O$
– melanterite	$Fe[SO_4] \cdot 7H_2O$
– coquimbite	$Fe_2^{3+}[SO_4]_3 \cdot 9H_2O$
– roemerite	$Fe^{2+}Fe_2^{3+}[SO_4]_4 \cdot 14H_2O$
– gypsum	$Ca[SO_4] \cdot 2H_2O$
– mirabilite	$Na_2[SO_4] \cdot 10H_2O$

Hydrous sulfates with foreign anions

– parabutlerite	$Fe^{3+}[OH \vert SO_4] \cdot 2H_2O$
– aluminite	$Al_2[(OH)_4 \vert SO_4] \cdot 7H_2O$
– copiapite	$(Fe^{2+}, Mg)\ Fe_4^{3+}[OH \vert (SO_4)_3]_2 \cdot 20H_2O$
– alumino-copiapite	$(Fe^{2+}, Mg)\ (Fe^{3+}, Al)_4\ [OH \vert SO_4)_3]_2 \cdot 20H_2O$
– sideronatrite	$Na_2Fe^{3+}[OH \vert (SO_4)_2] \cdot 3H_2O$
– metasideronatrite	$Na_2Fe^{3+}[OH \vert (SO_4)_2] \cdot 1^1/_2H_2O$

Thenardite (Figs. 431, 432) is a very common secondary mineral occurring as spherical particles and long fine fibres. The formation of Na-sulfates typically occurs in clays and bentonites containing exchangeable Na-ions. Here, thenardite fibres also occur in paragenesis with other secondary minerals, such as gypsum (Fig. 468). Thenardite fibres can also be seen in products manufactured from Na-clays.

Alunite (Figs. 433, 434) forms white earthy aggregates up to several centimeters in diameter, which are encountered in kaolins and clays. The alunite crystals are pseudocubic or rhombohedral.

Jarosite (Figs. 435–444) is a widespread mineral that occurs in sulfur-coloured botryoidal aggregates. Usually, jarosite crystals have a rhombohedral habit, but a few are elongate or platy. Occasionally, jarosite crystals form rosettes. There are no morphological differences observable between the potassium and sodium variety.

Szomolnokite (Figs. 445–453) occurs in clayey rocks, particularly in the vicinity of weathered pyrite, and usually in association with rozenite. The crystals occur in fibrous form, as long needles, as idiomorphic crystals (Figs. 451, 452) and in globular aggregates.

Rozenite (Figs. 453, 454, 459–464) occurs as white crusts on the surfaces and in the vicinity of weathered pyrites, and is commonly associated with other iron sulfates. The morphological habit comprises fibres and fine-grained crusts that may display contraction cracks (aged gels).

Melanterite (Figs. 465, 466) was identified as a constituent mineral of the so called blue stone of the Sedlec kaolin, Czechoslovakia. The mineral occurs in stalklike and platy crystals.

Coquimbite (Figs. 455–457) is found as white crusts on pyrite in kaolin. It has an acicular or fibrous habit and is associated with copiapite.

Roemerite (Figs. 467, 479) occurs as white needles and crusts on weathered pyrite, together with gypsum and rozenite.

Gypsum (Figs. 468–477) is encountered in kaolins as crystals up to 10 centimeters in length and some samples even longer. In white and coloured crusts it is frequently associated with other sulfates. Most crystals have a prismatic habit; twinning is common. Sometimes globular aggregates are seen, consisting of small idiomorphous crystals

(Fig. 470) or rosettes of stalklike crystals partly with skeletal growth (Fig. 471). Gypsum aggregates are often found at the exits of pore-water channels (Fig. 473).

Mirabilite (Fig. 478) occurs as translucent fibrous crystals on clays. Due to dehydration already at room temperature the mirabilite alters to thenardite, the fibres decomposing to a powdery mass.

Parabutlerite (Fig. 476) occurs together with gypsum in yellow and green crusts.

Aluminite (Figs. 477, 478) occurs as white nodules in the covering strata of the kaolin at Fuchsberg Hill near Halle, GDR, partly in association with alunite and gypsum. The aluminite of this locality was first described by DIECK (1859). The crystals have a tabular habit.

Copiapite (Figs. 457, 479—482) and alumino-copiapite (Fig. 482) are found in kaolins and clays in the immediate vicinity of weathered pyrites, where they occur in thin platelets which may form rosette-shaped intergrowths.

Sideronatrite and metasideronatrite (Figs. 483, 484) are associated with each other in white, yellow and brown efflorescences in kaolins. Morphologically, they occur as needles which may be arranged in tufts.

3.7.3. *Phosphates* — Figs. 485—490

Phosphates have primarily been described as major concretions in clayey rocks which, however, will not be discussed here. In clays and kaolins the relatively common vivianite ($Fe_3^{2+}[PO_4]_2 \cdot 8H_2O$) lines joints or impregnates the rock, which, when fresh, has a dark blue coloration which turns brown on exposure to air. This vivianite variety forms elongate and compact crystals (Figs. 485—488). Amorphous white, brownish or greenish phosphates occur locally as precipitates on joints or surfaces of various clayey rocks, e.g. in alum schist in the form of globular aggregates which may have shrinkage cracks (Figs. 489, 490). Such phosphates have a conchoidal fracture fabric and can easily be confused with allophane. Variscite ($Al[PO_4] \cdot 2H_2O$) occurs as globular aggregates along joints (Fig. 492). Crandallite ($CaAl_3[PO_4]_2(OH)_5$) occurs as platy idiomorphic crystals (Fig. 396) in the fine-grained fractions of kaolins. Recently, SZPILA (1981) has described several phosphates, many of them rather rare ones, which occur as trace minerals in the fine-grained fractions of kaolins. Using special enrichment techniques, SZPILA isolated some crystals and identified them as phosphates by means of microprobe analysis. These rare crystallites can easily be overlooked and certainly are more widespread than assumed earlier.

3.7.4. *Allophane* — Figs. 493—502

$Al_2O_3 \cdot SiO_2 \cdot H_2O$ (Al:Si $\approx$ 1:1)

Allophane occurs in zones of weathering and as secretions on joints and rock surfaces. Many of the substances originally referred to as allophane have been found to be clay minerals or phosphates (e.g. variscite). However, our samples from Gräfenthal, GDR (type locality, STROHMEIER & KOLKMANN 1796) and Zuckmantel, Czechoslovakia, definitely contained allophane, which was indicated by the botryoidal surface of the amorphous particles observable in the replicas and SEMs. The diameters of the spheres range from 0.02 µm up to some microns. Some of the Gräfenthal allophane has crystallized to hydrargillite, which has a somewhat different crystal habit (Figs. 495—498).

As an admixture in clay rocks allophane may easily be overlooked. A reliable identification technique is selective dissolution (after HARMAN 1969), i.e. the allophane is dissolved from the replicas while the clay minerals are left in place. In the oxidizing zones of ore deposits allophane may be easily confused with amorphous phosphates, which under these conditions precipitate as gels showing a morphology similar to that of allophane (Figs. 489—490).

3.7.5. *Opal-CT and amorphous silica* — Figs. 503—510

Opal-CT and amorphous silica occur in some clayey and carbonate-rich sediments and also in bentonites. Well known examples are the authigenic opal-CT lepispheres found in sediments (FLÖRKE et al. 1976) and in the anauxite of Bílina, Czechoslovakia (KELLER 1982). However, it is far more difficult to identify the aggregates of colloidal amorphous silica or opal-CT that has been found in the kaolin of Ledce, Czechoslovakia (Fig. 508) (BABUREK & STÖRR 1967), in the anauxite of Bilina (Figs. 503—505) (BEUTELSPACHER & VAN DER MAREL 1968), in the non-carbonates of the Cretaceous chalk of the Isle of Rügen (Figs. 509, 510, 581, 582) (STÖRR 1966), or in the Pacific deep sea clay (Fig. 591).

3.7.6. *Carbonates* — Figs. 511—514

Carbonates are well known as secondary minerals in clays and kaolins, and occur primarily in the form of siderite ($FeCO_3$), calcite ($CaCO_3$) and dolomite ($CaMg[CO_3]_2$) concretions and also as joint fillings. The idiomorphous siderite has already been discussed in the section on the iron minerals (Figs. 421—430). Calcitic joint fillings are usually hypidiomorphous (Fig. 511). Compact calcite can be recognized by its etching pits (Fig. 512). The calcitic concretions shown in Figure 512 were found on joint faces of phosphorite concretions in the Friedland clay.

3.8. Clay rocks

Natural occurrence of clay minerals in a mono-mineralic form is a rarity. In rocks we generally find mixtures of clay minerals and other minerals, and identification may be difficult. The following list includes those clay rocks whose mineral composition and texture are discussed:

— kaolins
— bentonitic and montmorillonitic clays
— kaolinitic clays
— illitic clays

mixed-layer clays and carbonates

— recent deep-sea clays
— clay minerals in coals
— diatomaceous clays

3.8.1. *Kaolins* — Figs. 515—538

The main constituent mineral of the kaolins is kaolinite, whereas the minor constituents include mica, halloysite, muscovite-montmorillonite mixed-layers, smectite, iron and titanium minerals, phosphates and sulfates as well as feldspar relics and quartz. Most of these minerals have been discussed in detail so that only some specific data are added here. In TEM the diagnostic criterion of halloysite is its tubular form (Figs. 515, 516). In the SEMs, however, the tubules are not always sufficiently resolved. They appear as elongate crystals that are difficult to differentiate from other elongate crystal forms, such as those of the muscovite-montmorillonite mixed-layers (Figs. 517, 518, 525—536). In the weathering profiles, particularly in the transition zones to the unaltered rock, there are aggregates of strongly disordered kaolinites, which have neither typical pseudohexagonal contours nor fully developed tubular forms. The halloysites develop from these particles. In such transition zones the halloysites may be considerably enriched (Figs. 533—538).

Muscovite-montmorillonite mixed-layers and illite can be recognized in the kaolins by their slat-shaped morphology. Because the slats are rather thin they may be easily overlooked in the TEM. Larger mixed-layer slats develop particularly in the pores of the kaolins, where these slats find favourable growth conditions so that they become as long as several microns and are easy to differentiate from the kaolinites in the SEM. The growth of the mixed-layer slats is particularly well observable in weathered pitchstones (Figs. 525–530).

Partly weathered feldspars are widespread in kaolins. The former can be recognized by the cleavage cracks and the texture of the decomposing feldspar (Figs. 531, 532). With increased weathering more and more clay minerals are formed until the original texture is lost (Figs. 533–538).

Quartz is a common constituent of the kaolins. It usually forms xenomorphic crystals, and only seldom idiomorphic ones. Both forms are easy to identify in the SEMs and the replica (Figs. 523, 524).

3.8.2. *Bentonites* — Figs. 539–562

Montmorillonite is usually the main constituent mineral of bentonites which may also contain minor volcanic glass, cristobalite and other primary minerals. In the TEM the montmorillonite is easily recognizable by the swelling-related features described earlier. The fabric of the bentonites varies considerably, reflecting the different genetic conditions that have given rise to different bentonite varieties.
These varieties in clude:

- bentonites (in the restricted sense)
- residual bentonites
- montmorillonitic clays
- hydrothermal bentonites.

In the restricted sense bentonites are tuffs that were altered in marine environments by montmorillonitization of the volcanic glass. These bentonites clearly reflect the fabric of the volcanic glass. The onset of montmorillonitization is indicated by the bent montmorillonite platelets, which may form rosette-shaped aggregates. With increasing montmorillonitization of the glass the original glass texture gradually disappears (Figs. 539–548), until textural differences are no longer observable. This stage of alteration has been reached in the Wyoming bentonite, in which glass particles are rarely seen (Figs. 549–552). The weakly oriented fabric may be of diagenetic origin. In the bentonite beds the fabric is much more weakly oriented than in the surrounding sediments, as e.g. in the Belle Fourche Shale (see below).

Residual bentonites, which represent montmorillonitized basic or intermediate igneous rocks, show, depending on their various parent rocks, great variations in mineral composition and often contain large amounts of Fe-rich montmorillonites and many relics of primary minerals, which frequently are only partially montmorillonitized (Figs. 557–562).

Montmorillonitic clays (Figs. 553–554) are characterized by a pronounced orientation of their constituent particles. Such fabric is a characteristic feature of the Belle Fourche Shale, which was sampled only 50 cm above the Clay Spur Bentonite.

Hydrothermal montmorillonites have a more homogeneous fabric. In the montmorillonite of Michalkovo, Bulgaria, slat-shaped crystals have grown into the pores (Figs. 204–210).

3.8.3. Clayey sediments

The mineral composition of sedimentary clays shows major variations which reflect the differing genetic conditions. Most of the clayey sediments are fine-grained, with grain diameters being only fractions of a micron. Reliable identification of such fine particles requires the use of a TEM.

Plastic kaolinitic clays (ball clays; Figs. 561–565) occur in idiomorphous pseudohexagonal crystals. According to their genesis (redeposited kaolinitic residuum) their mineral composition is relatively uniform, with only small amounts of clayey admixtures, such as illite, mica, montmorillonite or mixed-layers. In the lithified fire-clays, however, there are kaolinites, some of which have irregular borders (Fig. 565).

Most of the marine clayey sediments contain three-layer silicates. Dominant in the plastic sediments are the muscovite-montmorillonite mixed-layers, which can be recognized by their slat-like morphology in the TEMs (Figs. 566, 567, 579–582). These mixed-layers, however, do not everywhere occur in slats, but also in an irregular nepheloid shape indicative of illite (Figs. 571, 572). In the SEMs the slats have not been resolved and only the irregular scales (Figs. 575, 576) and their orientation are visible. The embedded quartz grains are clearly differentiable from the surrounding clay (Figs. 577, 578), though they are coated by clay minerals which also are clearly visible (Fig. 577).

In the micrographs of limestones the clay minerals are difficult to recognize among the much more coarse-grained carbonates. Such clay minerals are easier to identify if they occur in clay streaks, as can be seen in Figures 555 and 556, which show montmorillonite streaks in an organic clastite.

Usually the clay minerals can be recognized only after carbonate removal (Figs. 579 to 582).

Many Tertiary and Quaternary clays contain diatom tests (Fig. 583). In the kieselgurs such tests are the main constituent and can be recognized by their characteristic shape (Figs. 583–588). As decomposition proceeds, only aggregates of amorphous silica are left, as e.g. in the recent deep-sea clay (Fig. 591).

The mineral composition of recent deep-sea clays is rather complex. The TEMs primarily show irregularly shaped muscovite-montmorillonite mixed-layers and only sparse slats (Figs. 589–592). Kaolinite (Fig. 589), mica (Fig. 592) and possibly montmorillonite (Fig. 591) may also occur in such clays. A detailed investigation of the clay minerals in coals has started only recently, after GLUSKOTER (1965) developed the low-temperature ashing (LTA) technique. By this method the organic constituents are ashed in an oxygen plasma at $< 425\,^{\circ}K$ so that the unchanged clay minerals are enriched in the ash. Figures 593–598 show the minerals contained in the ashes of brown coals after LTA (STÖRR & ADOLPHI 1981, ADOLPHI & STÖRR 1985). These minerals include kaolinite, mica, pyrite, jarosite and artefactic sulfates from organic Ca and S represented by anhydrite (crystals in Figs. 594, 596) and amorphous Ca-sulfate or bassanite (grains in Figs. 594, 596).

Low-temperature ashing can also be used for "etching" SEM-specimens by exposing them for some minutes to the oxygen plasma, which oxidizes the organic constituents, while the minerals are left unchanged. Figures 599 and 600 show the texture of kaolinite in hard coal and Figures 601 and 602 that of bituminous shale. In the latter the bitumen particles have been oxidized so that the mineral texture (montmorillonite with interlayered pyrite) has become visible.

References

ADOLPHI, P.; STÖRR, M.: Glow discharge exited low temperature ashing — A new technique for separating mineral matter of coal. — Fuel, London 6 (1985) 2, pp. 151—155.

ARDENNE, M. v.; ENDELL, K.; HOFMANN, U.: Untersuchungen feinster Fraktionen von Bentoniten und Tonböden mit dem Universal-Elektronenmikroskop. — Ber. dt. keram. Gesellsch., Berlin 21 (1940), pp. 205—227.

BABUREK, J.; STÖRR, M.: Mineralogische und technologische Untersuchungen des Kaolins von Ledče, Plzeň Basin, Czechoslovakia. — Keram. Z., Freyburg/Br. 18 (1968).

BASSET, G. A.: Nucleation of evaporated metal layers on single crystal substrates. — In: IX. internat. Congr. Electr. Microsc. Berlin [West] 1958, Bd. I. — Berlin [West], 1960, pp. 512 to 515.

BATES, T. F.; COMER, J. J.: Electron microscopy of clay surfaces. — Proc. 3rd nation. Conf. Clays and Clay Minerals, Houston (1954), 1—9. — Washington, 1955.

BETHGE, H.: Über Grenzflächenreaktionen zur Entwicklung von üblicherweise im Abdruck nicht erkennbaren Oberflächenstrukturen. — In: IX. internat. Congr. Electr. Microsc. Berlin[West] 1958, Bd. I. — Berlin [West], 1960, pp. 409—415.

BETHGE, H.; KELLER, K. W.; STENZEL, H.: Zur elektronenmikroskopischen Sichtbarmachung unterschiedlicher Bildungs- und Adsorptionseigenschaften an Lamellenstufen auf NaCl-Kristallen. — Naturwissenschaften, Berlin [West] et al. 49 (1962), pp. 152—153.

BURST, J. F.: Mineral heterogenity in "Glauconite" pellets. — Amer. Mineral., Menasha 43 (1958) pp. 481—497.

BYSTRÖM, A. M.: Mineralogy of the Ordovician bentonite beds at Kinnekulle, Sweden. — SGU, Årsbok, Stockholm 48 (1954), pp. 1—62.

ČEKIN, S. S.; FINKO, V. I.: SAMOTOIN, N. D.: Nekotorye osobennosti preobrazovanija polevych špatov i muskovita v kaolinovych korach vyvetrivanija. — In: Kaoliny. — Moskva: Izd. Nauka, 1974, pp. 88—106.

BEUTELSPACHER, M.; VAN DER MAREL, M. W.: Atlas der Elektronenmikroskopie der Tonminerale und ihrer Beimengungen. — Amsterdam, 1968, 333 pp.

EITEL, W.; SCHUSTERIUS, L.: Die Auswertung übermikroskopischer Bilder zur Bestimmung der Kornverteilung von Tonen. — Naturwissenschaften, Berlin [West] et al. 28 (1940) 19, pp. 300—303.

FELDHAUS, D.: Zum Vorkommen von Alunit und Jarosit in mitteloligozänen Sedimenten am Fuchsberg bei Marl (nördlich von Halle/Saale). — Hall. Jb. Geowiss., Gotha/Leipzig 6 (1981), pp. 107—108.

FLÖRKE, O. W.; Hollmann, R.; VON RAD, U.; RÖSCH, H.: Intergrowth and twinning in opal-CT lepispheres. — Contrib. Mineral. Petrol. — Heidelberg, 1976.

FREYBERG, B. v.: Der Aluminit von Halle, eine an das Tertiär gebundene Mineralneubildung. — Steinbruch, Berlin 16 (1921) 6, pp. 98—100.

GARD, J. A.: Electron-optical investigations of clays. — Mineral. Soc., Monograph 3. — London, 1971.

GLAUERT, A. M. (Ed.): Practical methods in electron microscopy. — Amsterdam, New York, Oxford: North Holland Publ. Co., 1976.

GLUSKOTER, H. J.: Electronic low temperature ashing of bituminous coals. — Fuel, London 44 (1965), pp. 285—291.
GRIM, R. E.: Clay Mineralogy. — New York, 1953, 384 pp.
GRIM, R. E.; KULBICKI, G.: Montmorillonite: High temperature reactions and classification. — Amer. Mineral., Menasha 46 (1961), pp. 1329—1365.
GRITSAENKO, G. S.; RUDNITSKAYA, E. S.; GORSHKOV, A. I.: Electron Microscopy of Minerals. — Moscow: Izd. Akad. Nauk SSSR, 1961.
GRITSAENKO, G. S.; SAMOTOIN, N. D.: The decoration method applied to the study of clay minerals. — Proc. internat. Clay Conf., Jerusalem 1 (1966), pp. 391—400.
GÜVEN, N.; PEASE, R. W.: Electron-optical investigations on montmorillonites. II. Morphological variations in the intermediate members of the montmorillonite-beidellite series. — Clays and Clay Minerals, London 23 (1975), pp. 187—191.
HAMDY, S.; RADCZEWSKI, O. E.: Struktur und Bildung leistenförmiger Illite. — Keram. Z., Bad Honnef 26 (1974) 6, pp. 315—324.
HARMAN, M.: Die selektive Löslichkeit als neue Methode zur Identifizierung des Allophans unter dem Elektronenmikroskop. — Geol. zb. 20,1 (Bratislava), pp. 153—162.
HAUSER, E. A.: The colloid science of important clay minerals. — 3rd nation. Conf. Clays and Clay Minerals, Houston (1954), pp. 442—472. — Washington, 1955.
HENNING, K.-H.: Mineralogische Untersuchung des eozänen Tones der Lagerstätte Friedland (Bezirk Neubrandenburg). — Ber. dt. Gesellsch. geol. Wiss., B., Berlin 16 (1971) 5, pp. 5—39.
— Elektronenmikroskopische Teilchengrößenanalyse von Kaoliniten. — In: Tonminerale. Genese, Lagerstätten, industrielle Bedeutung und Nutzung. — Berlin: Akademie-Verlag, 1976, pp. 75—82, (Schriftenr. geol. Wiss.; 5).
— Ergebnisse morphometrischer Messungen mit dem Elektronenmikroskop an Kaoliniten von Podlesí (ČSSR). — 8. Conf. Clay Min. and Petr., Teplice 1979, pp. 145—151.
HENNING, K.-H.; LANDGRAF, K. F.: Über Beziehungen zwischen Morphologie und Kristallchemie der dioktaedrischen Dreischichtsilikate. — Beitr. Mineral. Technol. Tone. — Greifswald, 1978, pp. 43—64.
HENNING, K.-H.; REINHARDT, B.: Zur Morphologie von Papierstreichkaolinen. — Zellstoff und Papier, Leipzig 28 (1979) 5, pp. 210—215.
HENNING, K.-H.; STÖRR, M.; ZWAHR, H.: Ergebnisse der elektronenmikroskopischen Untersuchung mit der Golddekorationsmethode an Kaolinit- und Glimmermineralen der Kaolinlagerstätte Caminau. — Z. geol. Wiss., Berlin 12 (1984) 4, pp. 532—528.
HUFF, W. D.: Morphological effects on illite as a result of potassium depletion. — Clays and Clay Minerals, London 20 (1972), pp. 245—301.
JASMUND, K.: Die silikatischen Tonminerale. — Weinheim, 1955.
KAY, D. (Ed.): Techniques for electron microscopy. 2nd Ed. — Oxford: Blackwell Scientific Publications, 1965.
KELLER, W. D.: Scan electron micrographs of kaolins collected from diverse environments of origin. — Clays and Clay Mineral, London
I. 24 (1976), pp. 107—113;
II. 24 (1976), pp. 114—117;
III. 24 (1976), pp. 262—264;
IV. 25 (1977), pp. 311—345.
— Effects of micro-sized mixtures of kaolin minerals on properties of kaolinites. — Clays and Clay Minerals, London 26 (1978), pp. 384—386.
— Classification of kaolins exemplified by their textures in scan electron micrographs. — Clays and Clay Minerals, London 26 (1978), pp. 1—20.
KÖSTER, H. M.: High resolution electronmicroscopy of clay minerals. — 5th Euroclay Conference, Praha 1983.
MERING, J.; OBERLIN, A.: The smectites. — In: The Electron-optical investigations of clays, ed. by J. A. GARD, Mineral. Soc., Monograph 3. — London, 1971, pp. 193—229.
MÜLLER, W.; BEHNE, W.: Zur Morphologie des Glaukonits. — Naturwissenschaften, Berlin [West] et al. 24 (1954), 575 pp.
MURRAY, H. H.: Genesis of the Georgia kaolin. — Proc. 6th internat. Symp. Genesis Kaolin, Tokyo 1976.
PICARD, E.: Erläuterungen zur geologischen Karte von Preußen und benachbarten Bundesstaaten, Lief. 5, Bl. 28 Halle/Saale (Nord). — Berlin: Preuß. geol. Landesanst., 1922, 59 pp.

PICHT, J.; HEYDENREICH, J.: Einführung in die Elektronenmikroskopie. — Berlin: Verlag Technik, 1966, 400 pp.

REKSHINSKAYA, L. G.: Atlas der Elektronen-Mikrophotographien von Tonmineralen und ihren natürlichen Assoziationen in Sedimentgesteinen. — Moscow, 1966, 232 pp.

REIMER, L.: Elektronenmikroskopische Untersuchungs- und Präparationsmethoden. — Berlin [West], Heidelberg, New York: Springer-Verlag, 1967, 598 pp.

ROSS, C. S.; HENDRICKS, S. B.: Minerals of the montmorillonite group. — Geol. Survey, Prof. Paper, Washington **205** B (1945), 1—77 pp.

RUSKO, J. A.: Kaolinisierung und Kaoline des ukrainischen Schilds. — Kiev: Naukova dumka, 1976, 158 pp.

SCHIMMEL, G.; VOGELL, W.: Methodensammlung der Elektronenmikroskopie. — Stuttgart: Wiss. Verlagsgesellsch., 1975, (aperiodical series).

STÖRR, M.; ADOLPHI, P.: Zur plasmachemischen Niedertemperatur-Veraschung von Braunkohlen. — Z. angew. Geol., Berlin **29** (1983) 6, pp. 272—277.

STÖRR, M.; BECK, H.; ULLRICH, B.; ZWAHR, H.: Sekundärminerale in Kaolinen und Tonen. — Z. geol. Wiss., Berlin **11** (1983) 9, pp. 1101—1124.

SUDO, T.; SHIMODA, S.; YOTSUMOTO, H.; AITA, S.: Electron micrographs of clay minerals. — Kodansha-Elsevier, 1981, 203 pp.

SUNAGAWA, J.; KOSHINO, Y.: Growth spirals of the kaolin minerals. — Amer. Mineral., Menasha **60** (1975), pp. 407—412.

SZPILA, K.: Some problems on the geochemistry of clays. — Tonmineralkonferenz, Vortrag 5, Greifswald 1981.

SZPILA, K.; WIEWIORA, A.; GAODMSKI, 00: Preliminary Investigation of Kaolinite-Smectite from Jeglowa, Lower Silesia. — Bull. Acad. Pol. d. Sci., Sev. d. Sci La tere XX (1972) 1, pp. 15 to 24.

WEAVER, C. E.; POLLARD, D.: The chemistry of clay minerals. — Amsterdam, 1973, 213 pp.

WIEWIORA, A.: A mixed-layer kaolinite-smectite from Lower Silesia, Poland. — Clays and Clay Minerals, London **19** (1971).

ZWAHR, H.; STÖRR, M.: Zur Ausbildung von Schadstoffen in Kaolinen der DDR und technologische Möglichkeiten ihrer Abreicherung durch Hochgradientmagnetseparation (HGMS). — Silikattechnik, Berlin **32** (1981) 9, pp. 262—265.

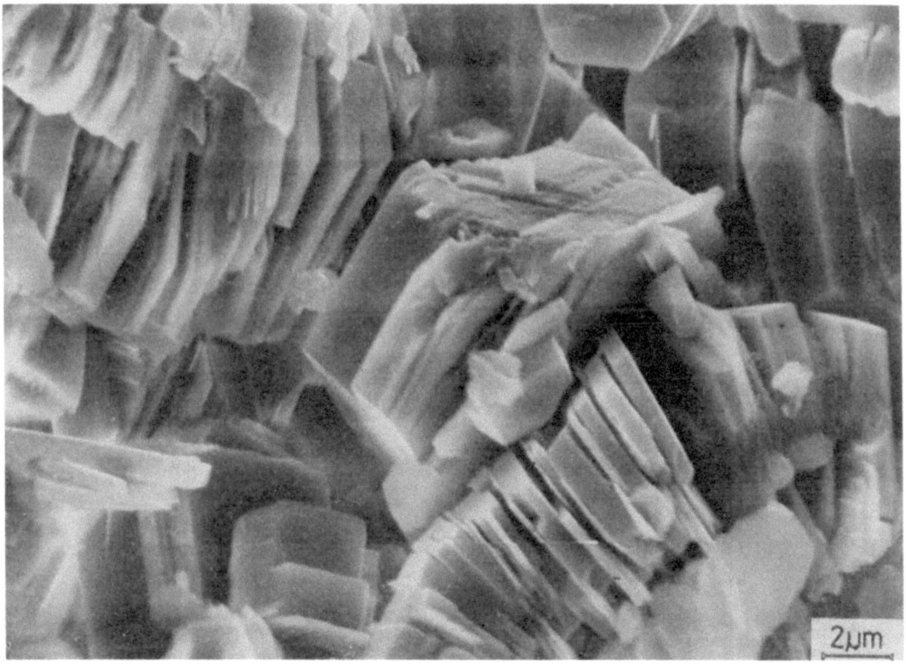

Figs. 7, 8. SEMs: *Kaolinite* from cracks in a siderite concretion in the shale of the Brazil Formation, Pennsylvanian, Brazil Coal Mine, Indiana, USA

Fig. 8.

Figs, 9, 10. Replicas: *Kaolinite* from cracks in a siderite concretion in the shale of the Brazil Formation, Pennsylvanian, Brazil Coal Mine, Indiana, USA

Fig. 10.

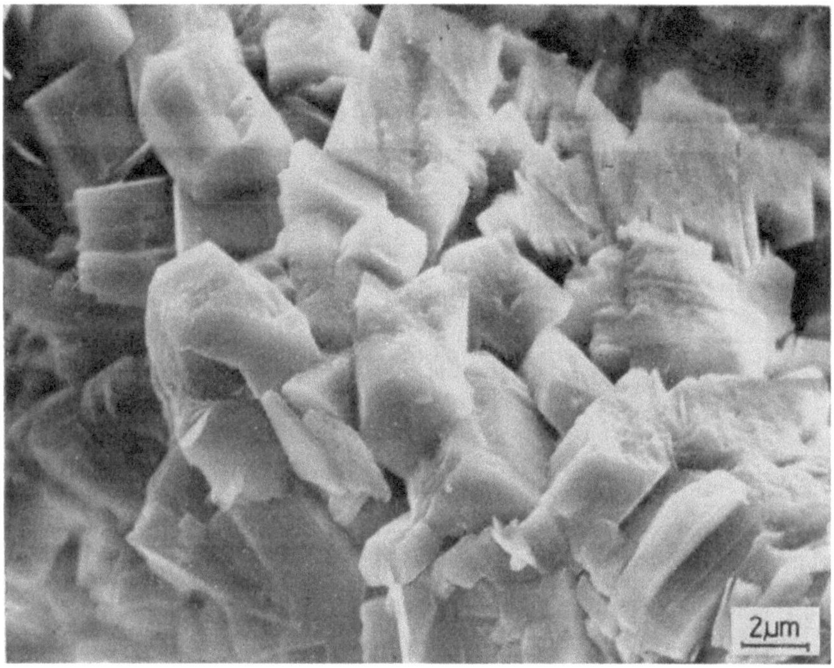

Figs. 11, 12. SEMs: *Kaolinite* from cracks in the rock coal of the Brazil Formation, Pennsylvanian, Brazil Coal Mine, Indiana, USA

Fig. 12.

Figs. 13, 14. SEMs: *Kaolinite* from cracks in a siderite concretion in the Belle Fourche Shale Crook County, Wyoming, USA

Fig. 14.

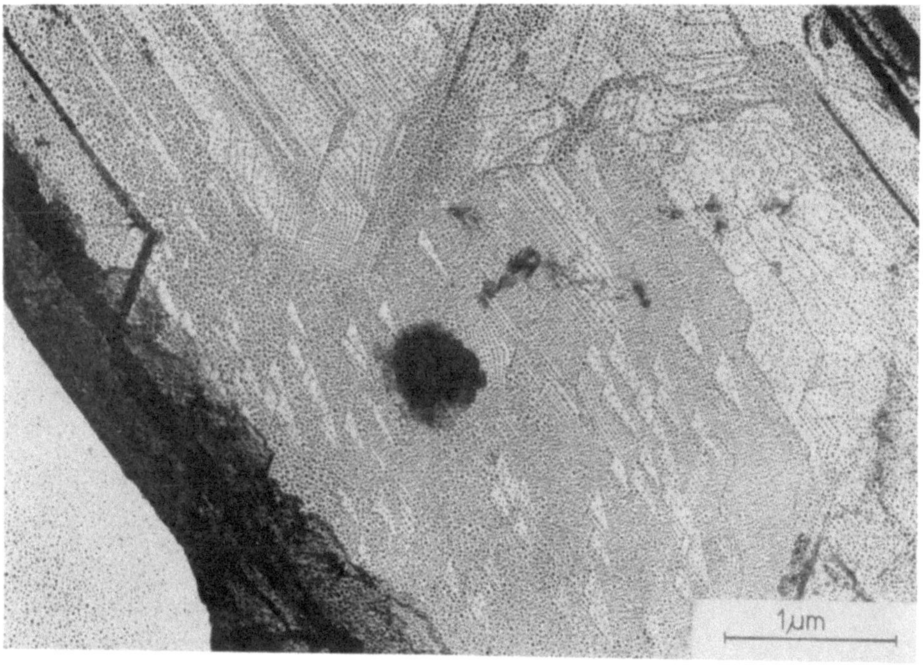

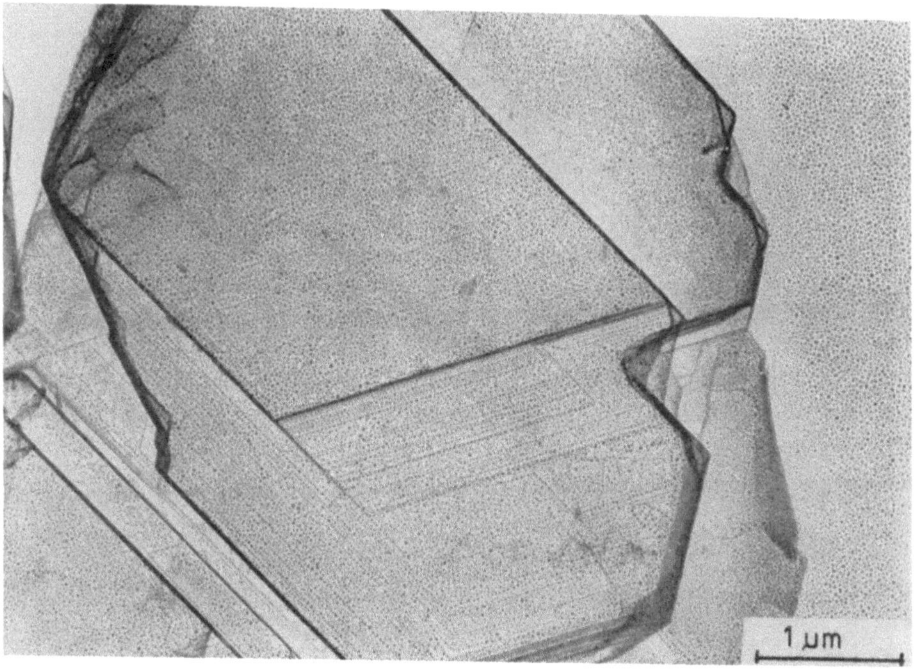

Fig. 15. Replica: *Kaolinite* with etching figures, gold decorated, from cracks in a siderite concretion in the Belle Fourche Shale, Crook County, Wyoming, USA

Fig. 16. Replica: *Kaolinite*, gold decorated, from cracks in a siderite concretion in the Belle Fourche Shale, Crook County, Wyoming, USA

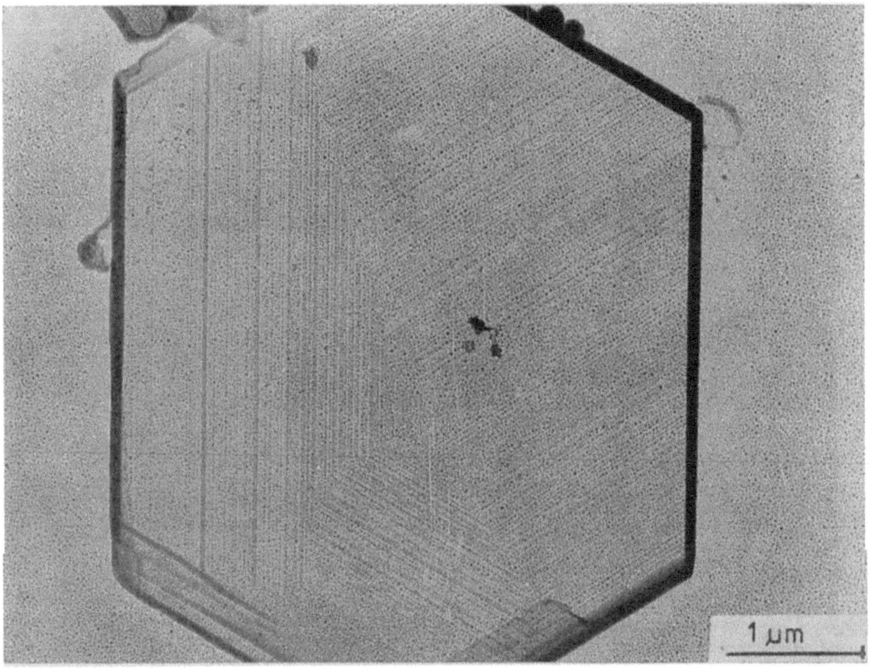

Fig. 17. Replica: *Kaolinite* from geodes in Mississipian Limestone, Keokuk, Iowa, USA

Figs. 18, 19, 20. Replicas: *Kaolinite*, gold decorated, from geodes in Mississippian Limestone, Keokuk, Iowa, USA

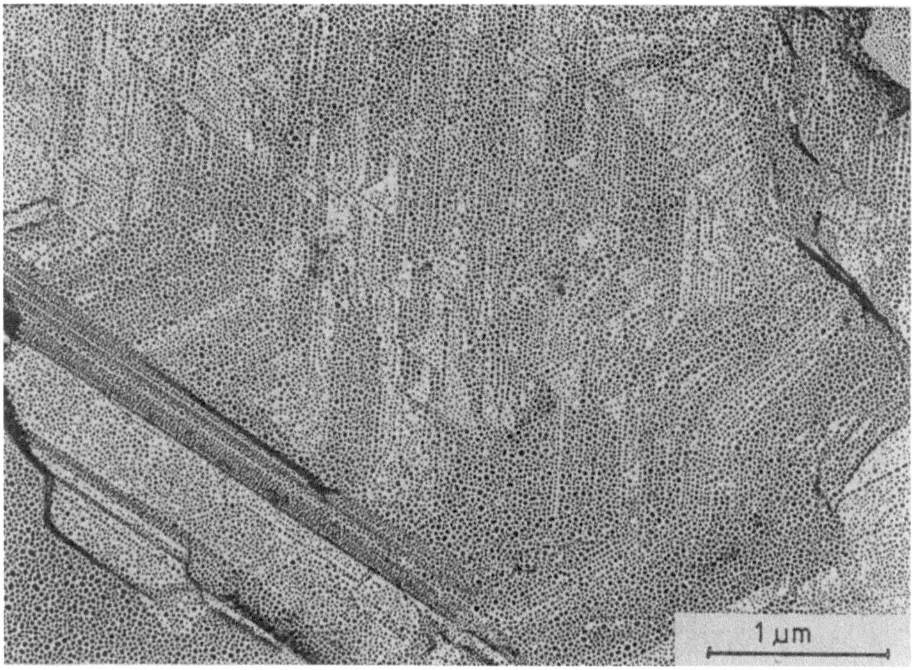

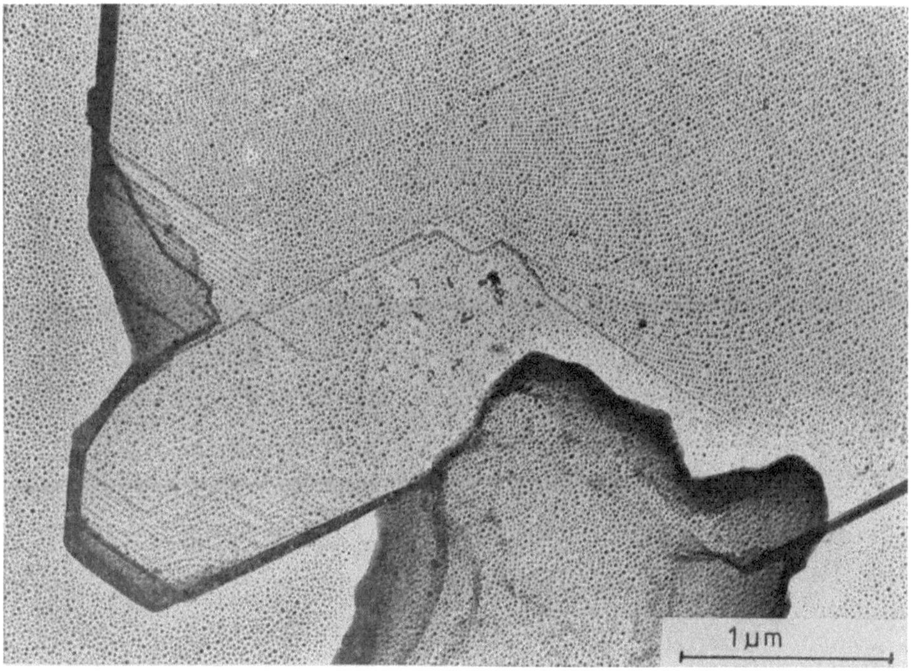

Fig. 19.

Fig. 20.

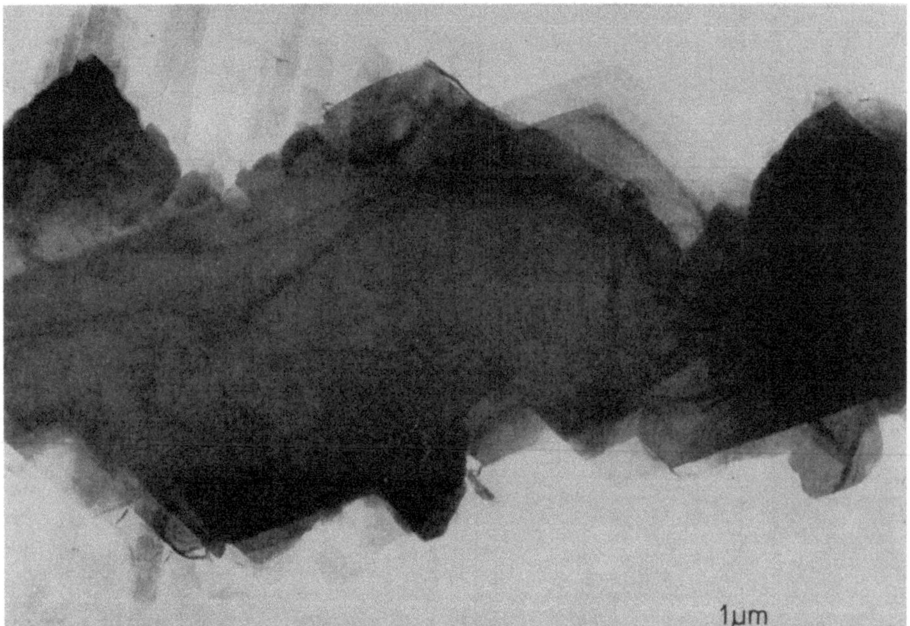

Fig. 21. TEM: *Kaolinite* from cracks in a siderite concretion in the Belle Fourche Shale, Crook County, Wyoming, USA

Fig. 22. TEM: *Kaolinite* and very thin slats of three-layer silicate from cracks in a siderite concretion in the Belle Fourche Shale, Crook County, Wyoming, USA

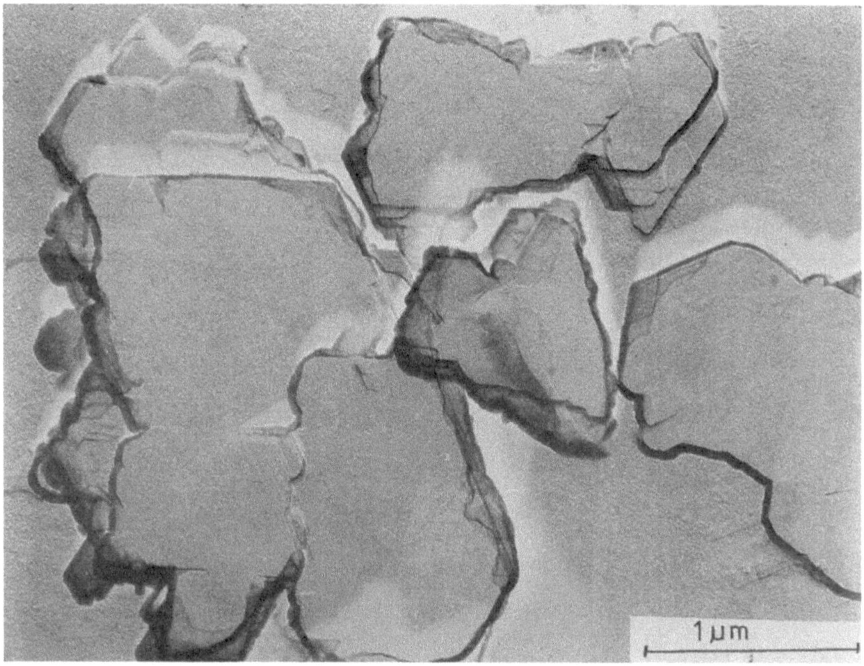

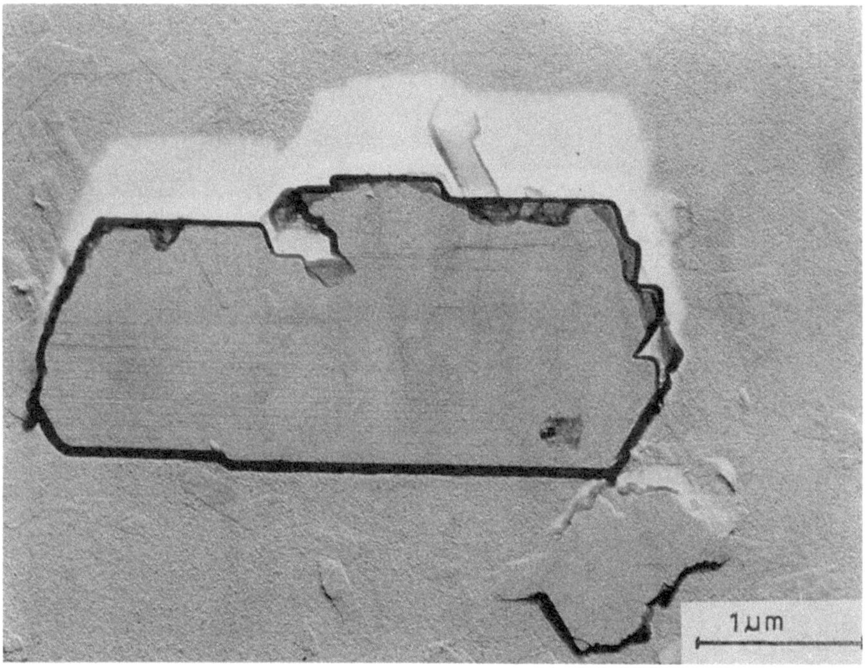

Fig. 23. Replica: *Kaolinite* from cracks in a siderite concretion in the Belle Fourche Shale, Crook County, Wyoming, USA

Fig. 24. Replica: *Kaolinite* and very thin slats of three-layer silicate from cracks in a siderite concretion in the Belle Fourche Shale, Crook County, Wyoming, USA

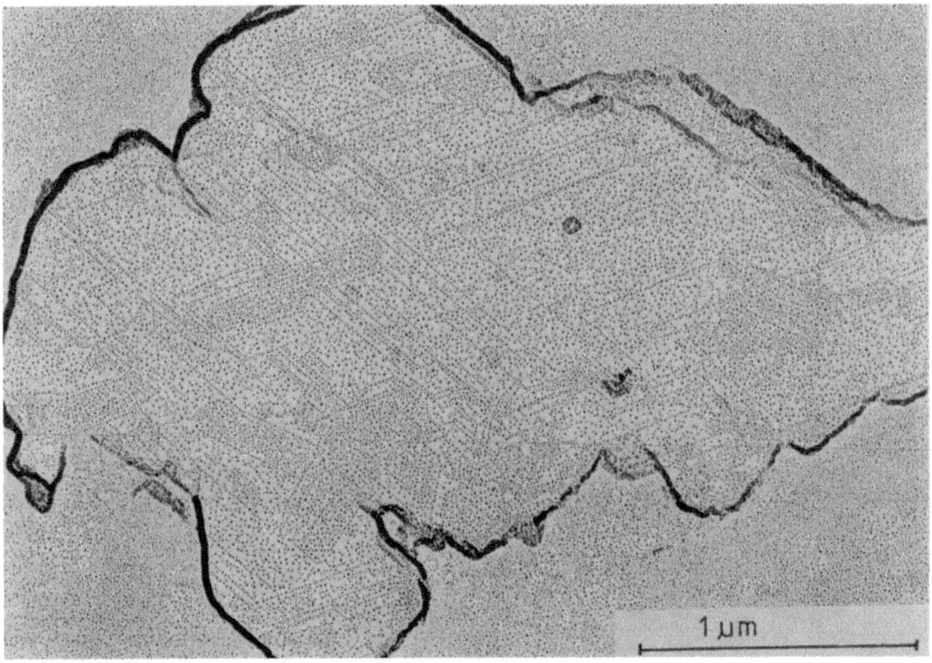

Figs. 25, 26, 27, 28. Replicas: *Kaolinite* and slats of three-layer silicate, gold decorated, from cracks in a siderite concretion in the Belle Fourche Shale, Crook County, Wyoming, USA

Fig. 26.

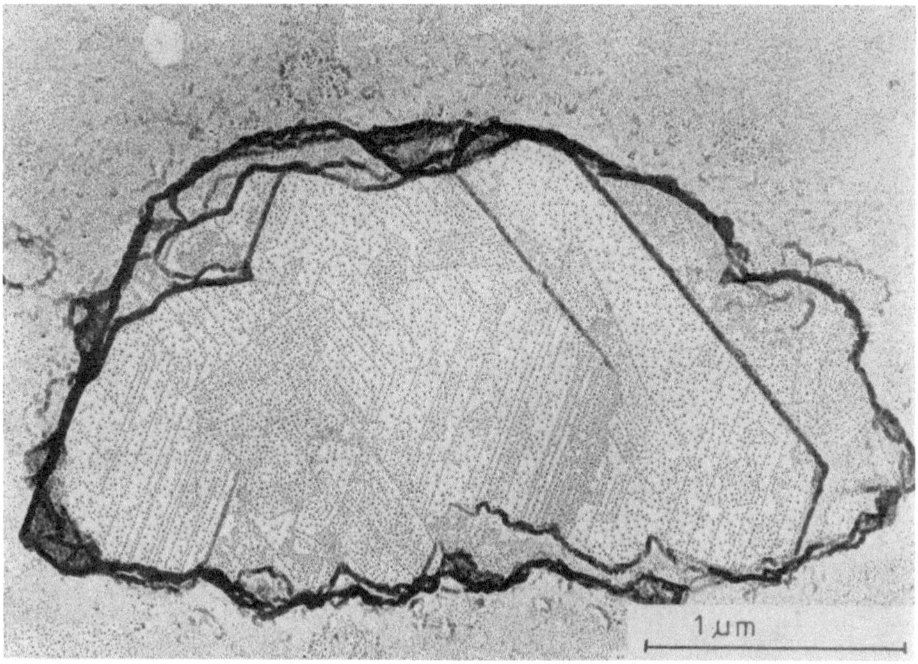

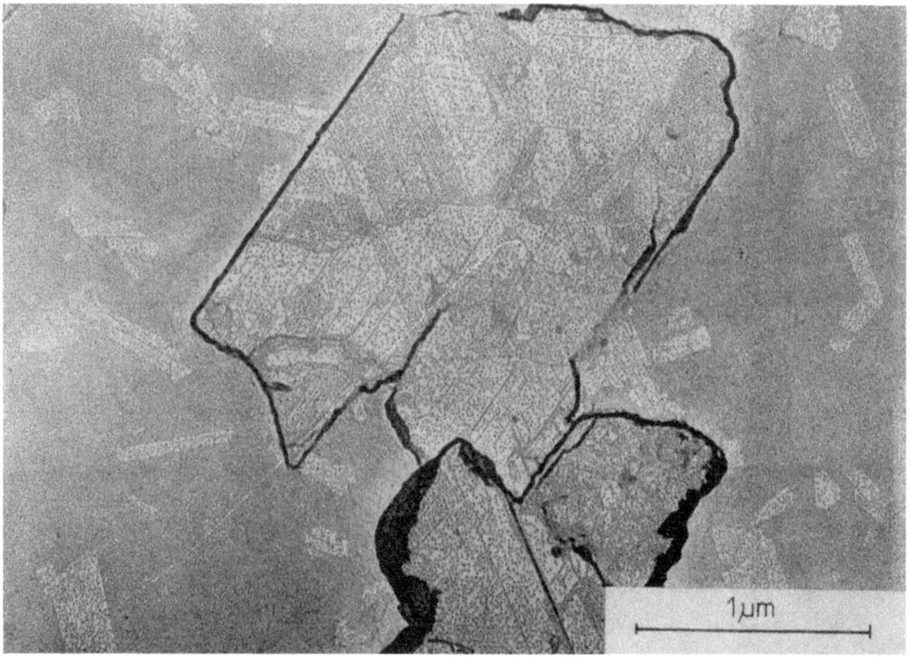

Fig. 27.

Fig. 28.

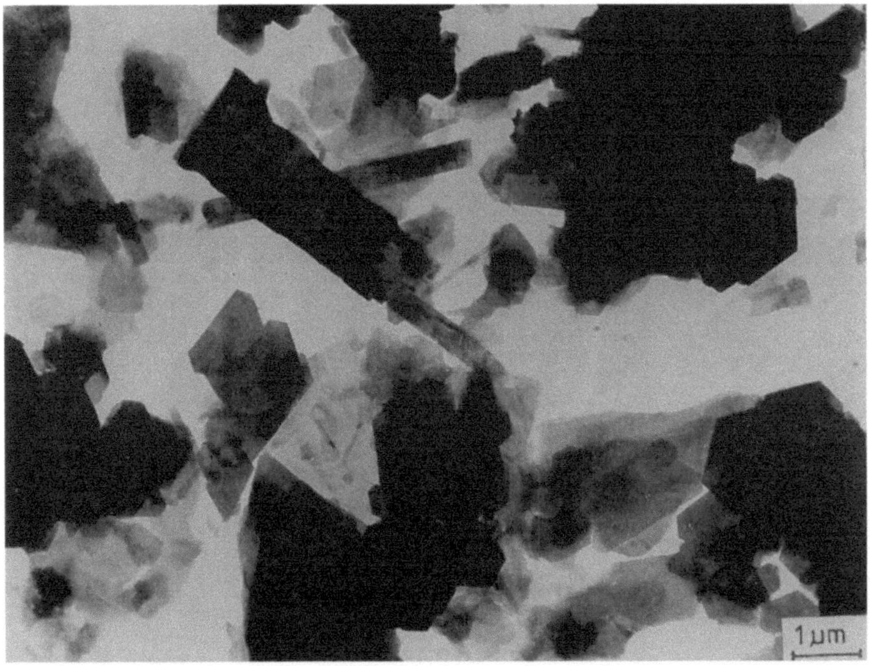

Fig. 29. TEM: *Kaolinite* from kaolinized pegmatite, Aue, GDR

Fig. 30. Replica: *Kaolinite* and *mica* (slats) from kaolinized pegmatite, Aue, GDR

Fig. 31. Replica: *Kaolinite* from kaolinized pegmatite, Aue, GDR

Fig. 32. SEM: *Kaolinite* from kaolinized pegmatite, Aue, GDR

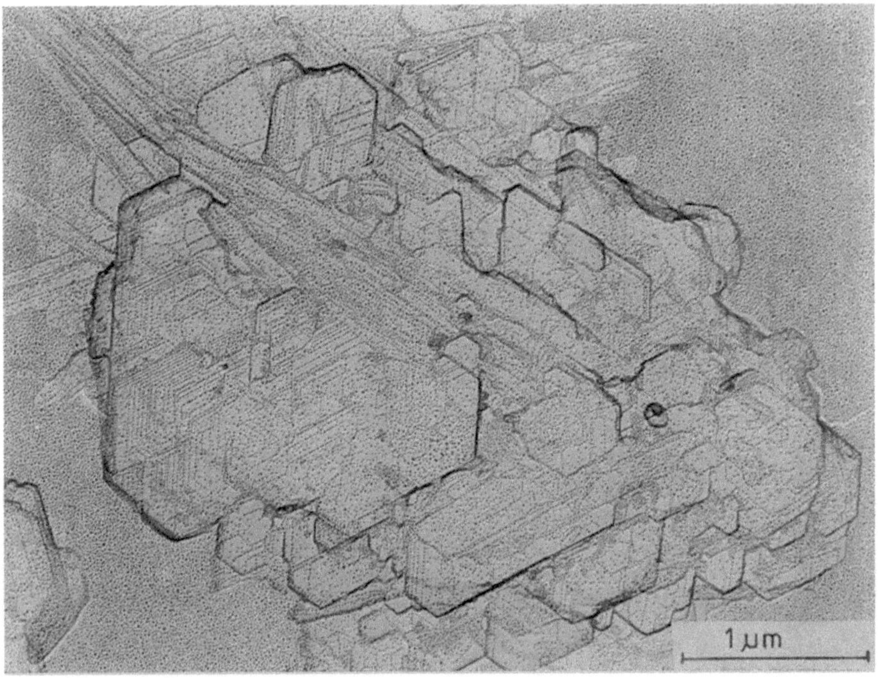

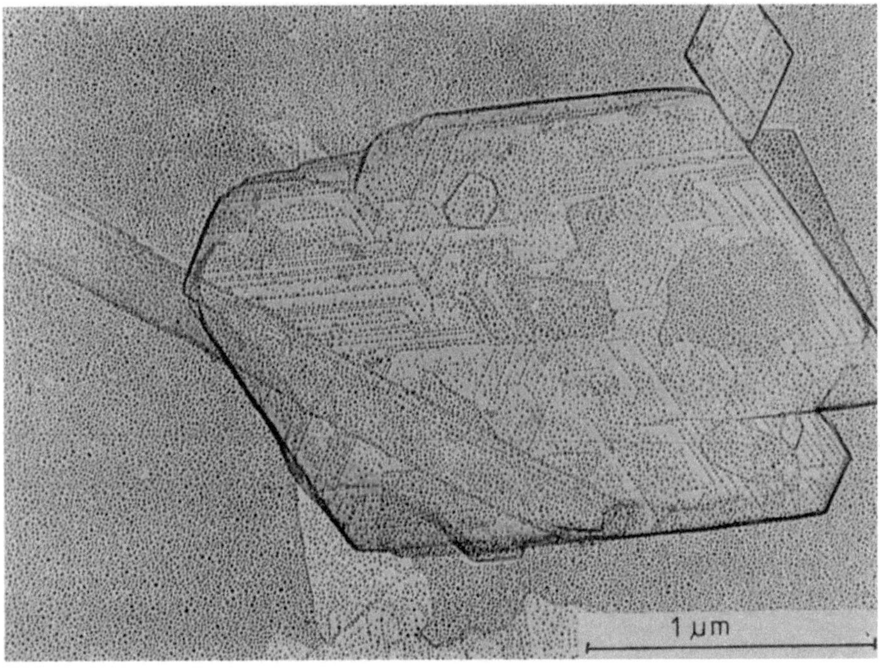

Figs. 33, 34. Replicas: *Kaolinite* and *mica* (slats), gold decorated, Aue, GDR

Fig. 34.

Figs. 35, 36, 37. Replicas: Diagenetic *kaolinite* from Carboniferous sandstone, drill core, Rügen, GDR

Fig. 36.

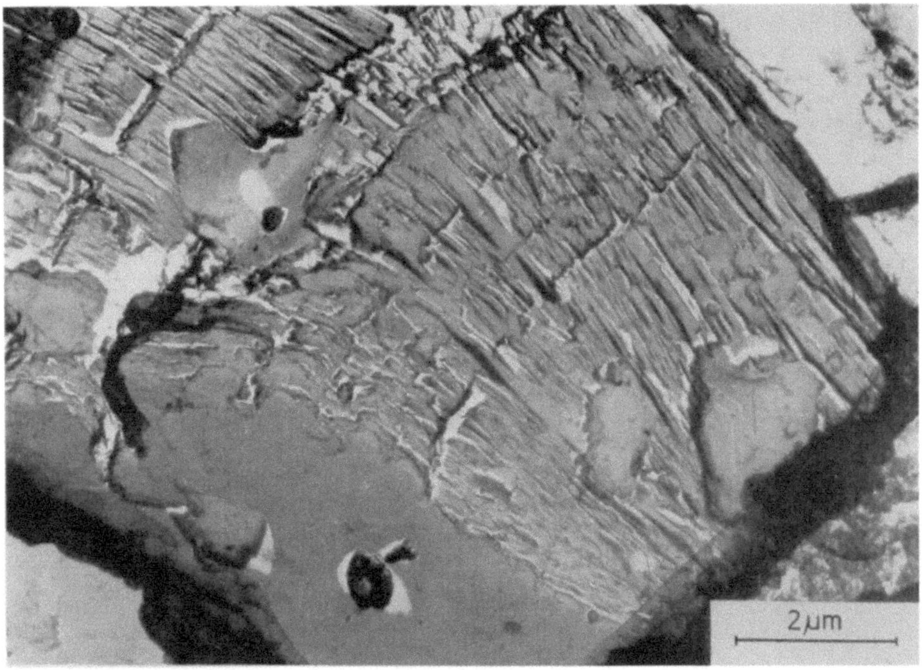

Fig. 37. (*explained at p. 59*)

Fig. 38. SEM: *Kaolinite* from hydrothermally altered pegmatites, tin deposit, Ehrenfriedersdorf, GDR

Figs. 39, 40. TEMs: *Synthetic kaolinite* (575 K, 48 h) from amorphous silica and amorphous aluminium hydroxide

Fig. 40.

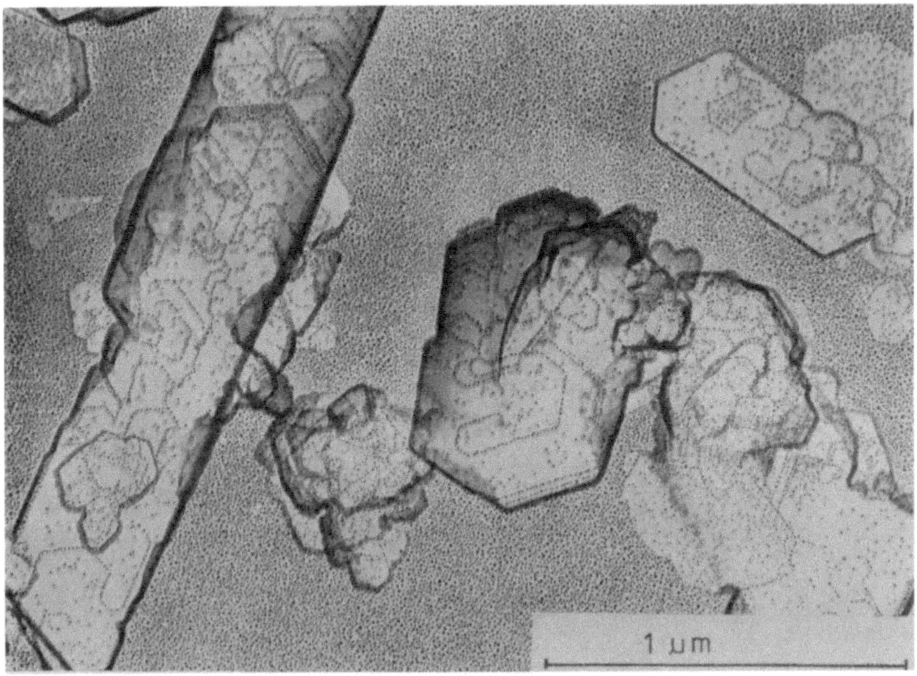

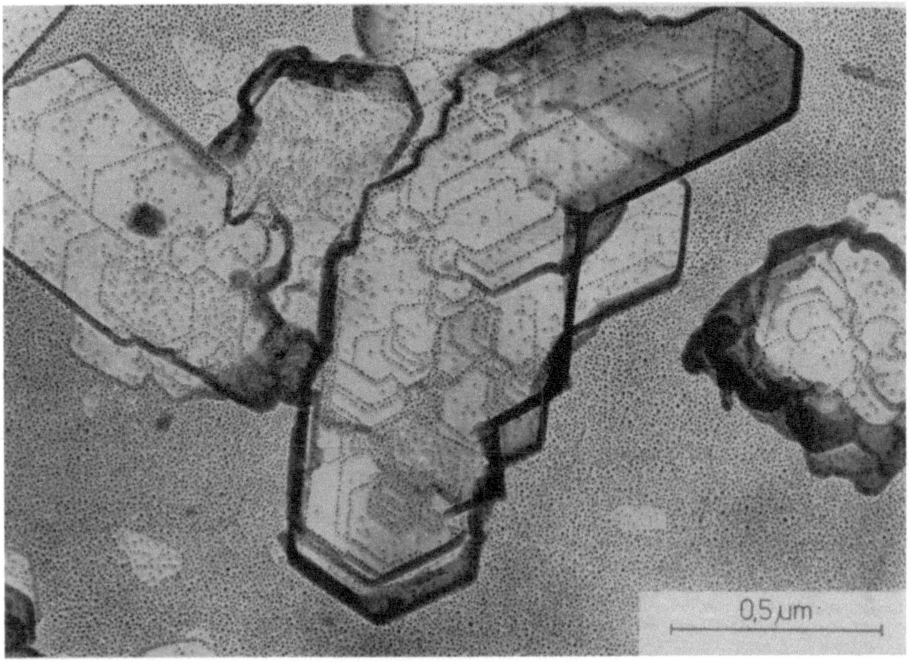

Figs. 41, 42, 43, 44. Replicas: *Synthetic kaolinite* (575 K, 48 h) from amorphous silica and amorphous aluminium hydroxide, gold decorated

Fig. 42.

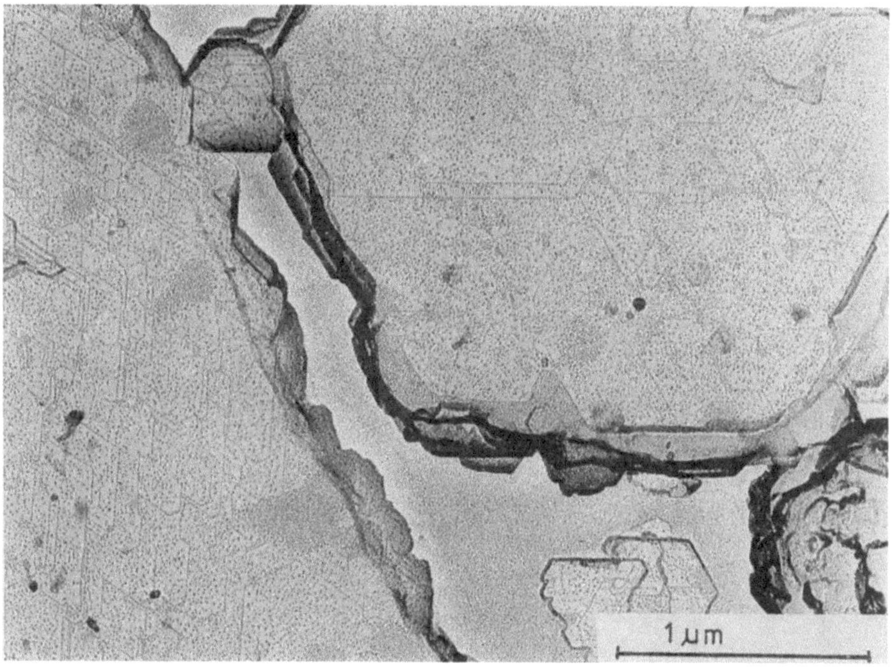

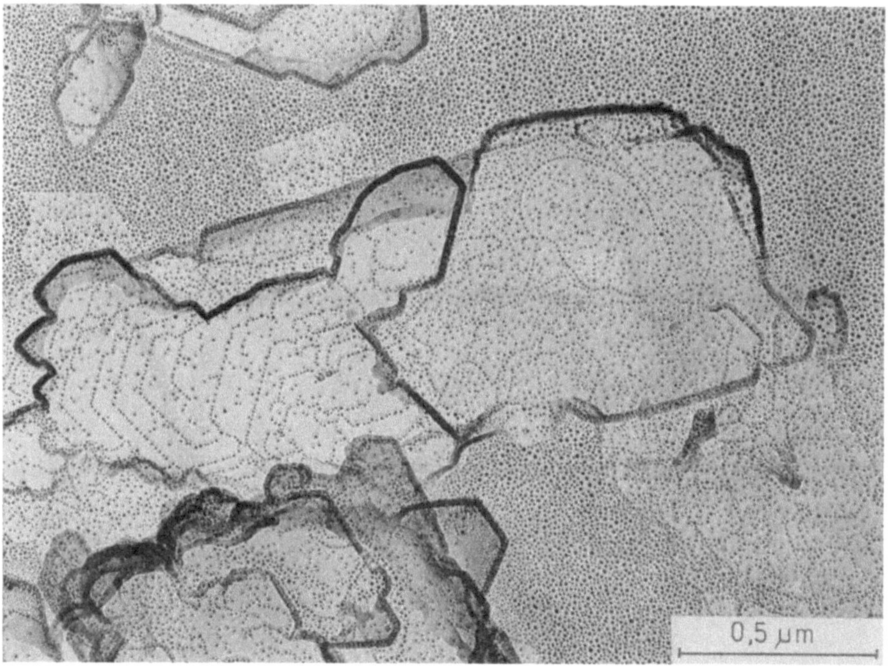

Fig. 43.

Fig. 44.

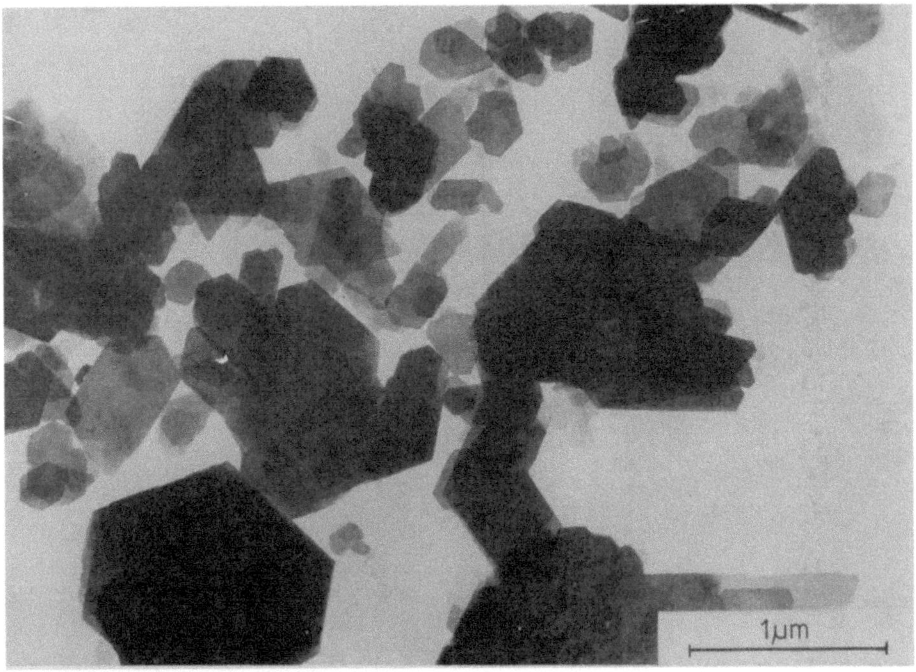

Figs. 45, 46. TEMs: *Kaolinite* from kaolinitic sandstone, Cretaceous, Pit Maria near Boleslawiec, Poland

Fig. 46.

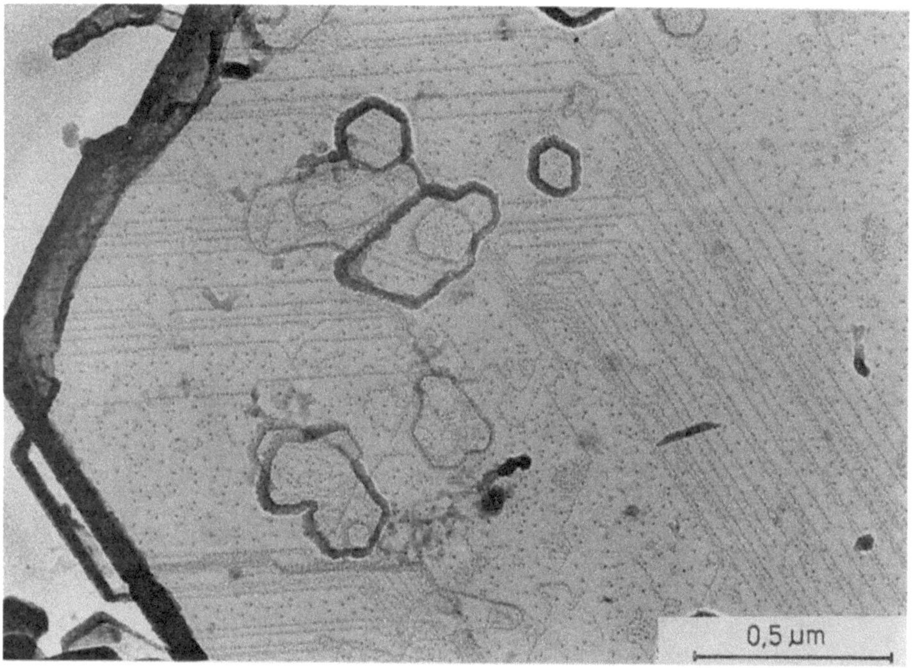

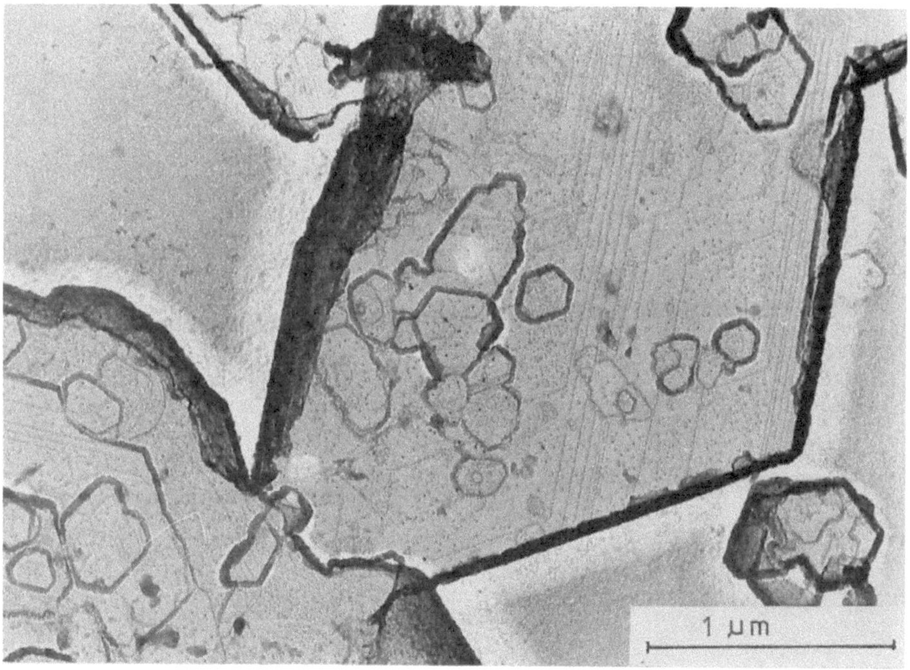

Figs. 47, 48. Replicas: *Kaolinite*, gold decorated, from kaolinitic sandstone, Cretaceous, Pit Maria near Boleslawiec, Poland

Fig. 48.

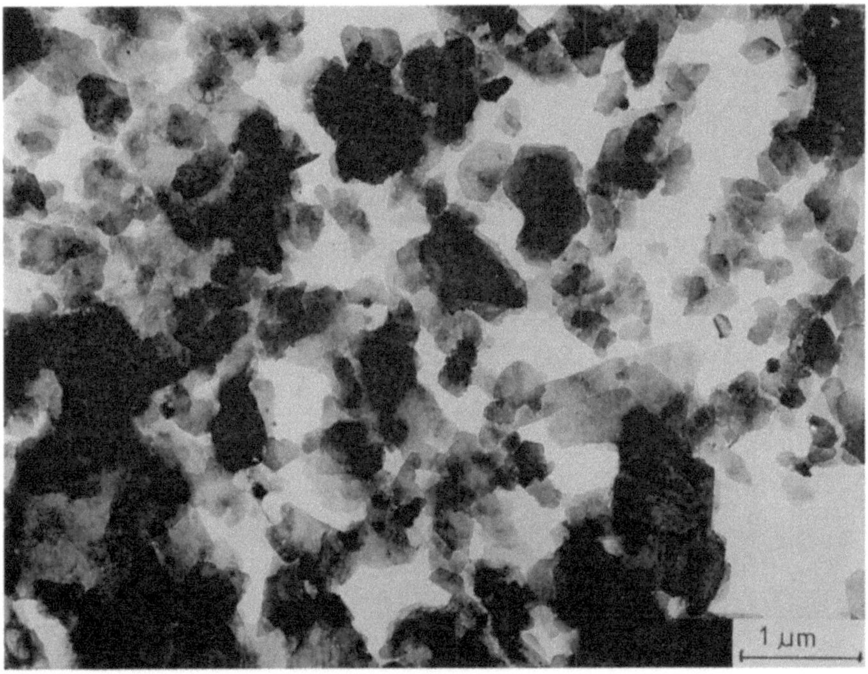

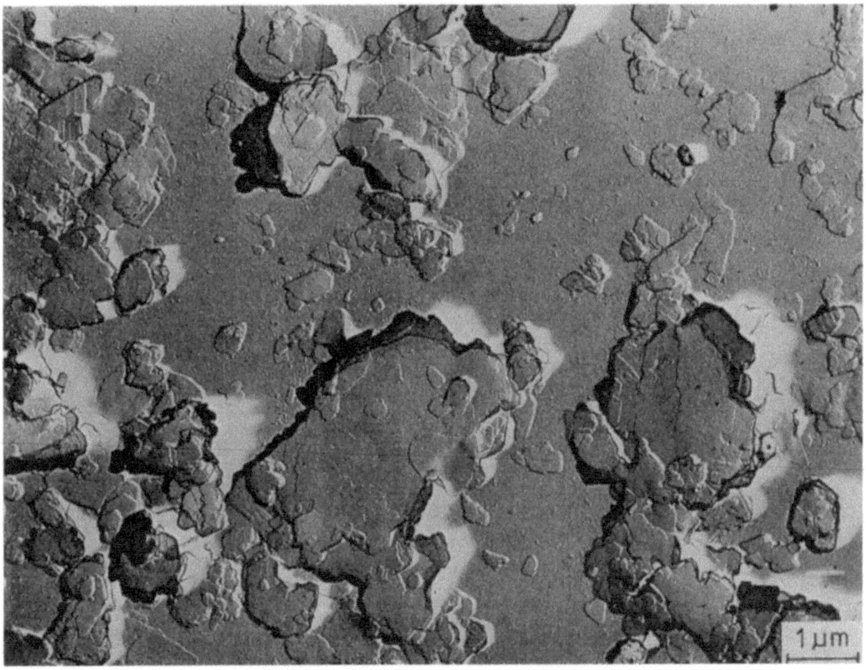

Fig. 49. TEM: *Kaolinite* from granodiorite kaolin, Caminau, GDR

Fig. 50. Replica: *Kaolinite* from granodiorite kaolin, Caminau, GDR

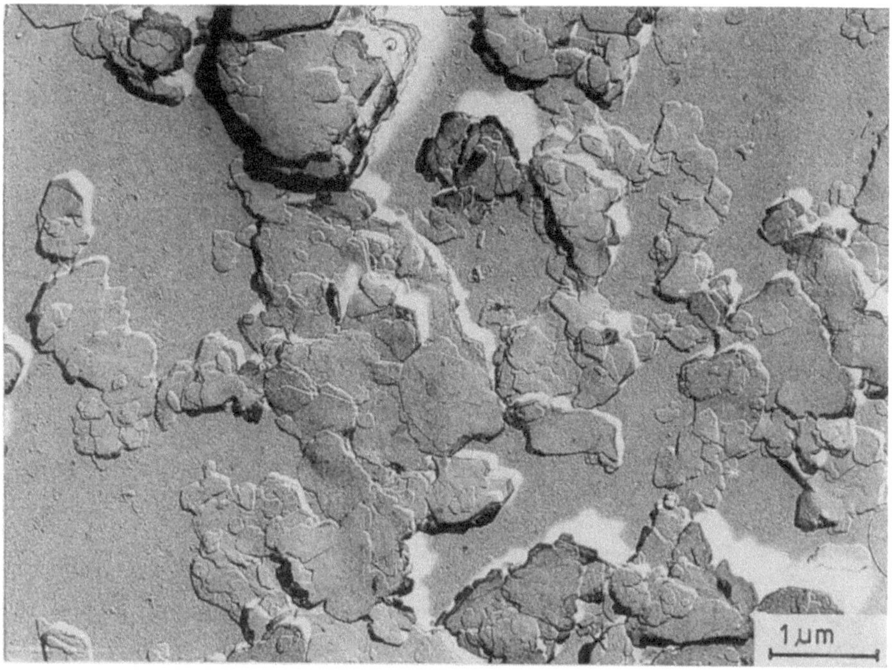

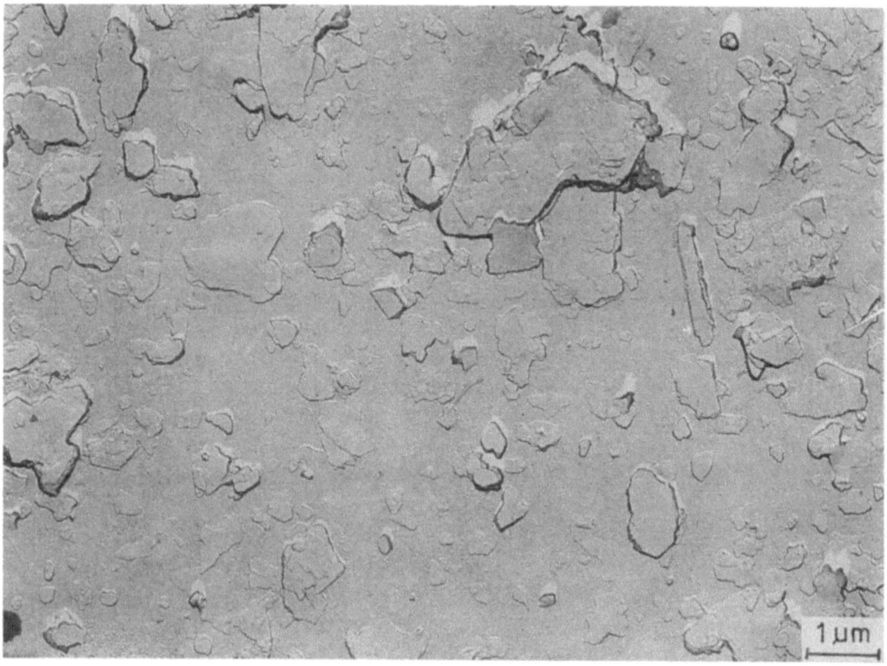

Fig. 51. Replica: *Kaolinite* aggregates from granodiorite kaolin, Caminau, GDR

Fig. 52. Replica: *Delaminated granodiorite kaolin,* Caminau, GDR

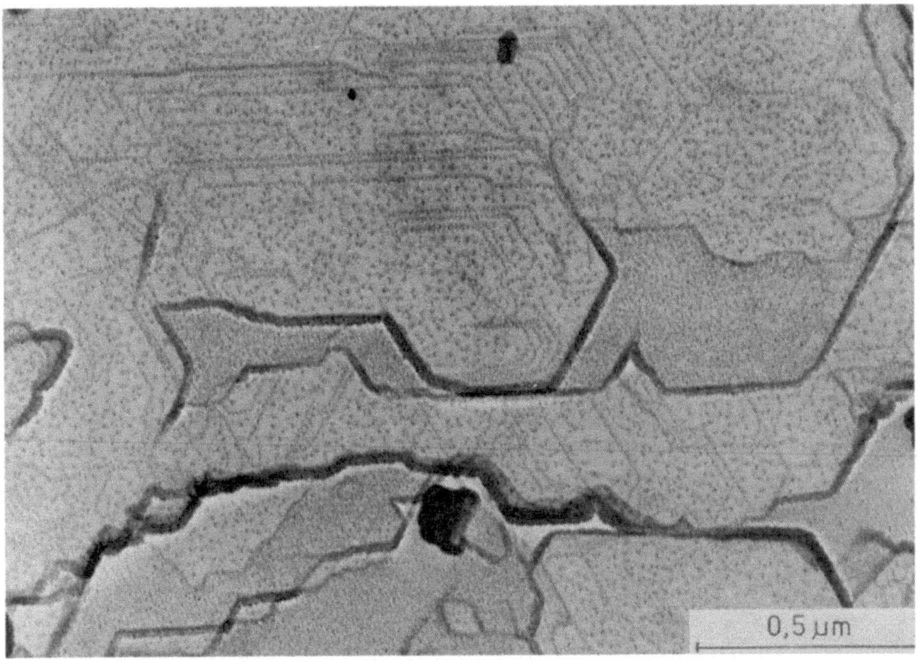

Figs. 53, 54. Replicas: Epitaxy *kaolinite/mica*, gold decorated, from granodiorite kaolin, Caminau, GDR

Fig. 54.

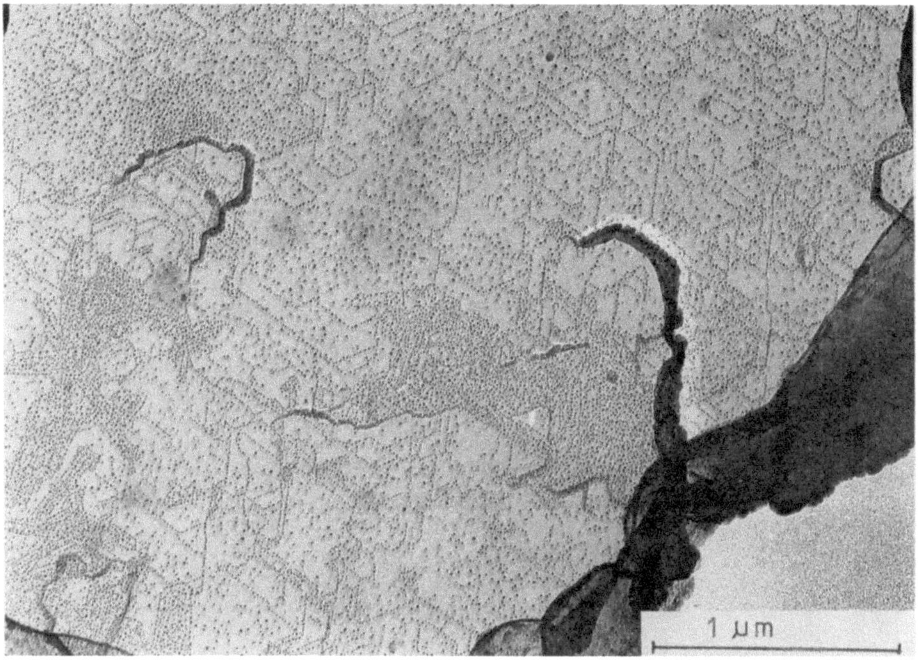

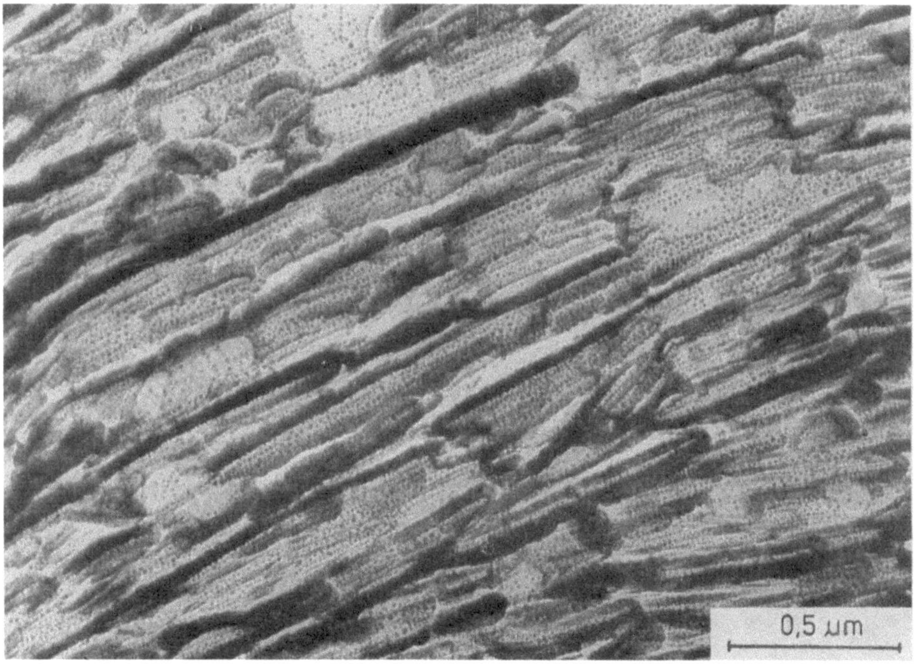

Fig. 55. Replica: *Kaolinite*, gold decorated, from granodiorite kaolin, Caminau, GDR

Fig. 56. Replica: *Kaolinite* book, gold decorated at the prism faces, from granodiorite kaolin, Caminau, GDR

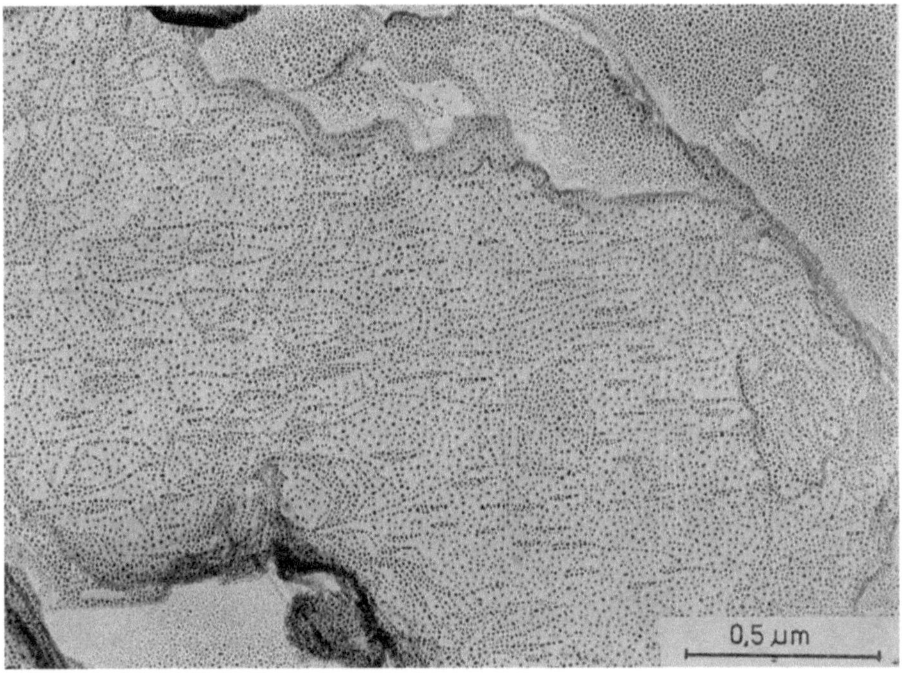

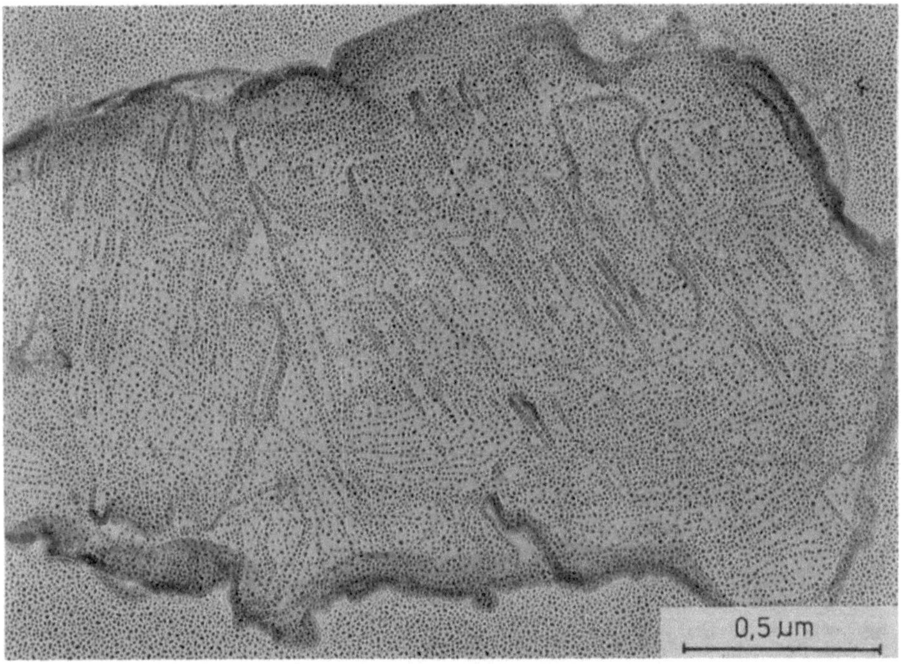

Figs. 57, 58. Replicas: *Kaolinite*, NH_4F-etched, gold decorated, from granodiorite kaolin, Caminau, GDR

Fig. 58.

Figs. 59, 60. SEMs: *Granodiorite kaolin* from Caminau, GDR

Fig. 60.

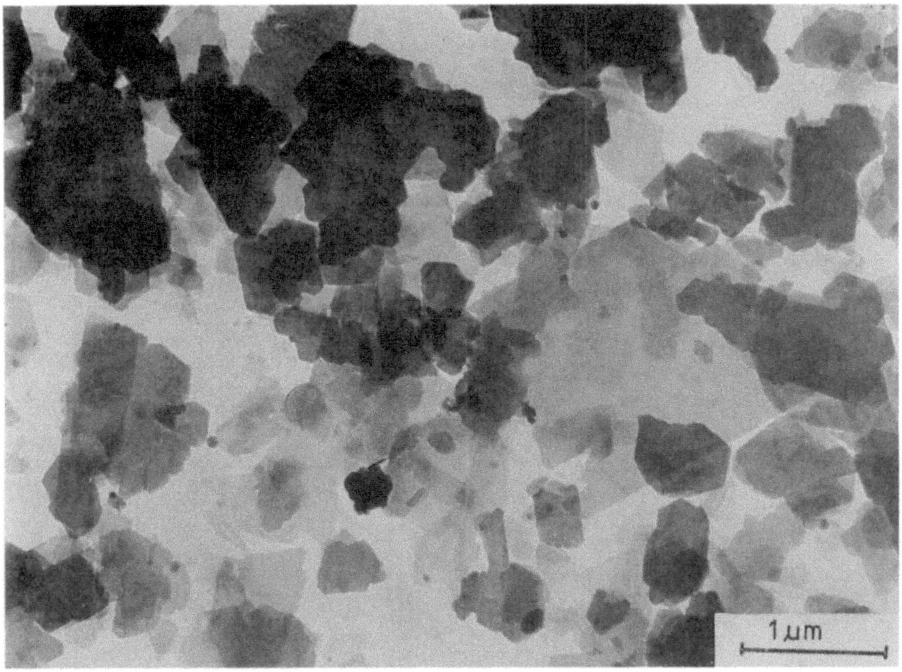

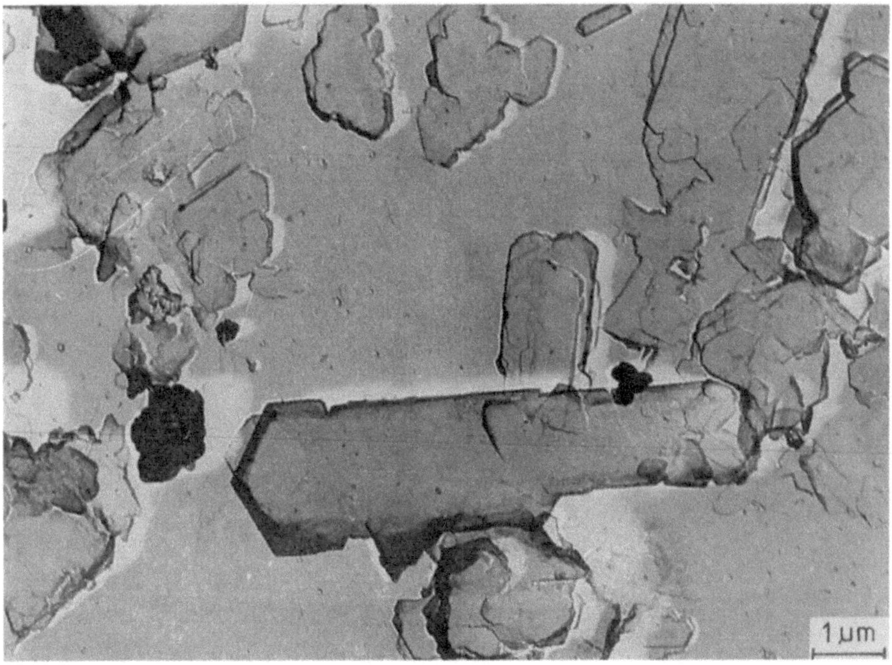

Fig. 61. TEM: *Kaolinite* from porphyry kaolin, Fuchsberg near Halle, GDR

Fig. 62. Replica: *Kaolinite* elongates from porphyry kaolin, Fuchsberg near Halle, GDR

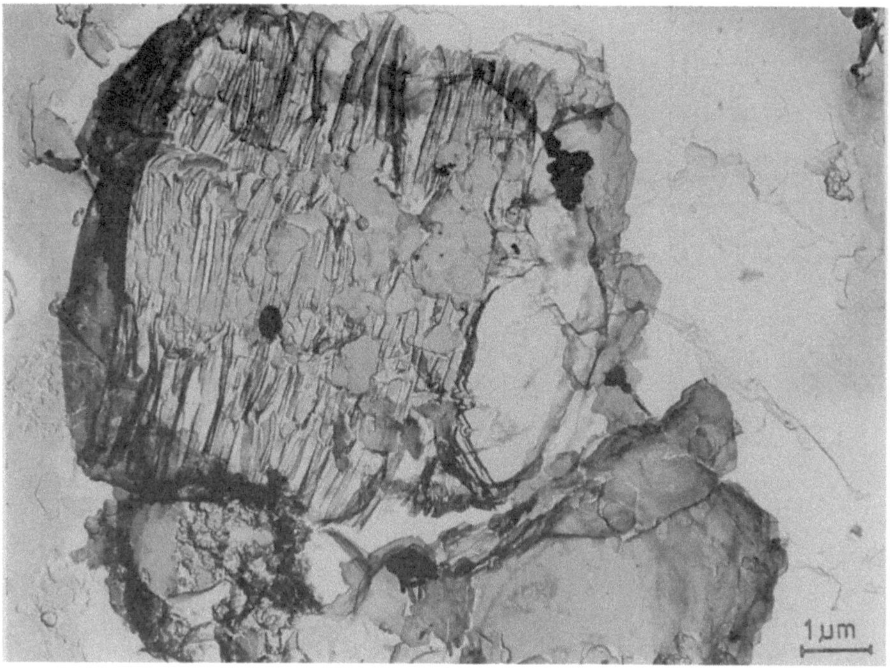

Fig. 63. Replica: *Kaolinite* book from porphyry kaolin, Fuchsberg near Halle, GDR

Fig. 64. SEM: *Kaolinite* from porphyry kaolin, Fuchsberg near Halle, GDR

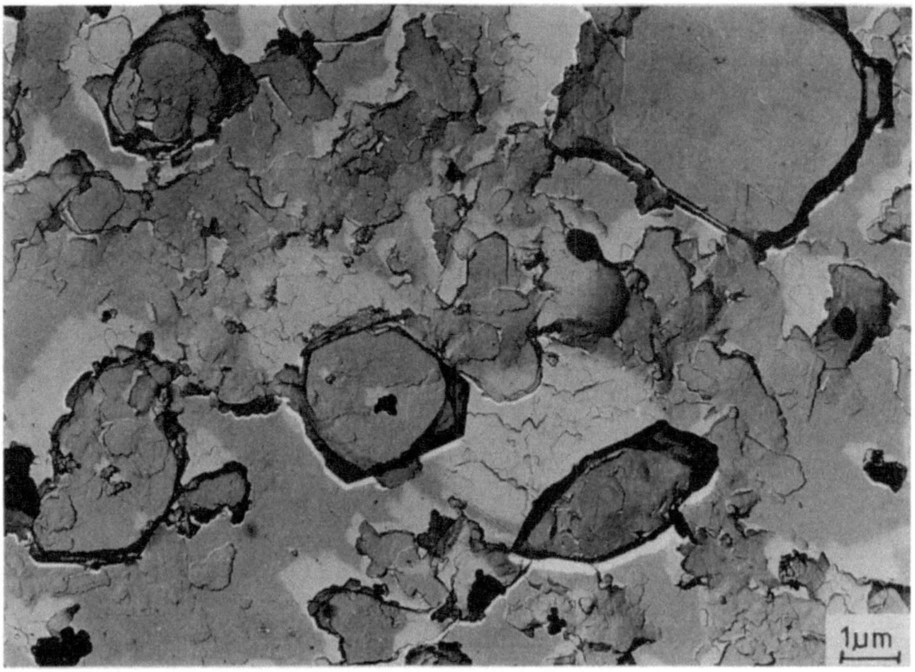

Fig. 65. TEM: *Kaolinite* from porphyry kaolin, Kemmlitz, GDR

Fig. 66. Replica: *Kaolinite* from porphyry kaolin, Kemmlitz, GDR

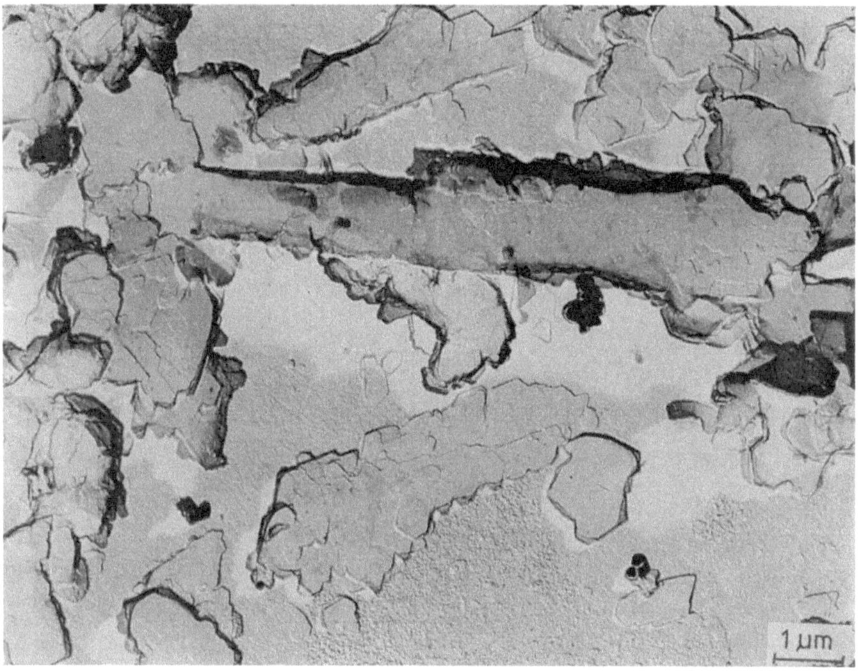

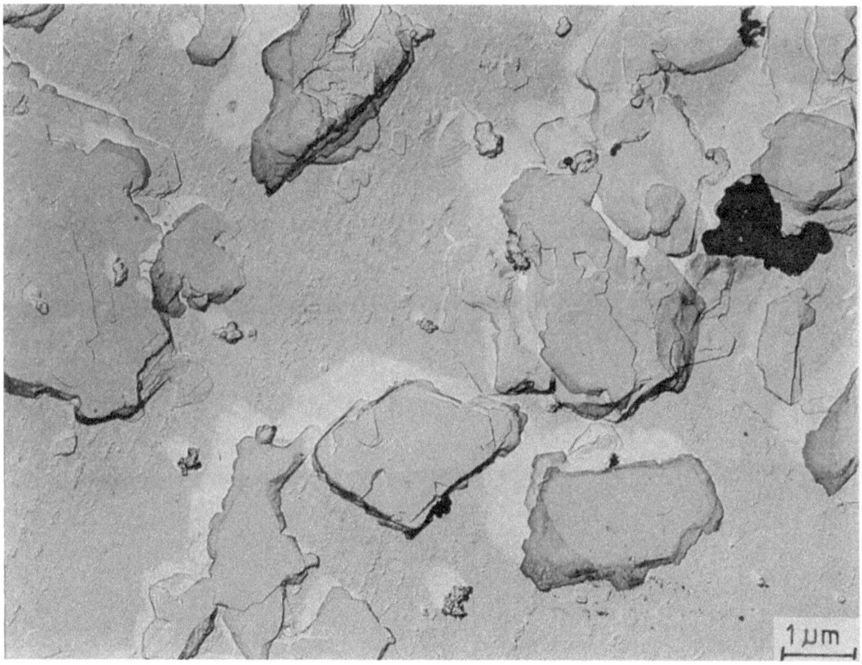

Fig. 67. Replica: *Kaolinite* from kaolinized Bunter Sandstone arkoses, Spergau near Halle, GDR

Fig. 68. Replica: *Kaolinite* from kaolinized greywacke, Espenhain near Leipzig, GDR

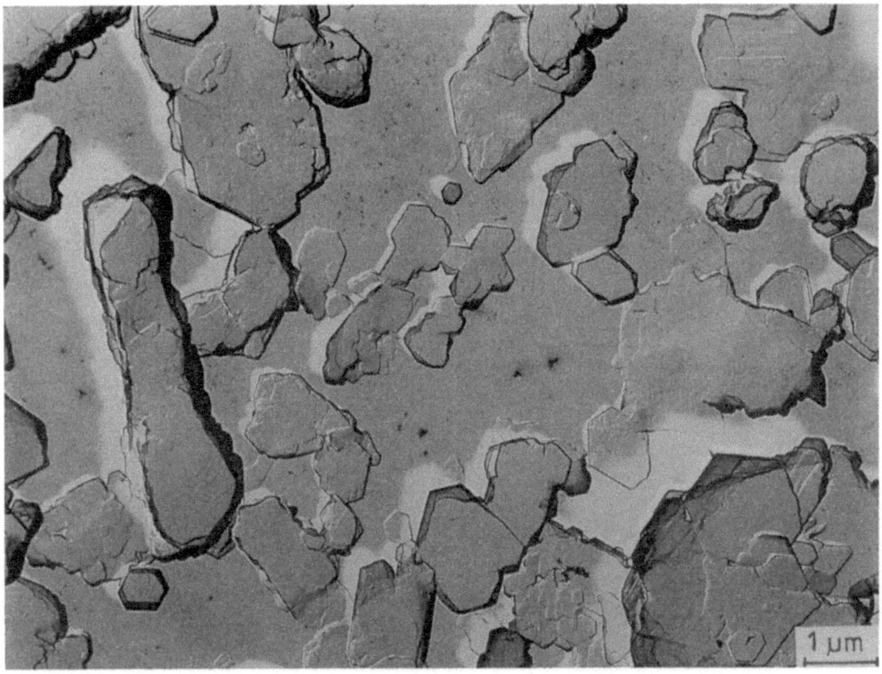

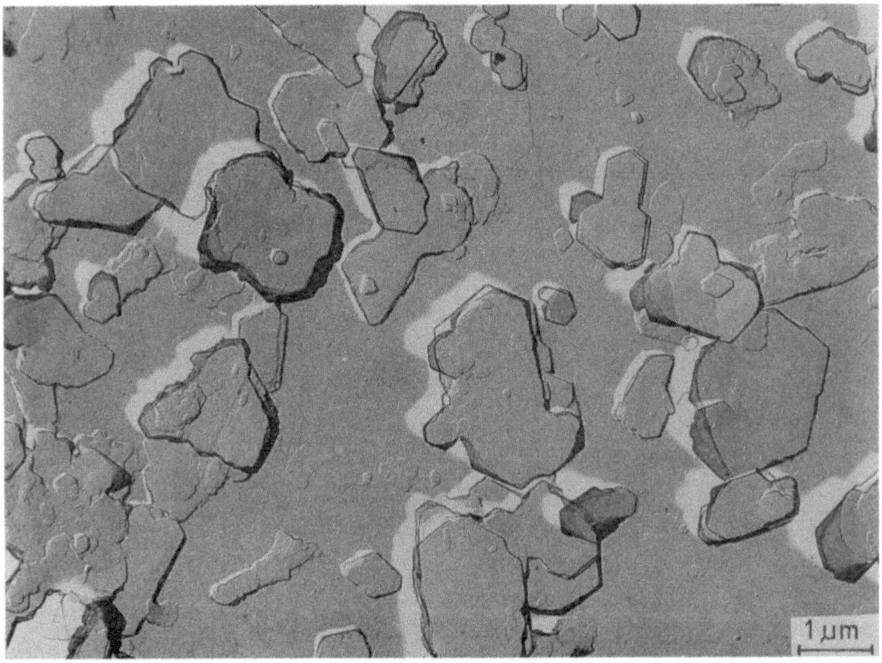

Figs. 69, 70. Replicas: *Kaolinite* from kaolinized arkoses, Kažnejov, Czechoslovakia

Fig. 70.

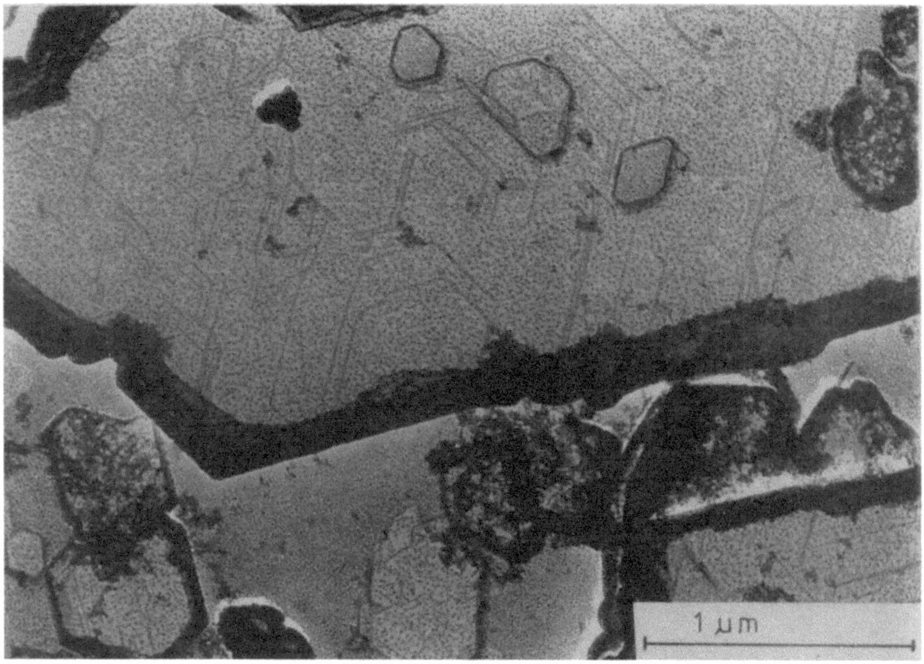

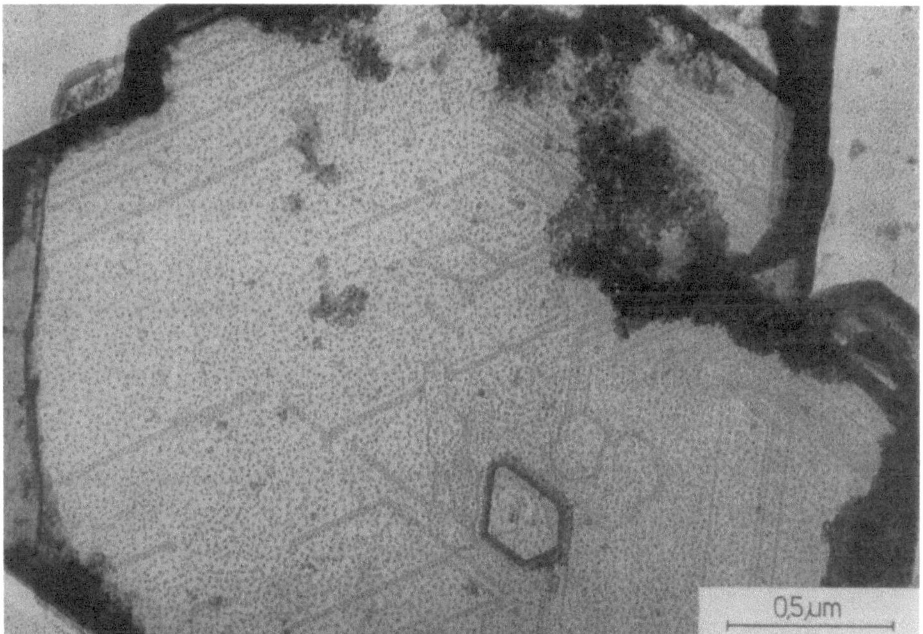

Figs. 71, 72. Replicas: *Kaolinite*, gold decorated, from kaolinized arkoses, Kažnejov, Czechoslovakia

Fig. 72.

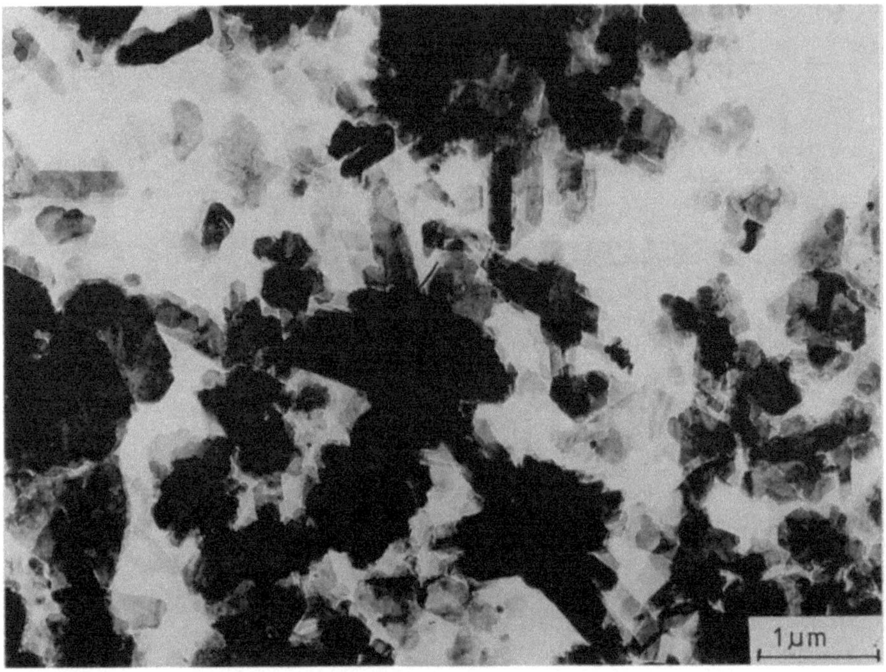

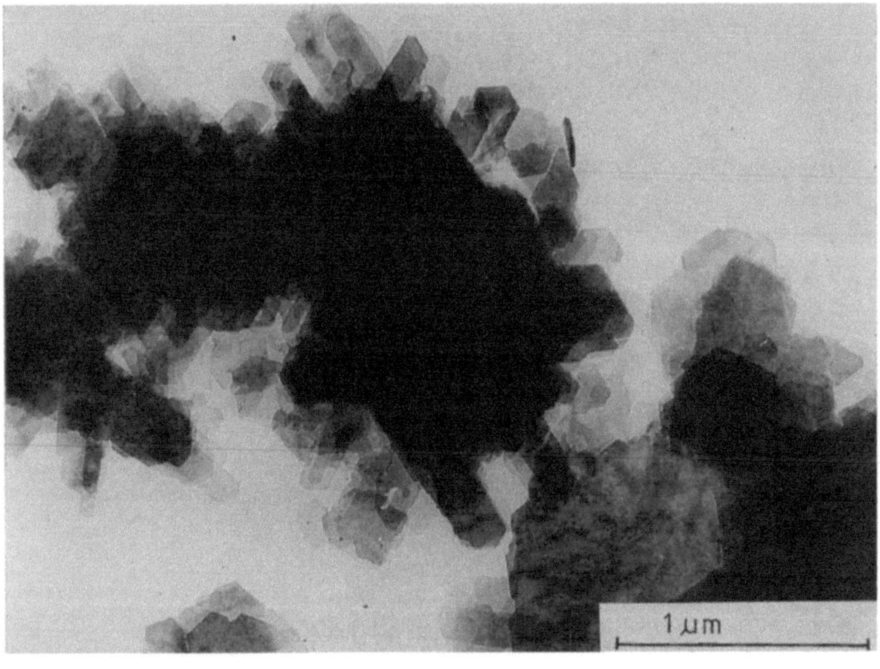

Figs. 73, 74. TEMs: *Kaolinite* from kaolin sand, Senovo, Bulgaria

Fig. 74.

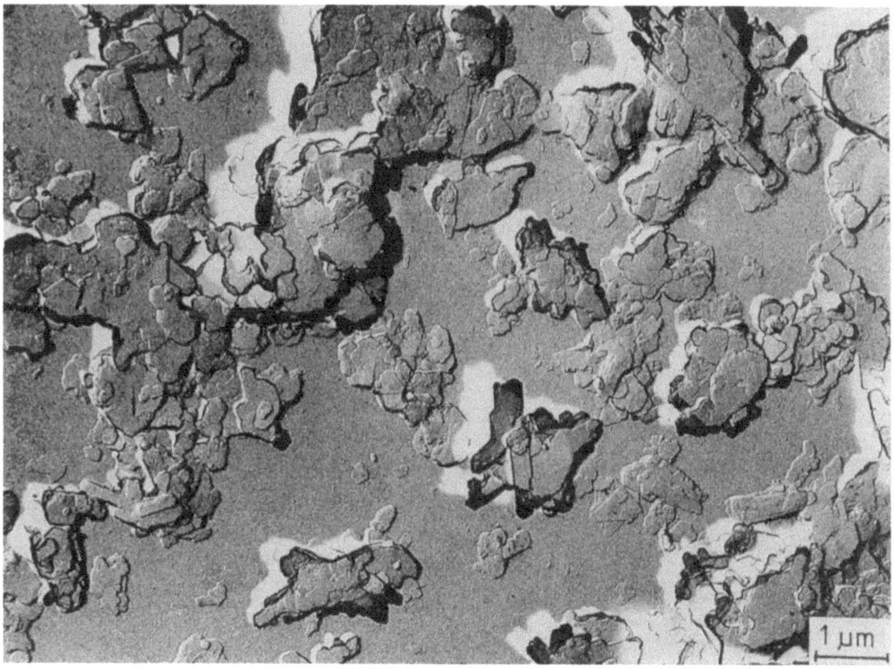

Figs. 75, 76. Replicas: *Kaolinite* from kaolin sand, Senovo, Bulgaria

Fig. 76

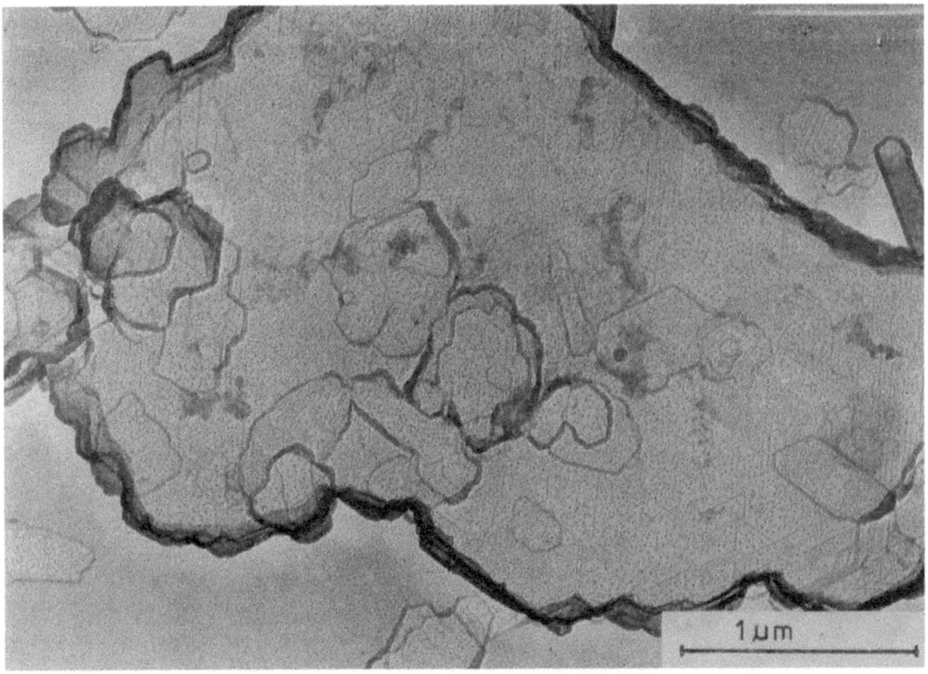

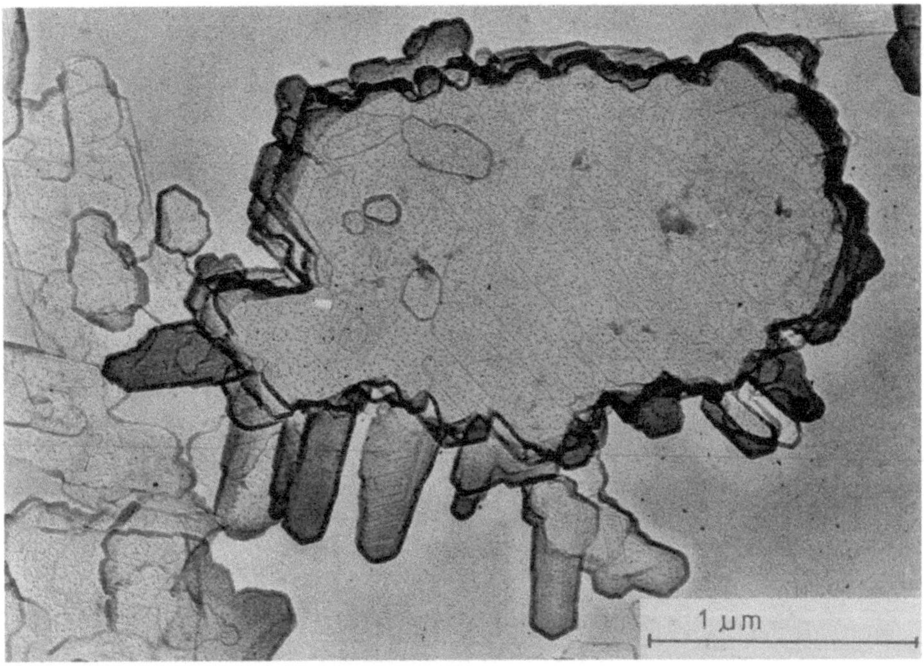

Figs. 77, 78. Replicas: *Kaolinite,* gold decorated, from kaolin sand, Šenovo, Bulgaria

Fig. 78.

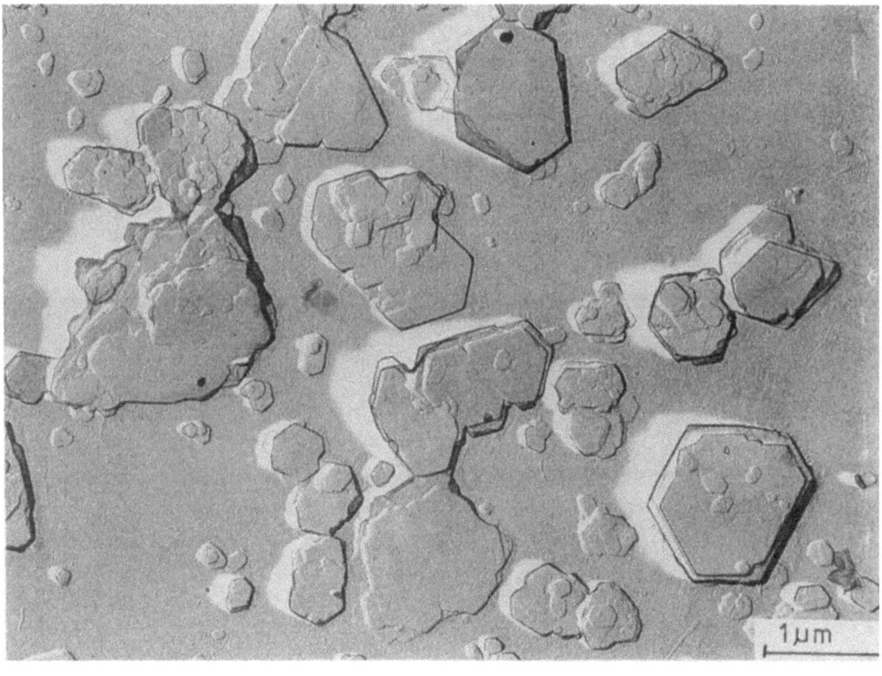

Fig. 79. TEM: *Kaolinite* from Georgia Kaolin (soft clay), Macon, Georgia, USA

Figs. 80, 81, 82. Replicas: *Kaolinite* from Georgia Kaolin (soft clay), Macon, Georgia, USA

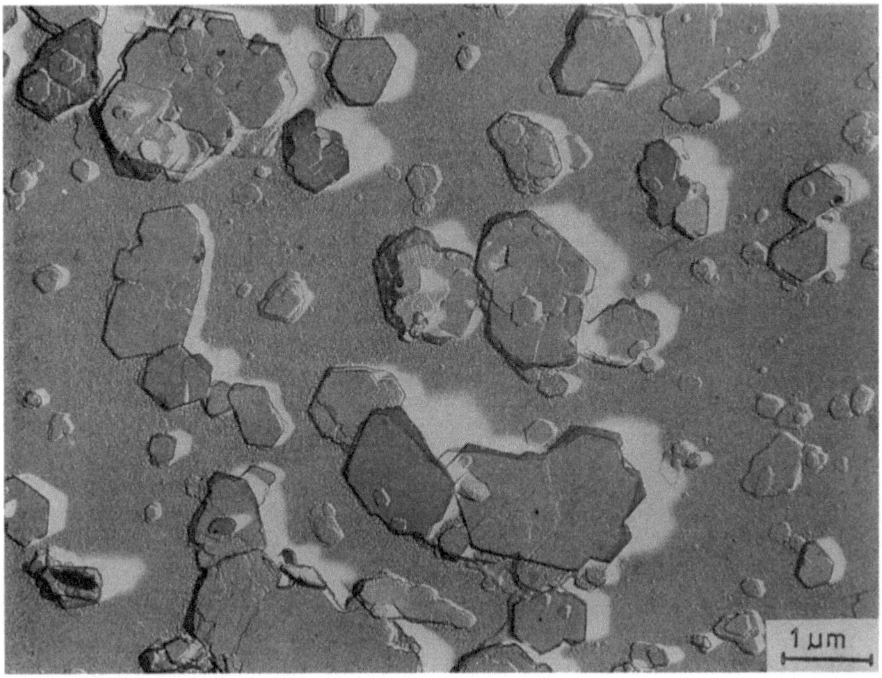

Fig. 81.

(*explained at p. 81*)

Fig. 82.

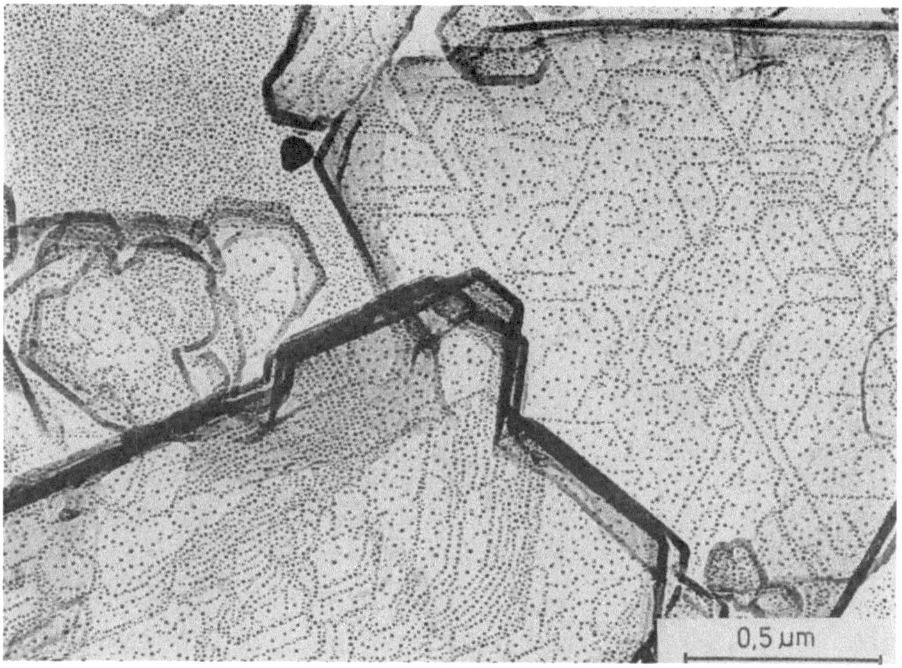

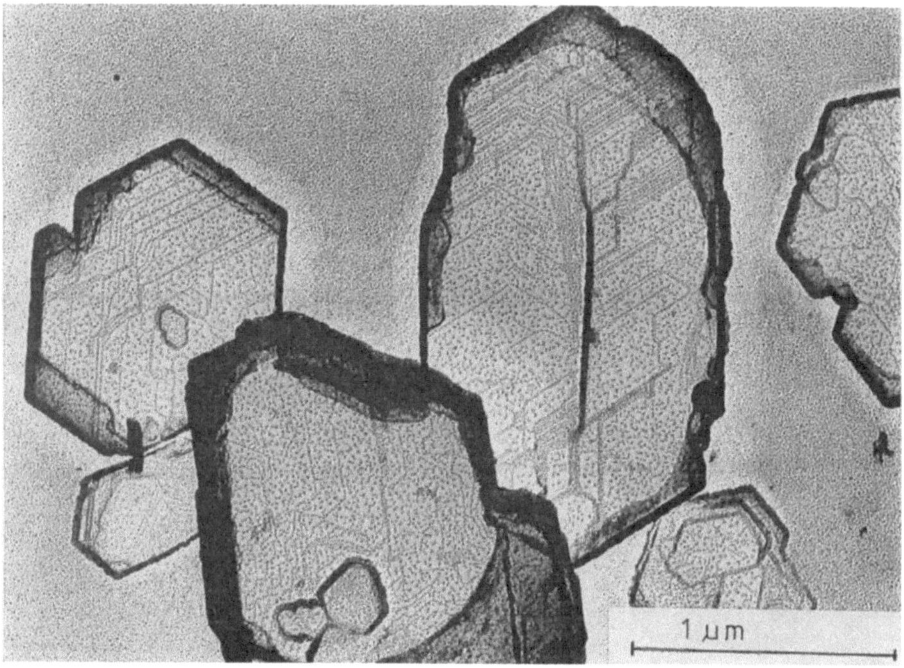

Figs. 83, 84. Replicas: *Kaolinite,* gold decorated, from Georgia Kaolin (soft clay), Macon, Georgia, USA

Fig. 84.

Figs. 85, 86. SEMs: *Kaolinite*, vermicular aggregates in Georgia Kaolin (soft clay), Macon, Georgia, USA

Fig. 86.

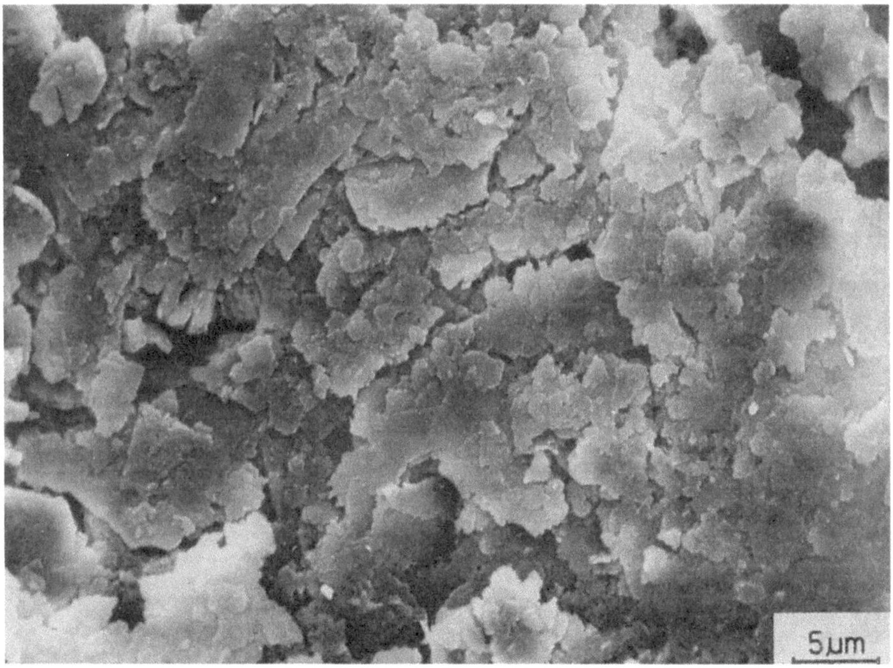

Fig. 87. Replica: *Kaolinite*, Georgia Kaolin (hard clay), Macon, Georgia, USA

Fig. 88. SEM: *Kaolinite* from Georgia Kaolin (hard clay), Macon, Georgia, USA

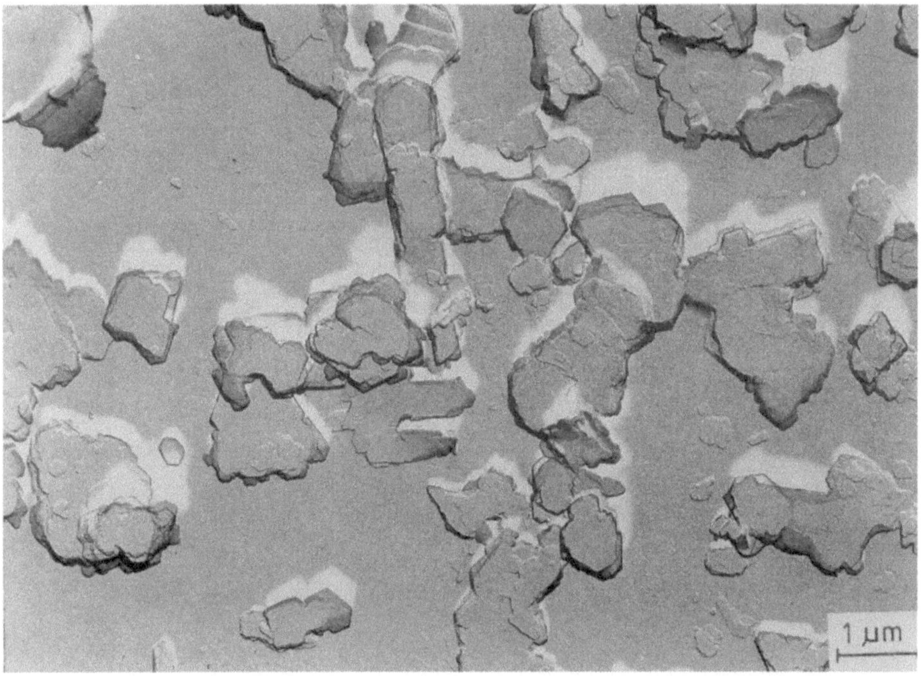

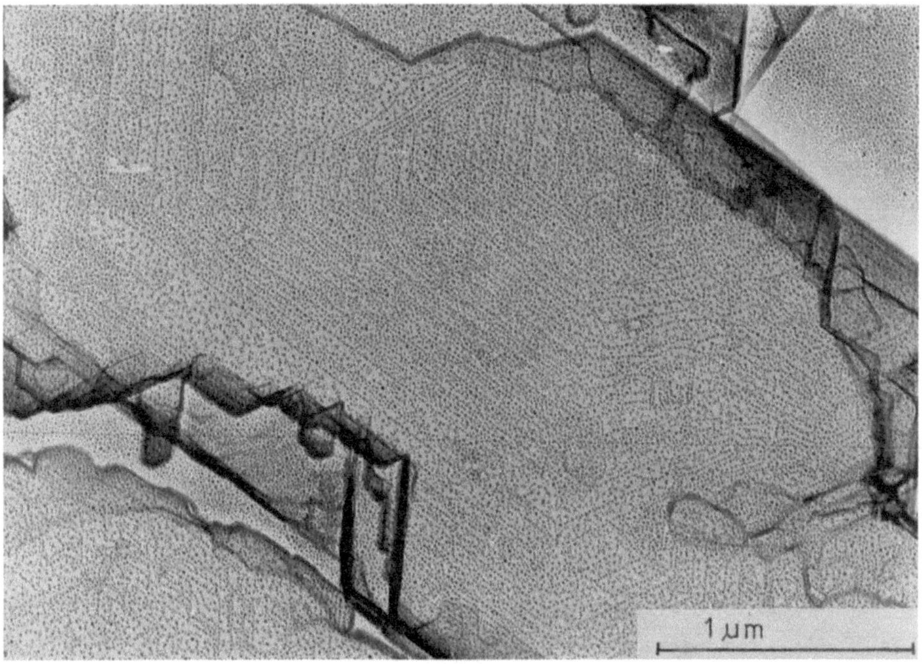

Fig. 89. Replica: *Kaolinite* from kaolinized granite, Alexeyevsk, USSR

Fig. 90. Replica: *Kaolinite*, gold decorated, from kaolinized granite, Alexeyevsk, USSR

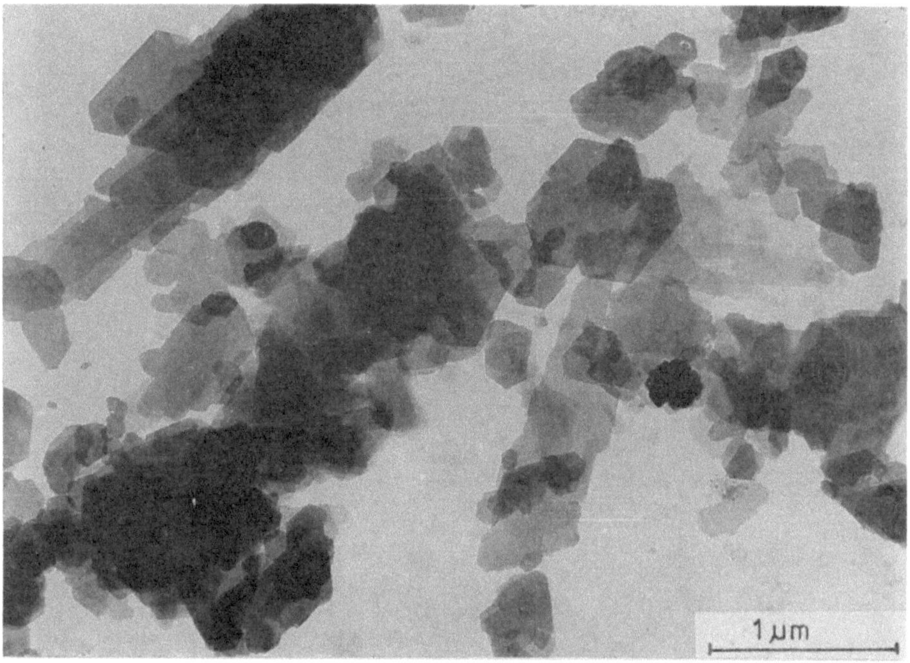

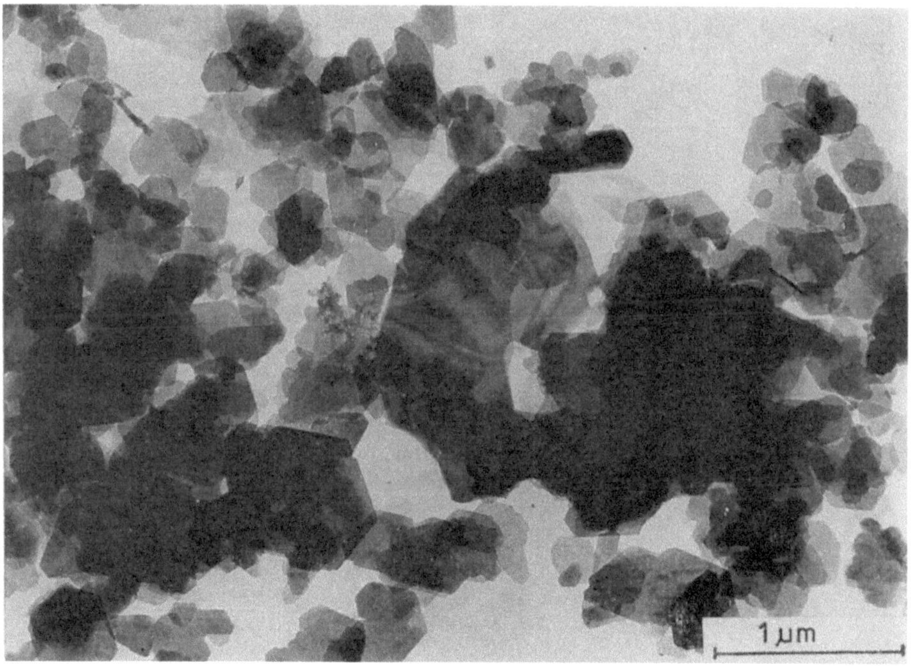

Figs. 91, 92. TEMs: *Kaolinite* from kaolinized granite, Podlesí, Czechoslovakia

Fig. 92.

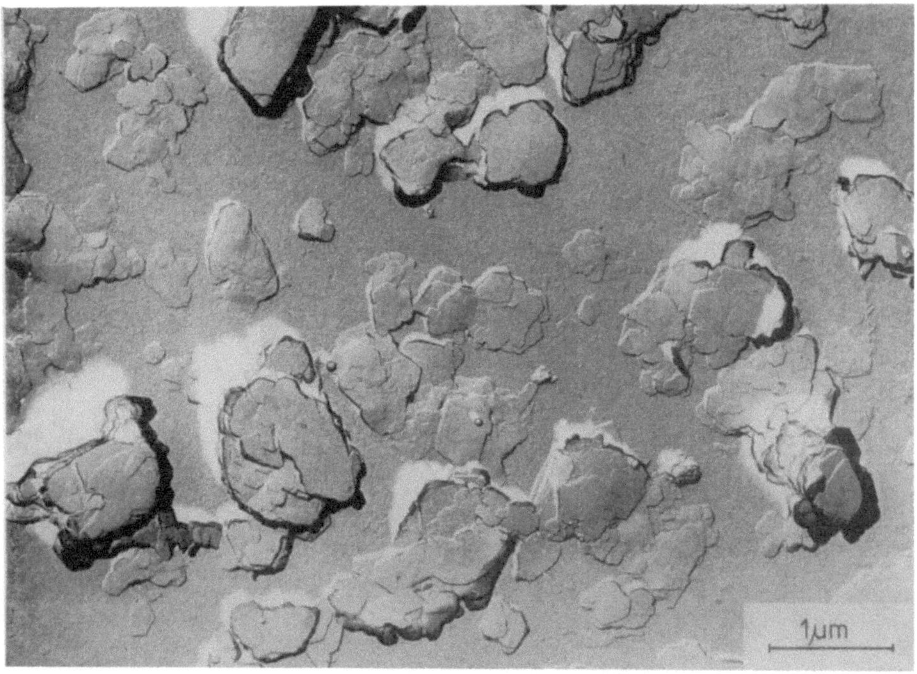

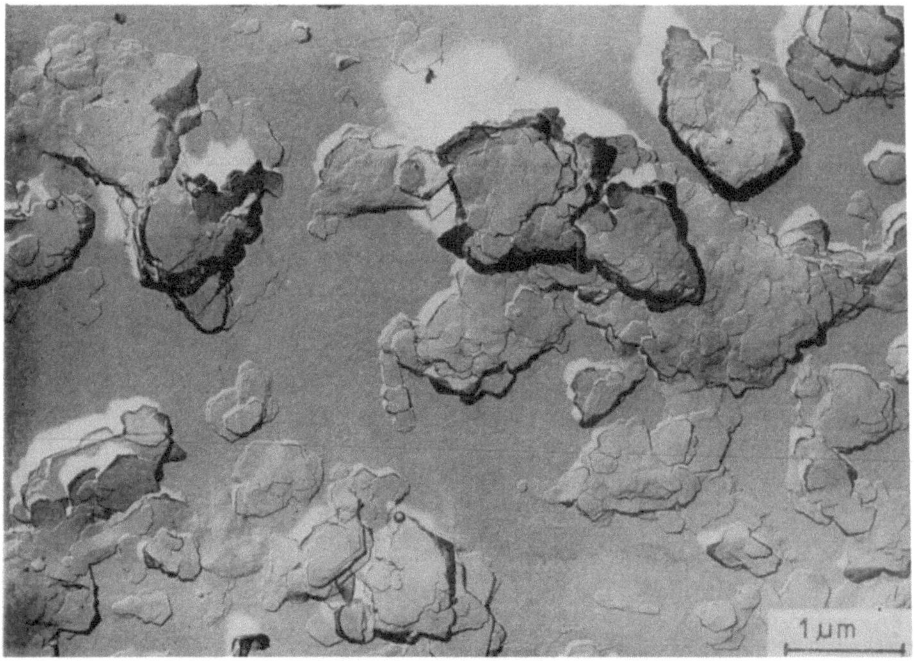

Figs. 93, 94. Replicas: *Kaolinite* from kaolinized granite, Podlesí, Czechoslovakia

Fig. 94.

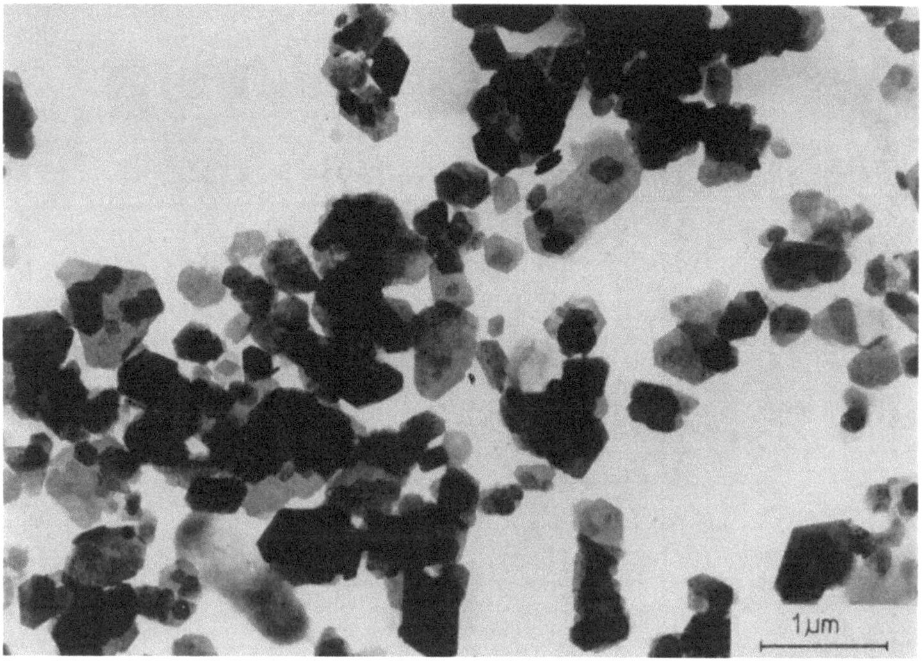

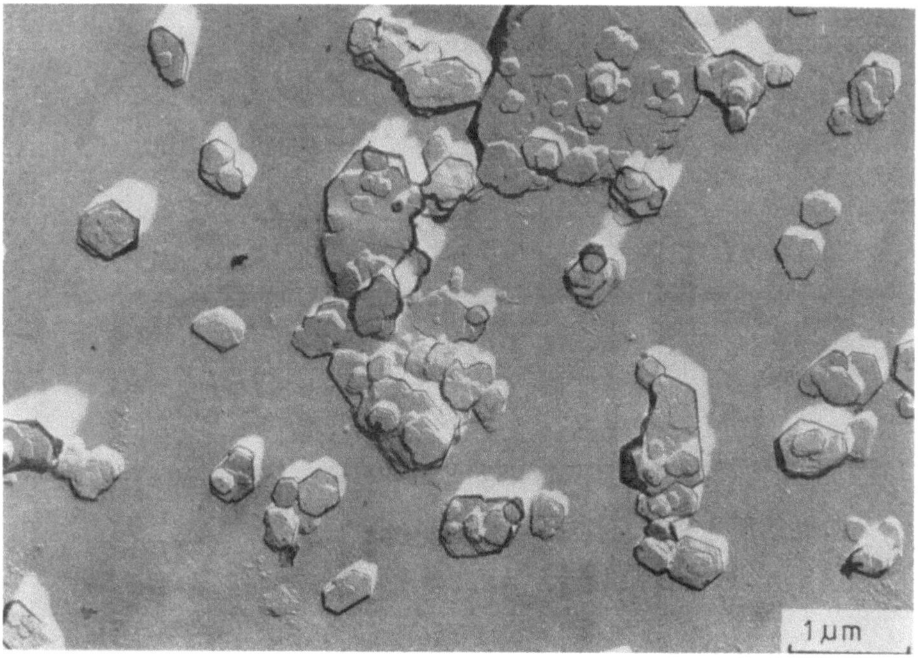

Fig. 95. TEM: *Kaolinite*, St. Felipe, Brazil

Fig. 96. Replica: *Kaolinite*, St. Felipe, Brazil

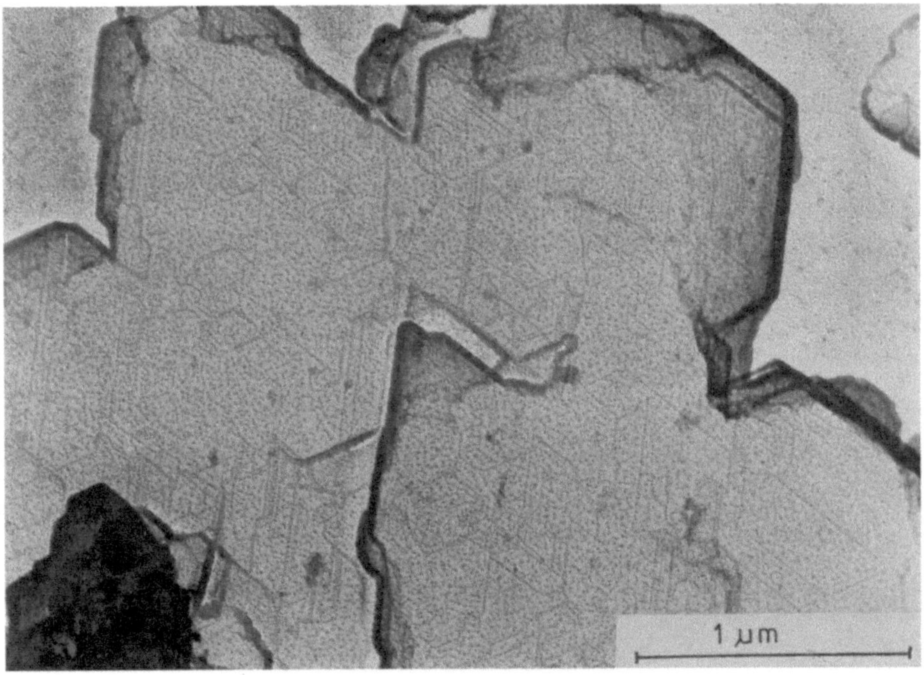

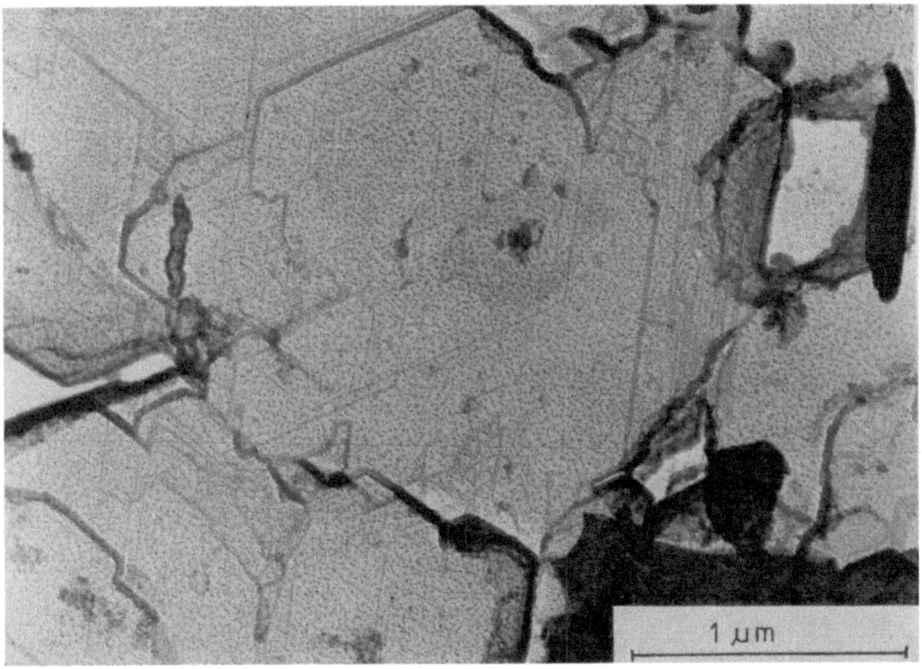

Figs. 97, 98. Replicas: *Kaolinite*, gold decorated, St. Felipe, Brazil

Fig. 98.

Fig. 99. TEM: *Kaolinite* (grade SPS) from altered granite, Cornwall, UK

Fig. 100. Replica: *Kaolinite* (grade SPS) from altered granite, Cornwall, UK

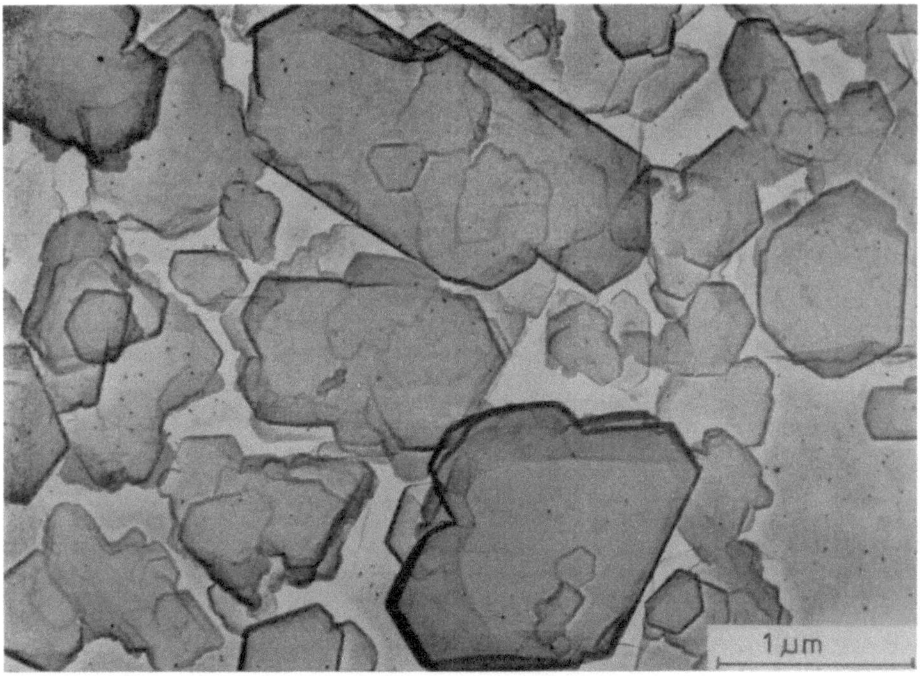

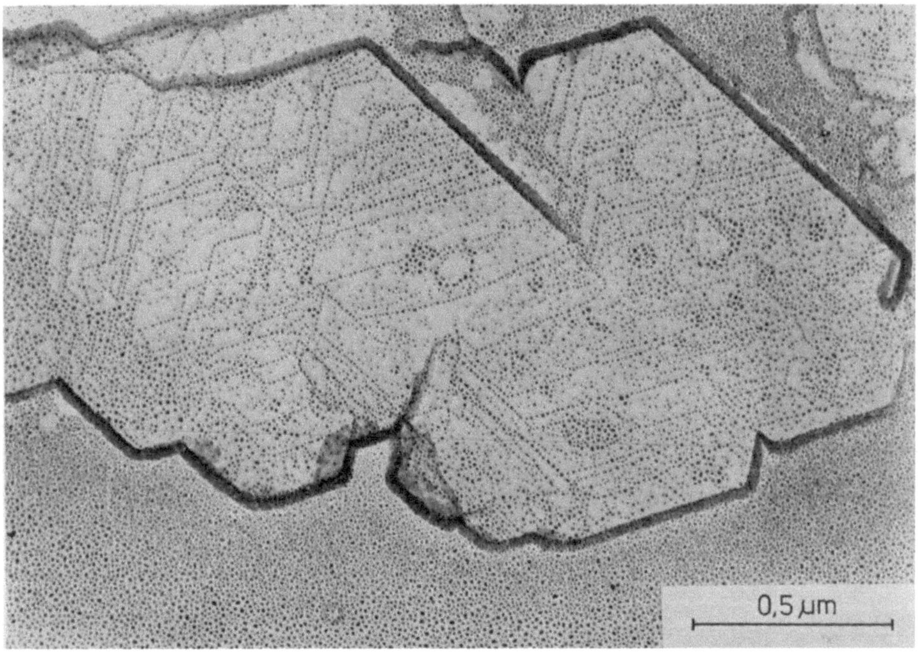

Fig. 101. Replica: *Kaolinite* (grade SPS) from altered granite, Cornwall, UK

Fig. 102. Replica: *Kaolinite*, gold decorated (grade SPS), Cornwall, UK

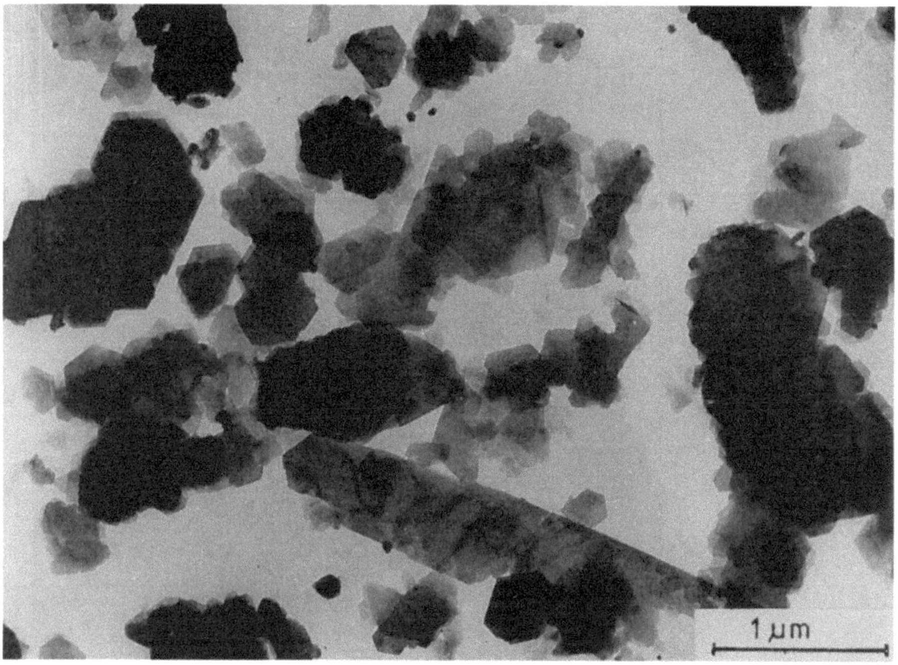

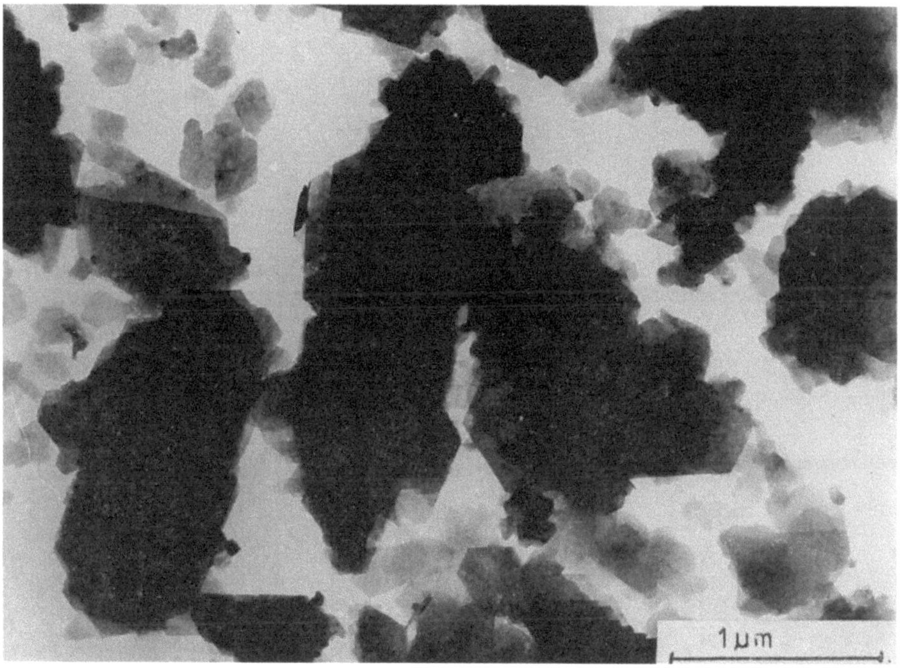

Figs. 103, 104. TEMs: *Kaolinite* from weathered vein granite, Callenberg, GDR

Fig. 104.

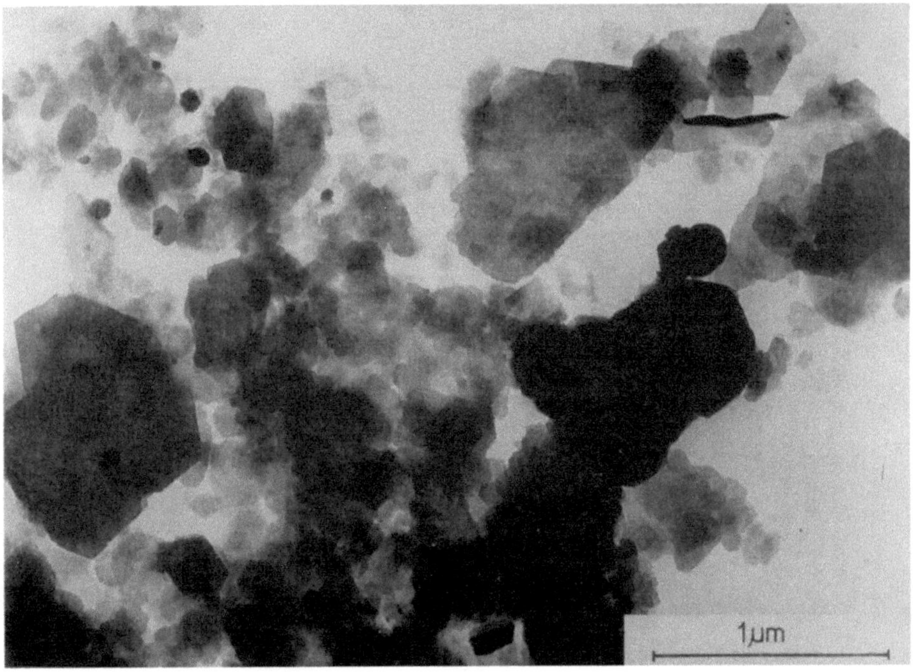

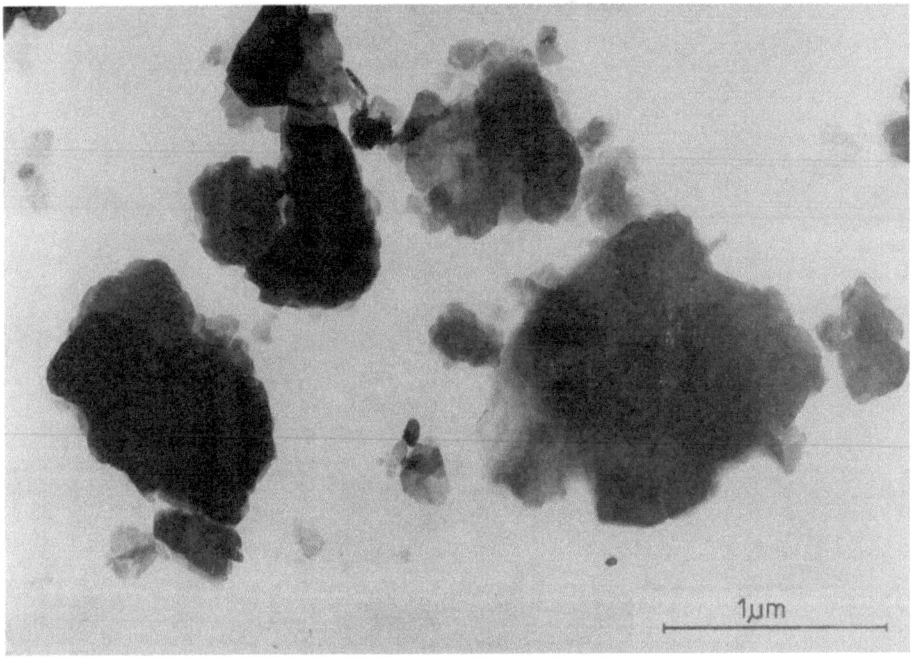

Fig. 105. TEM: *Disordered kaolinite* (fire clay), Czersegtomai, Hungary

Figs. 106, 107. TEMs: *Disordered kaolinite* (fire clay), Golden, Colorado, USA

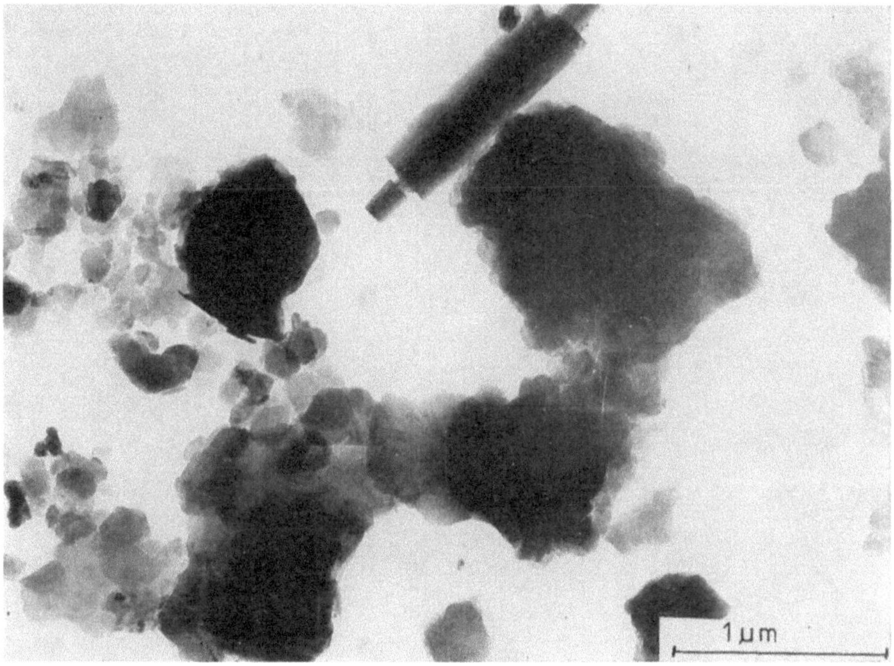

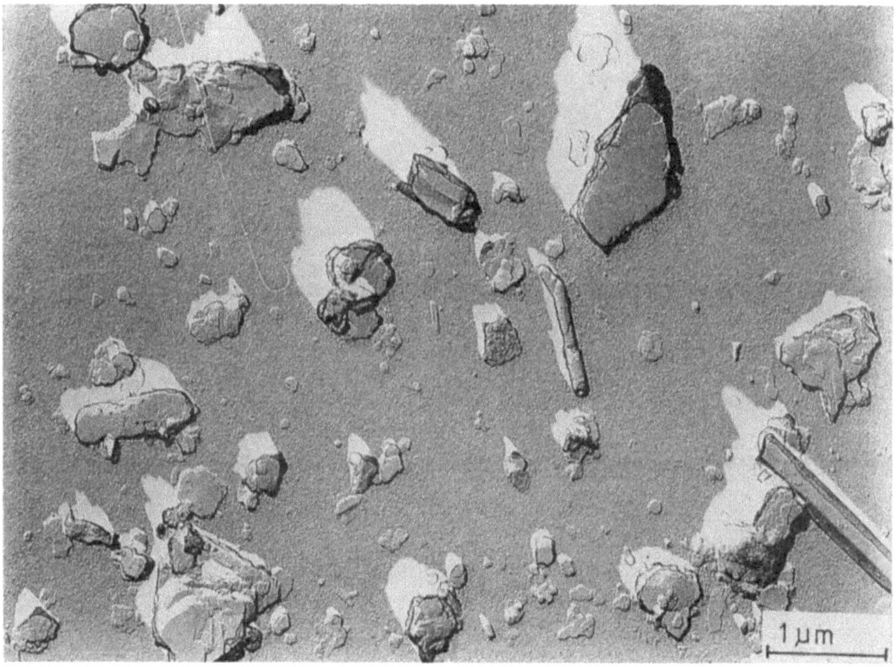

Fig. 107.

Fig. 108. Replica: *Disordered kaolinite* (fire clay) and halloysite, Golden, Colorado, USA

Fig. 109. TEM: *Dickite*, San Juanito, Chihuahua, Mexico

Fig. 110. Replica: *Dickite*, San Juanito, Chihuahua, Mexico

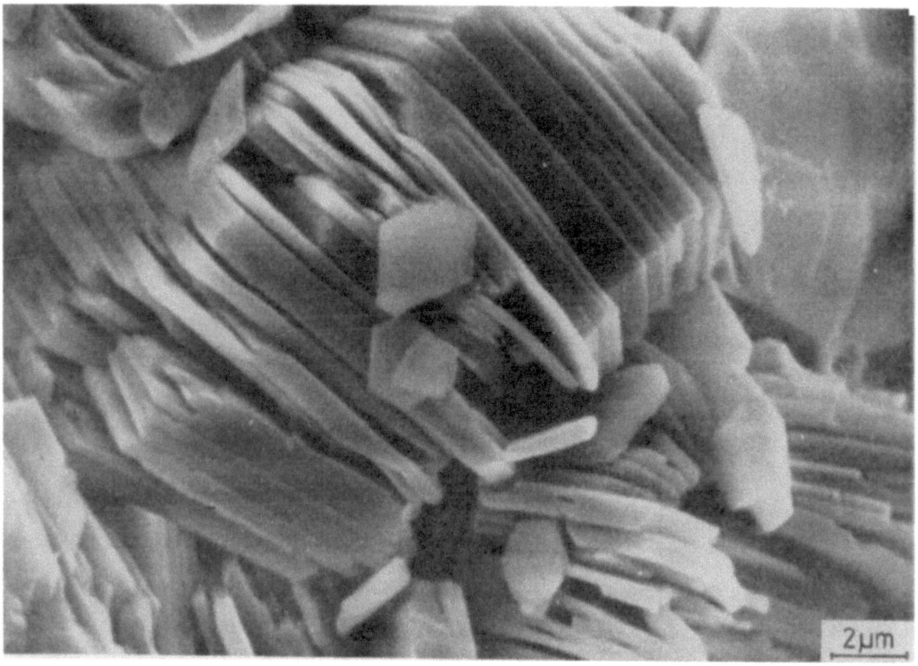

Figs. 111, 112, 113, 114. SEMs: *Dickite*, San Juanito, Chihuahua, Mexico

Fig. 112.

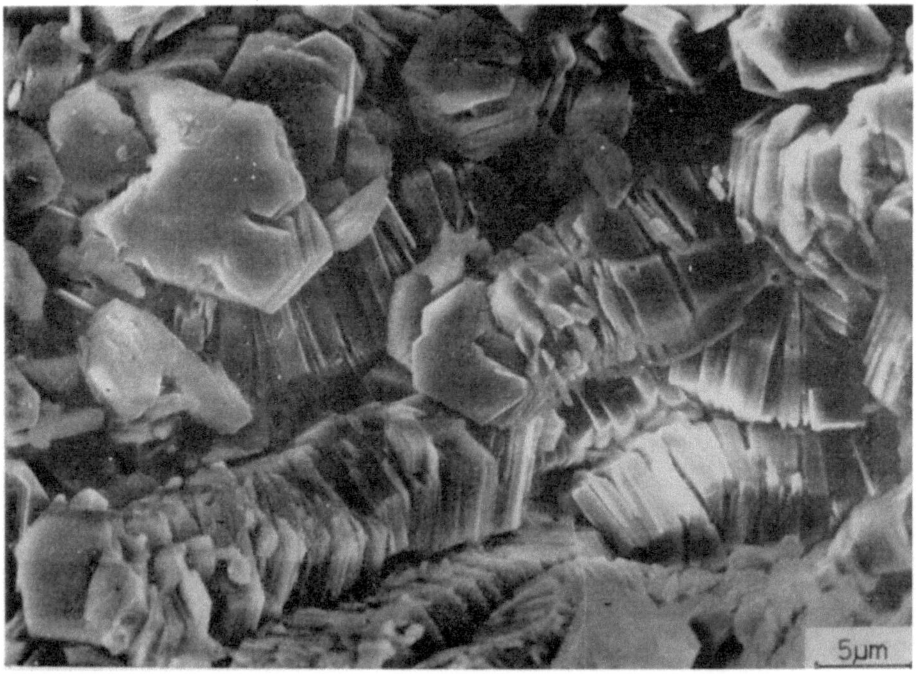

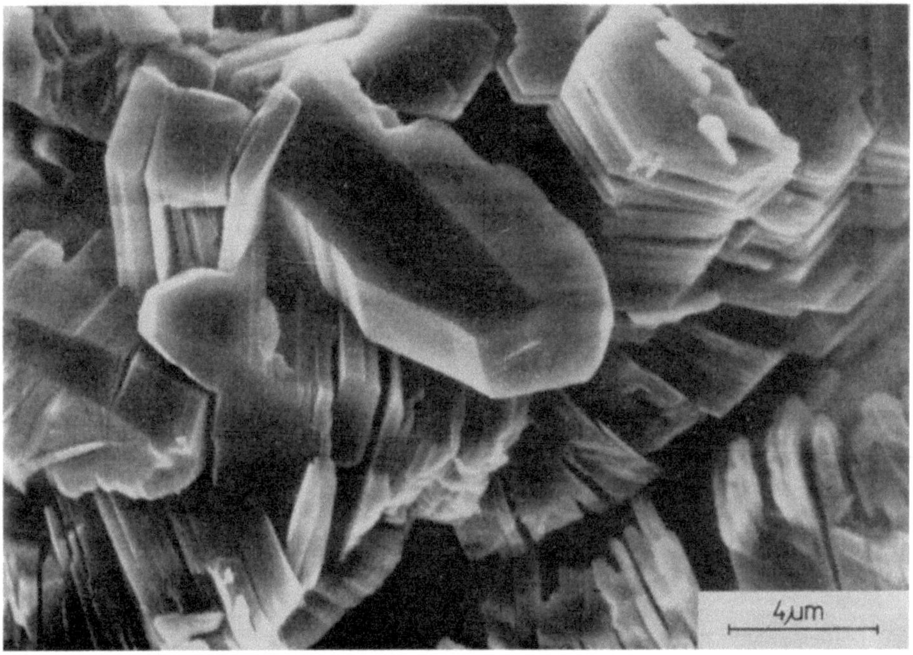

Fig. 113.

(explained at p. 97)

Fig. 114

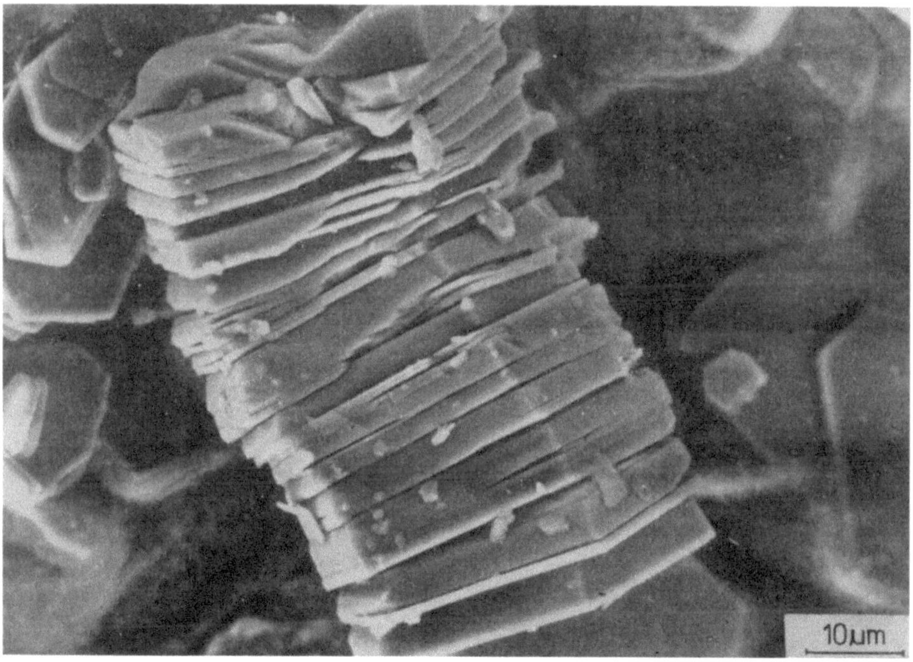

Figs. 115, 116. SEMs: *Dickite*, API No. 14, Red Mountain, Colorado, USA

Fig. 116.

Fig. 117. TEM: *Dickite* from a hydrothermal vein, tin deposit Ehrenfriedersdorf, GDR

Fig. 118. Replica: *Dickite* from a hydrothermal vein, tin deposit Ehrenfriedersdorf, GDR

Figs. 119, 120. SEMs: *Dickite* from a hydrothermal vein, tin deposit Ehrenfriedersdorf, GDR

Fig. 120.

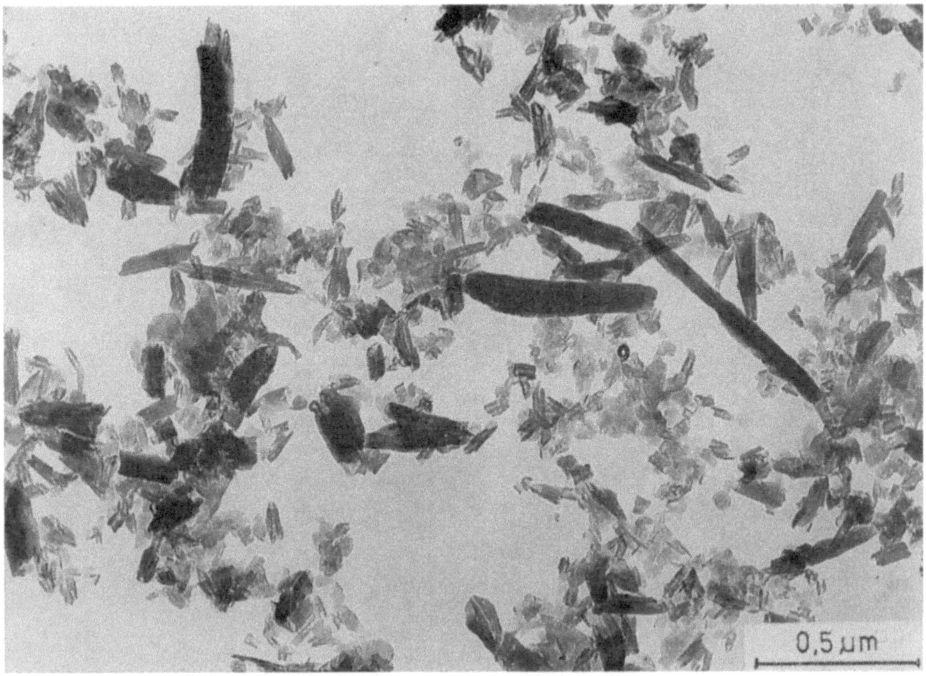

Fig. 121. TEM: *Halloysite*, Eureka, Utah, USA

Fig. 122. Replica: *Halloysite*, Eureka, Utah, USA

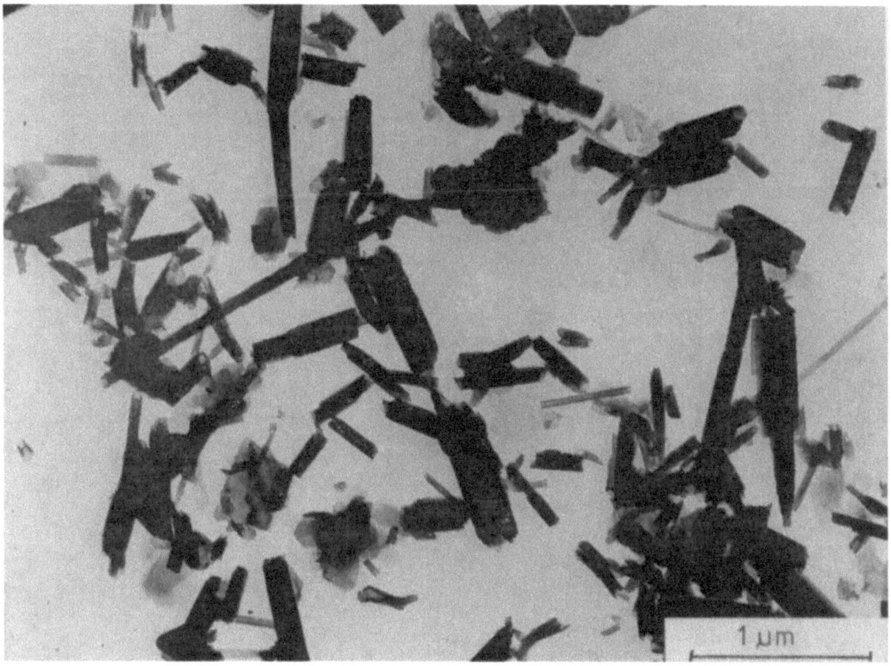

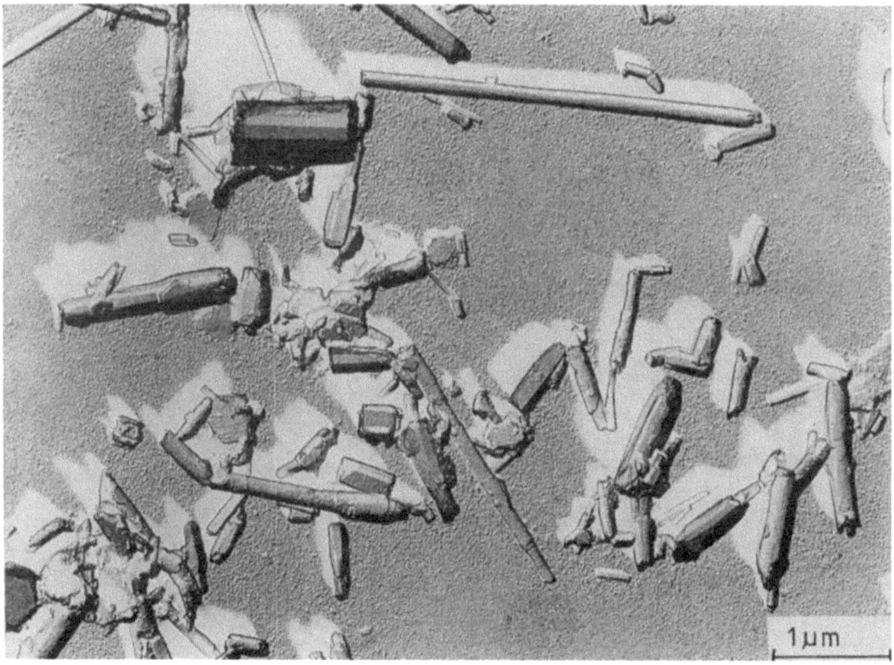

Fig. 123. TEM: *Halloysite*, API No. 29, Wagon Wheel Gap, Colorado, USA

Fig. 124. Replica: *Halloysite*, API No. 29, Wagon Wheel Gap, Colorado, USA

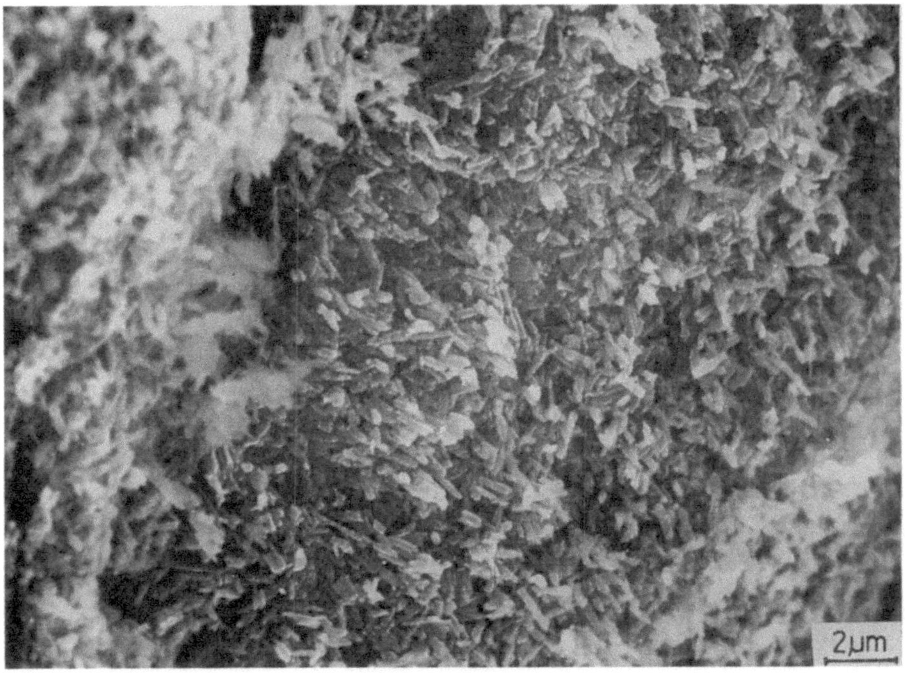

Figs. 125, 126. SEMs: *Halloysite*, API No. 29, Wagon Wheel Gap, Colorado, USA

Fig. 126.

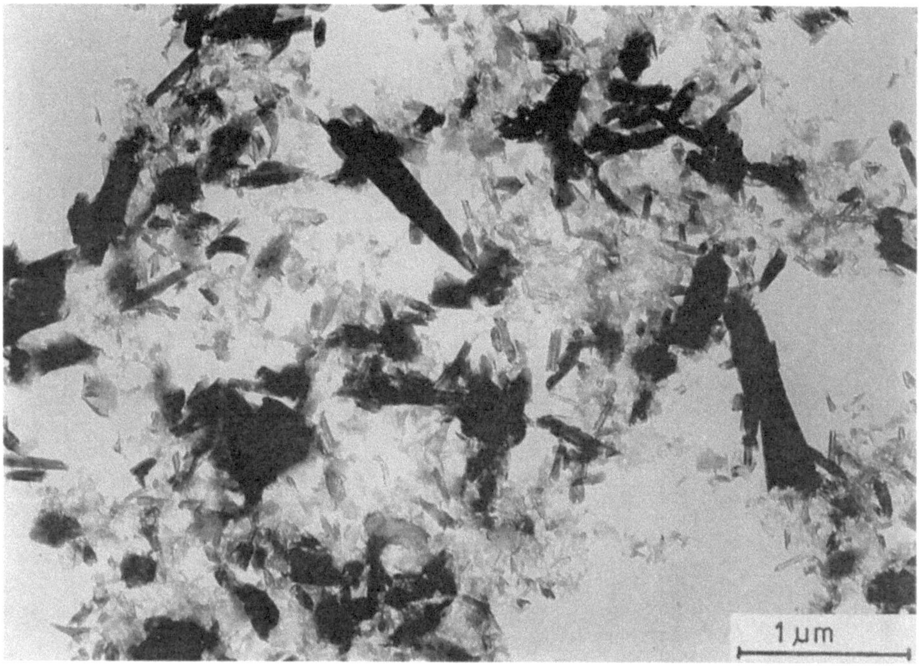

Fig. 127. TEM: *Halloysite*, API No. 12, Bedford, Indiana, USA

Fig. 128. Replica: *Halloysite*, API No. 12, Bedford, Indiana, USA

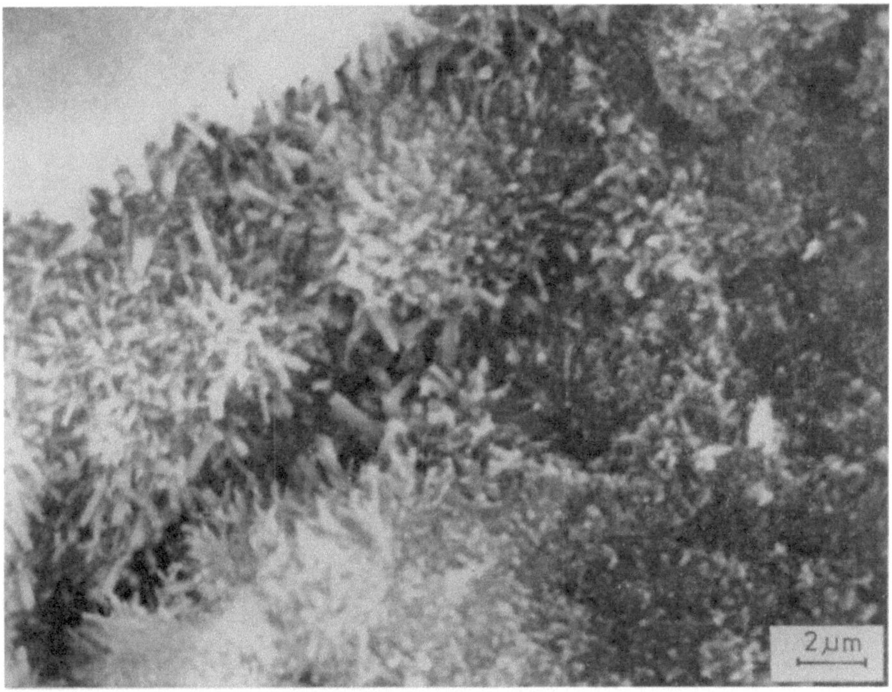

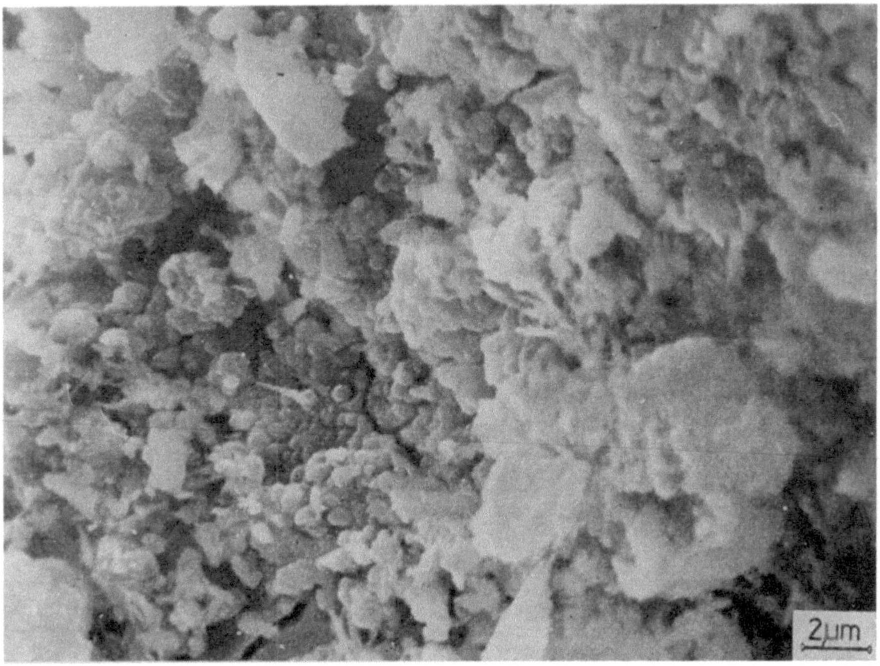

Figs. 129, 130, 131, 132. SEMs: *Halloysite*, API No. 12, Bedford, Indiana, USA

Fig. 130.

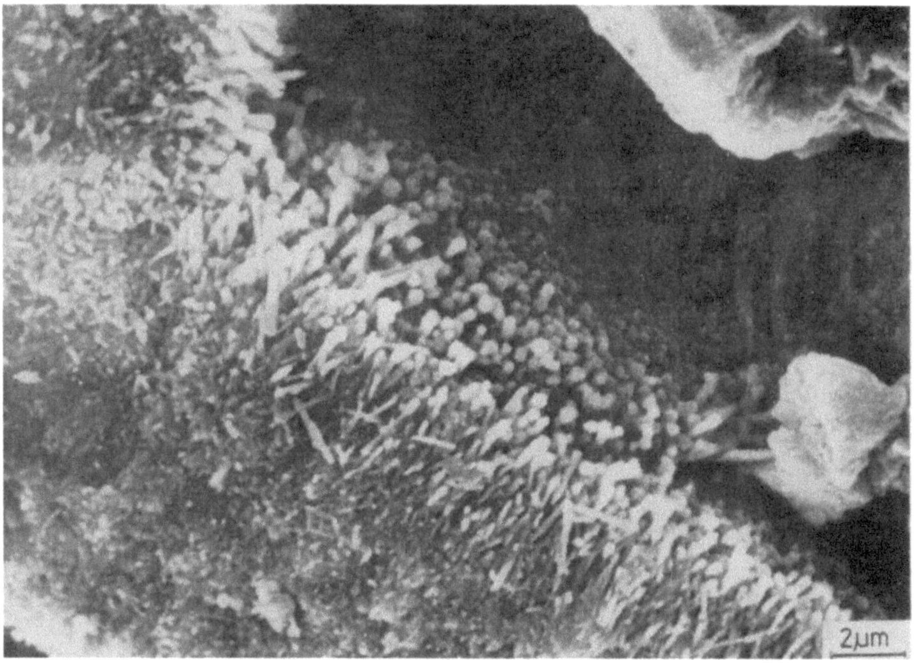

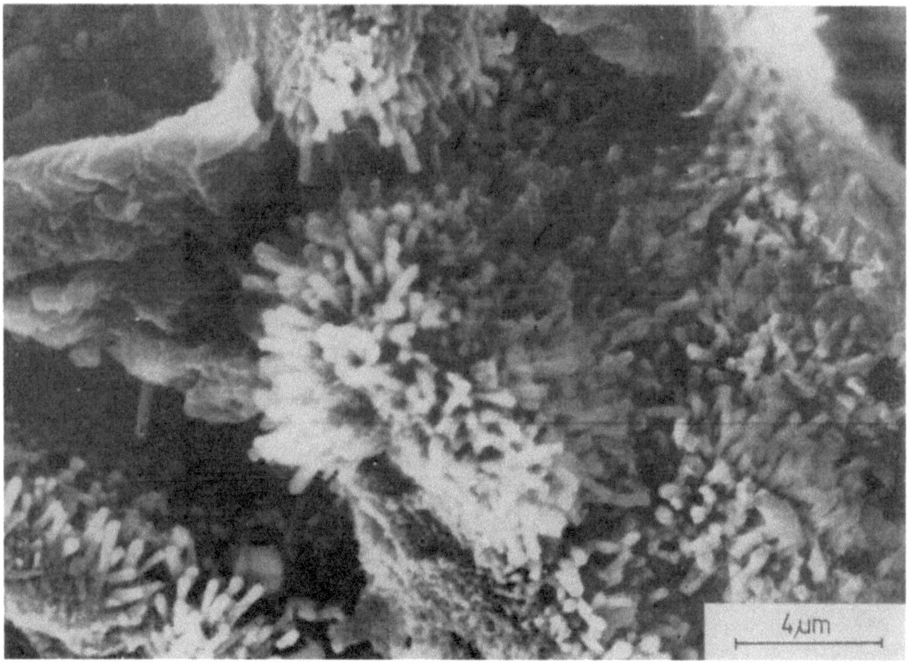

Fig. 131.

Fig. 132.

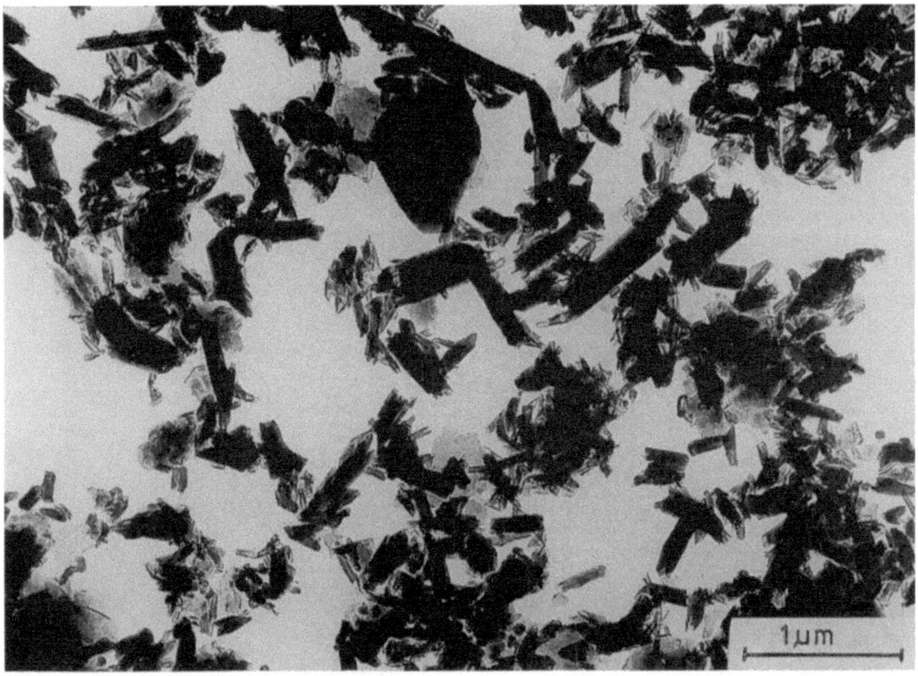

Figs. 133, 134. TEMs: *Halloysite* from weathered gabbro, Callenberg, GDR

Fig. 134.

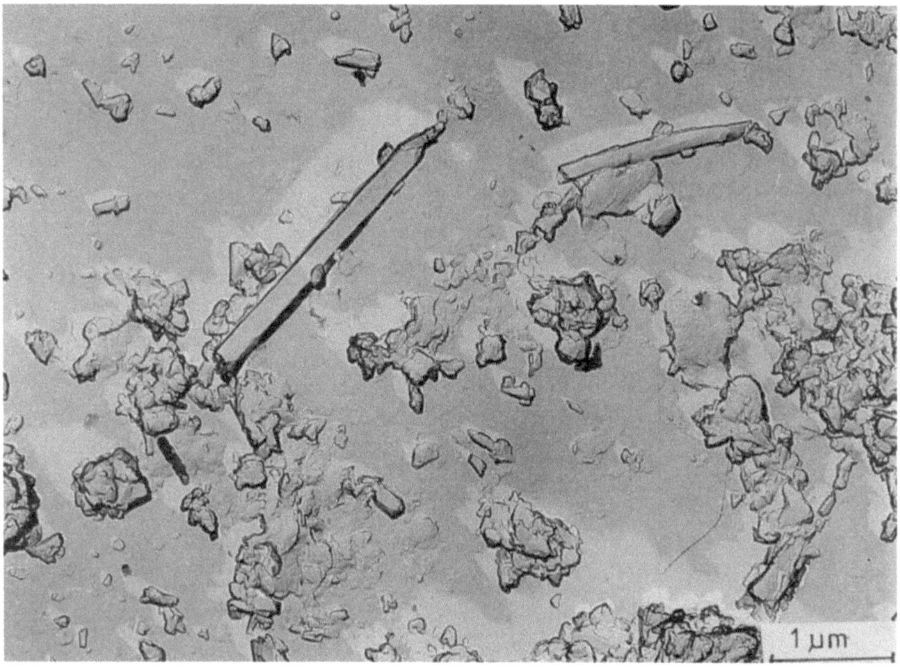

Fig. 135. Replica: *Halloysite* from weathered gabbro, Callenberg, GDR

Fig. 136. SEM: *Halloysite* from weathered gabbro, Callenberg, GDR

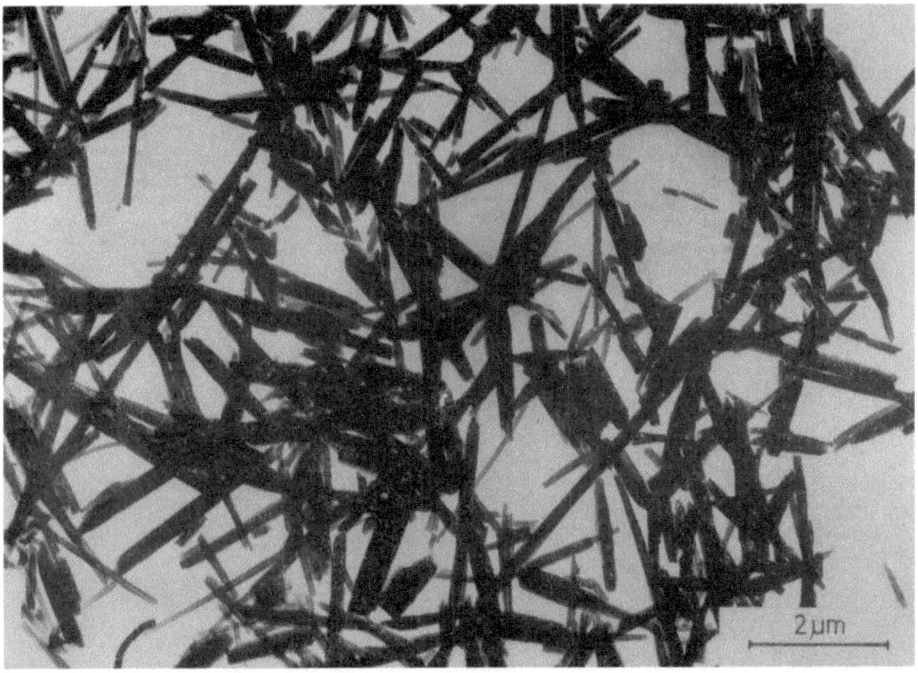

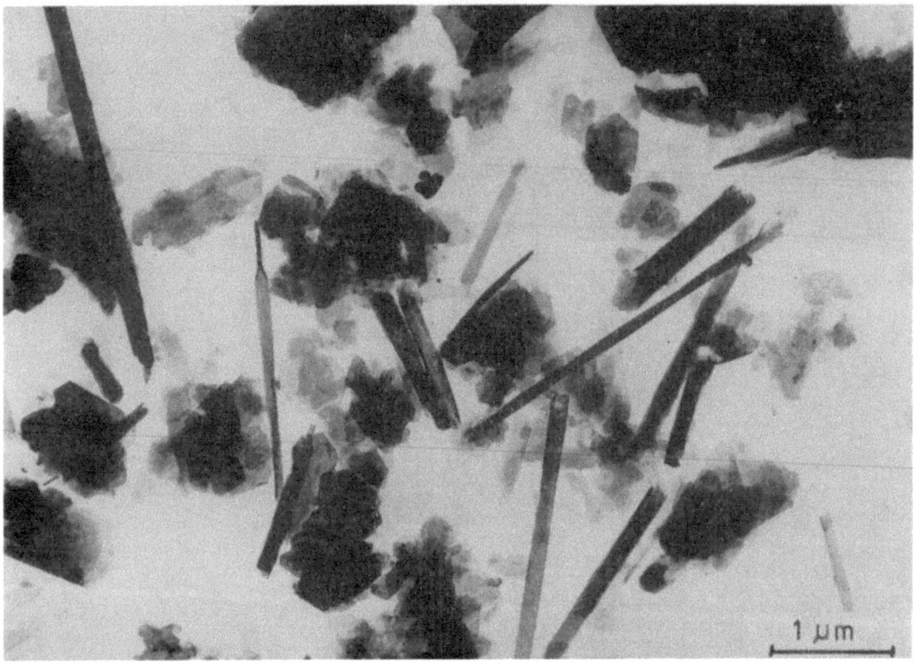

Fig. 137. TEM: *Halloysite* from the oxidizing zone of the fluorite deposit, Schönbrunn, GDR

Fig. 138. TEM: *Halloysite* from porphyry kaolin, Hohburg, GDR

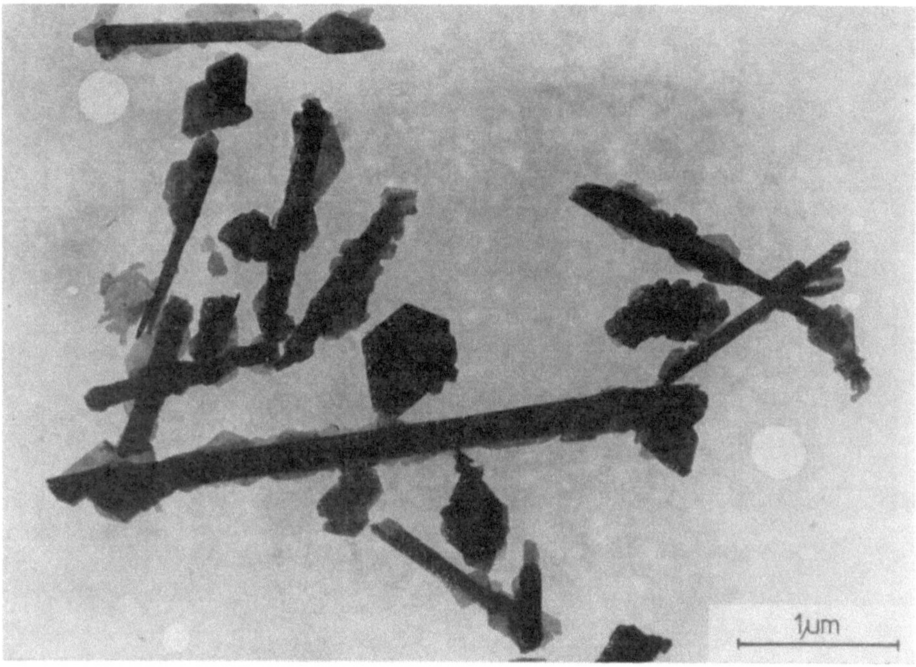

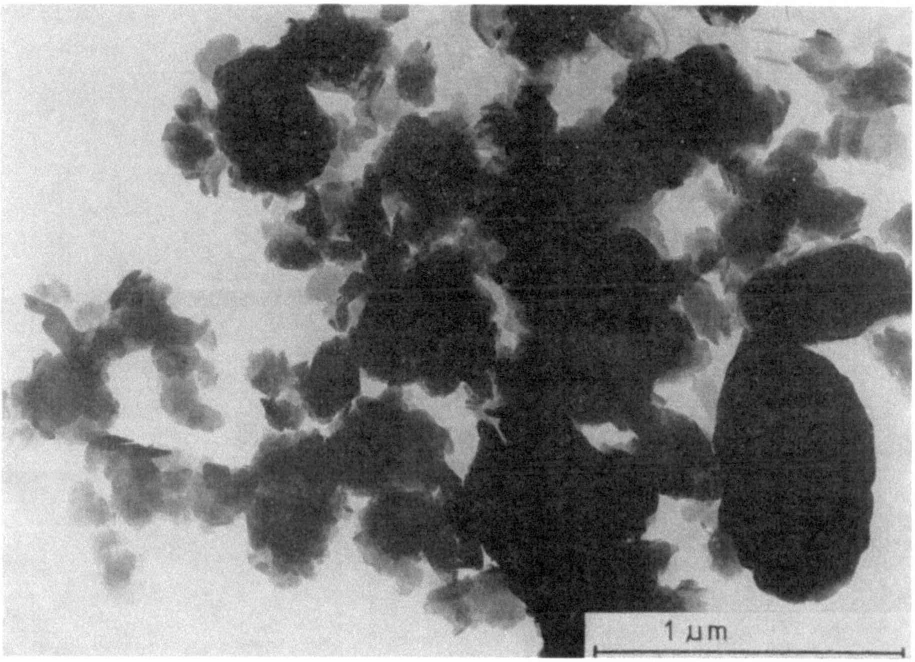

Fig. 139. TEM: *Halloysite* and kaolinite from granodiorite kaolin, Caminau, GDR

Fig. 140. TEM: *Halloysite*, Michalovce, Czechoslovakia

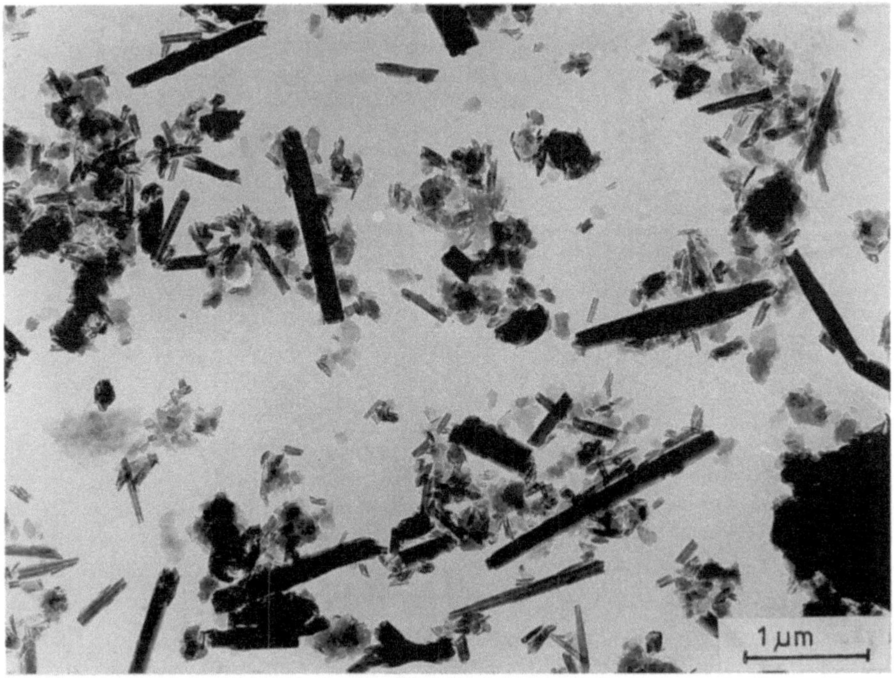

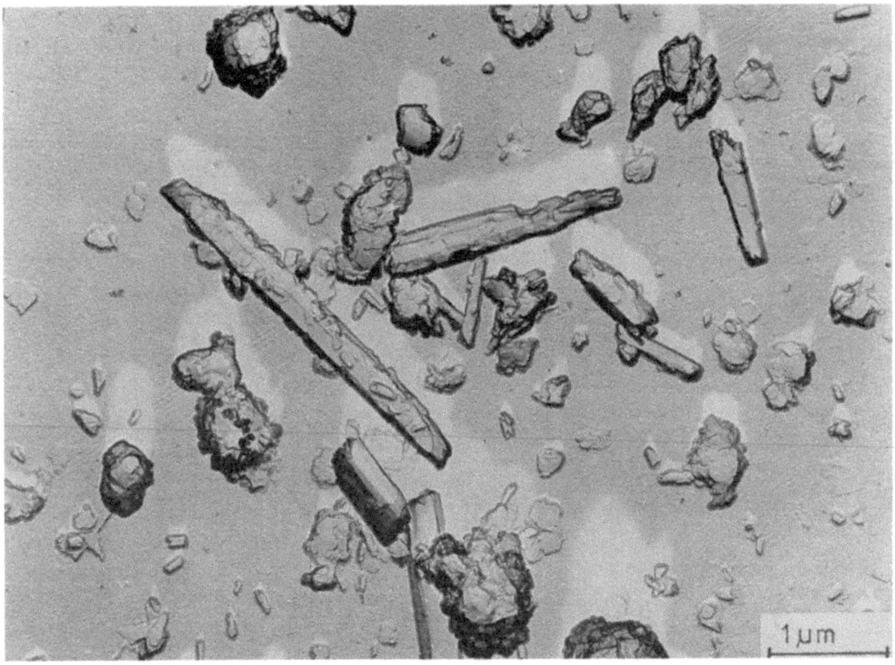

Fig. 141. TEM: *Halloysite*, Djebel Djebar, Algeria

Fig. 142. Replica: *Halloysite*, Djebel Djebar, Algeria

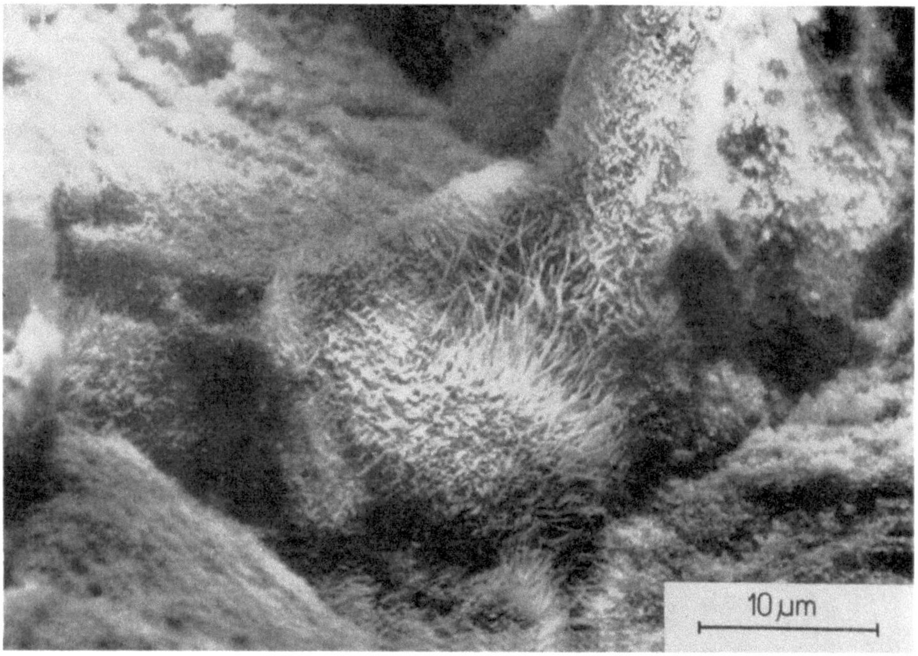

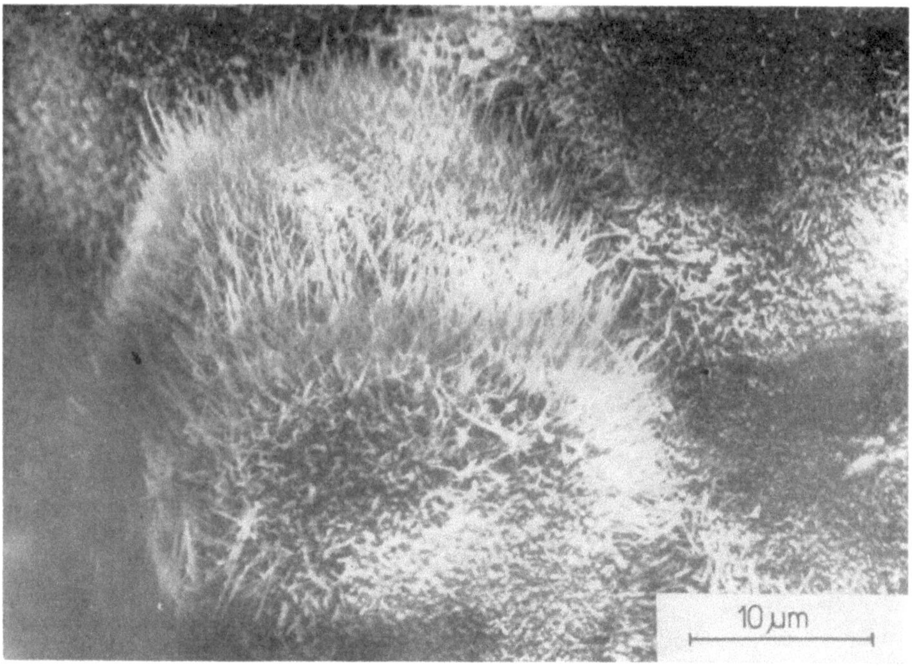

Figs. 143, 144, 145, 146. SEMs: *Halloysite*, Djebel Djebar, Algeria

Fig. 144.

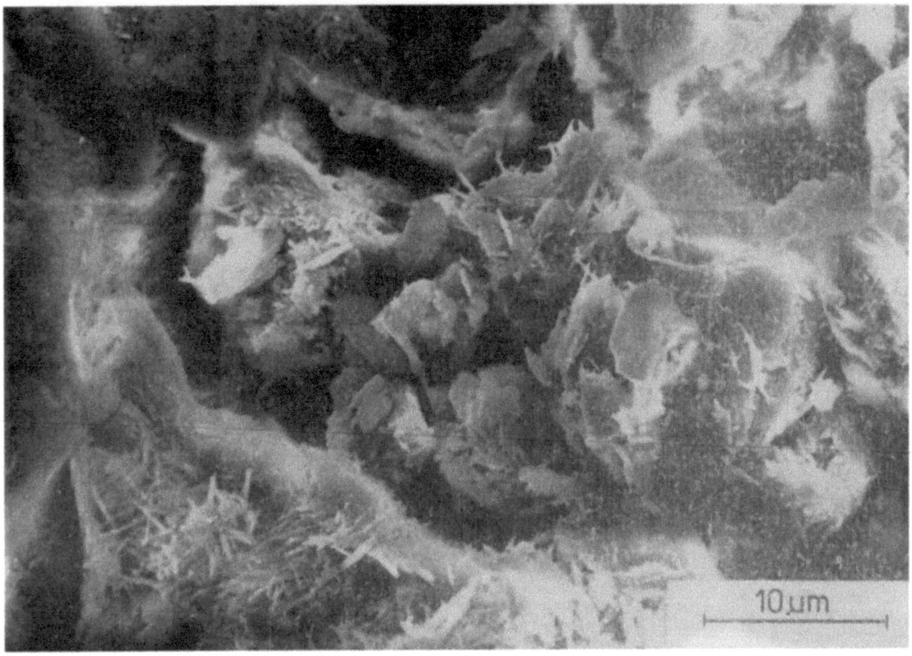

Fig. 145.

(*explained at p. 113*)

Fig. 146.

Figs. 147, 148. TEMs: *Clinochrysotile* from weathered serpentinite, Callenberg, GDR

Fig. 148.

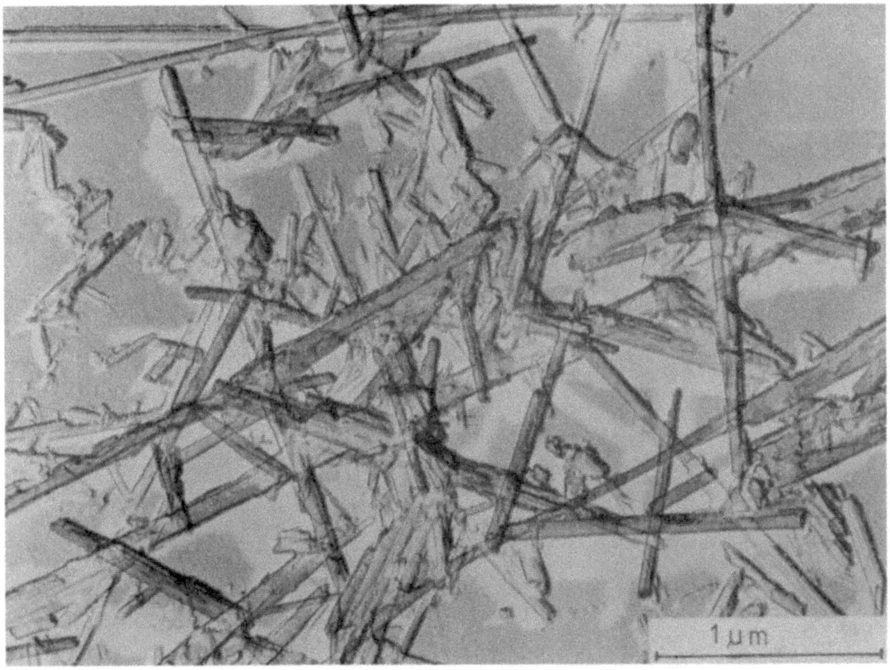

Figs. 149, 150. Replicas: *Clinochrysotile* from weathered serpentinite, Callenberg, GDR

Fig. 150.

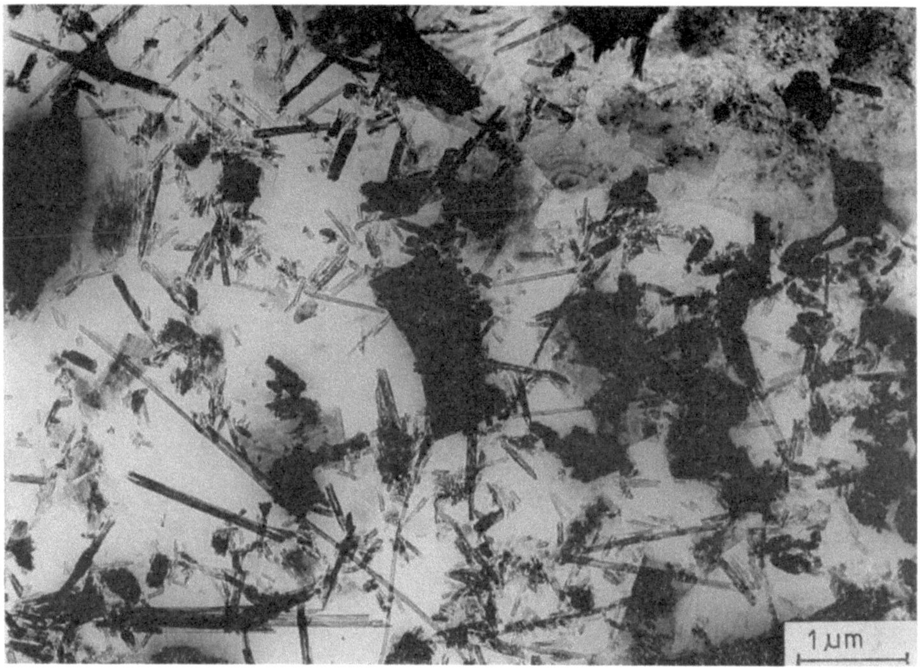

Fig. 151. TEM: *Orthochrysotile* from weathered serpentinite, Callenberg, GDR

Fig. 152. TEM: *Asbestos*, Orenburg, Urals, USSR

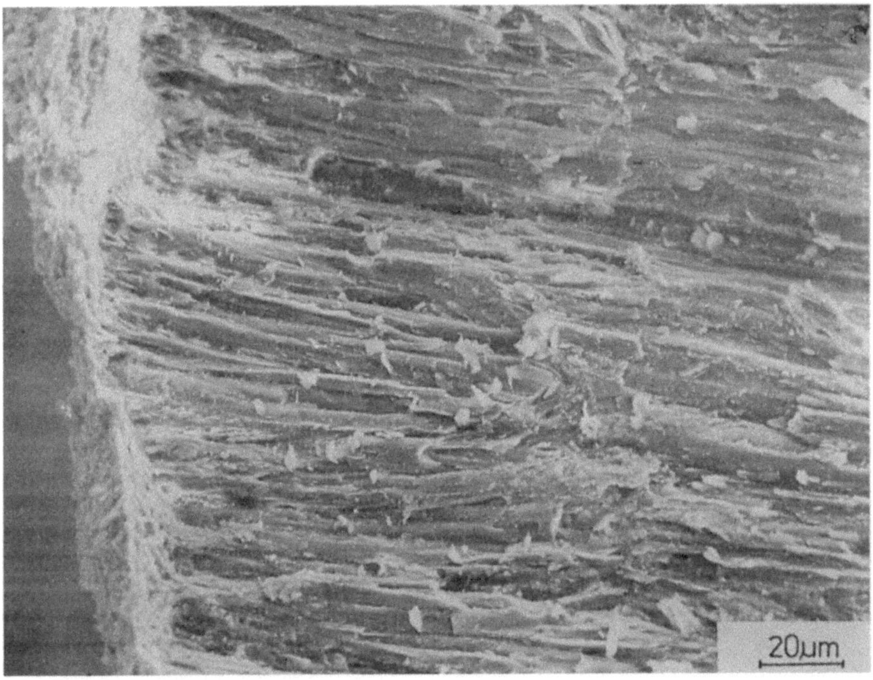

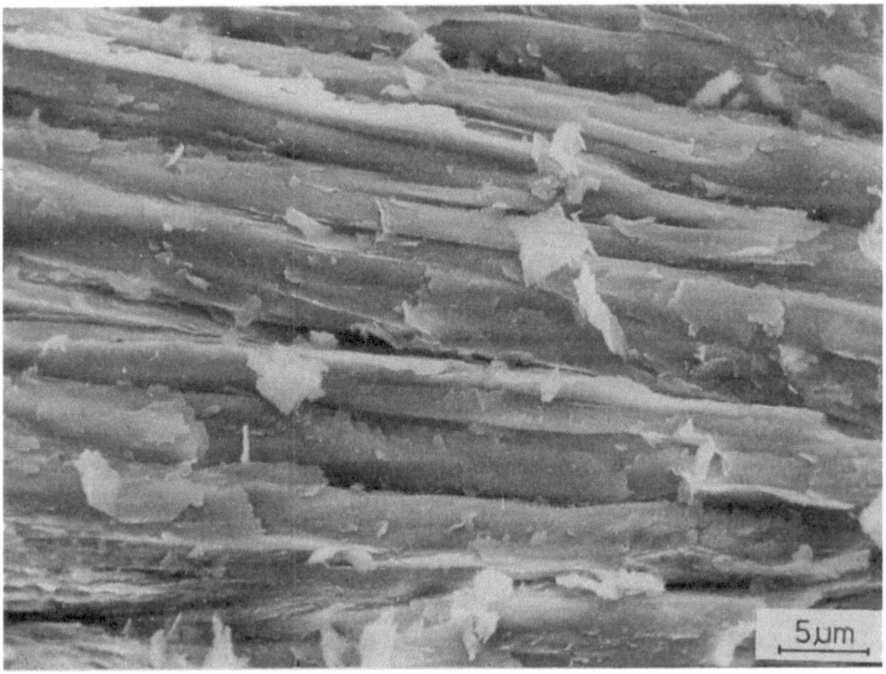

Fig. 153. SEM: *Chrysotile* from serpentinite, Ansprung, GDR

Fig. 154. SEM: *Chrysotile* from serpentinite, detail of Fig. 153, Ansprung, GDR

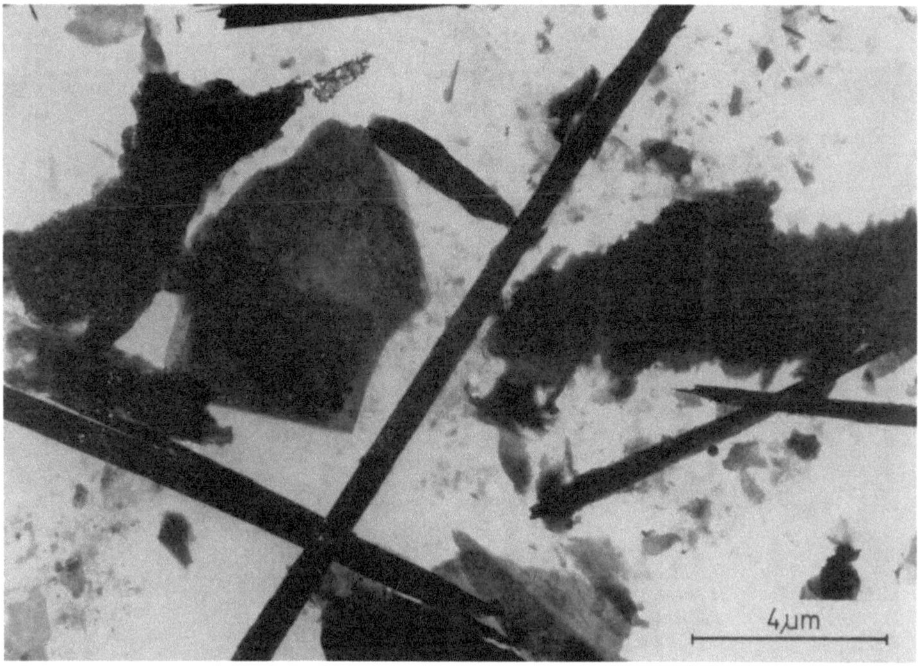

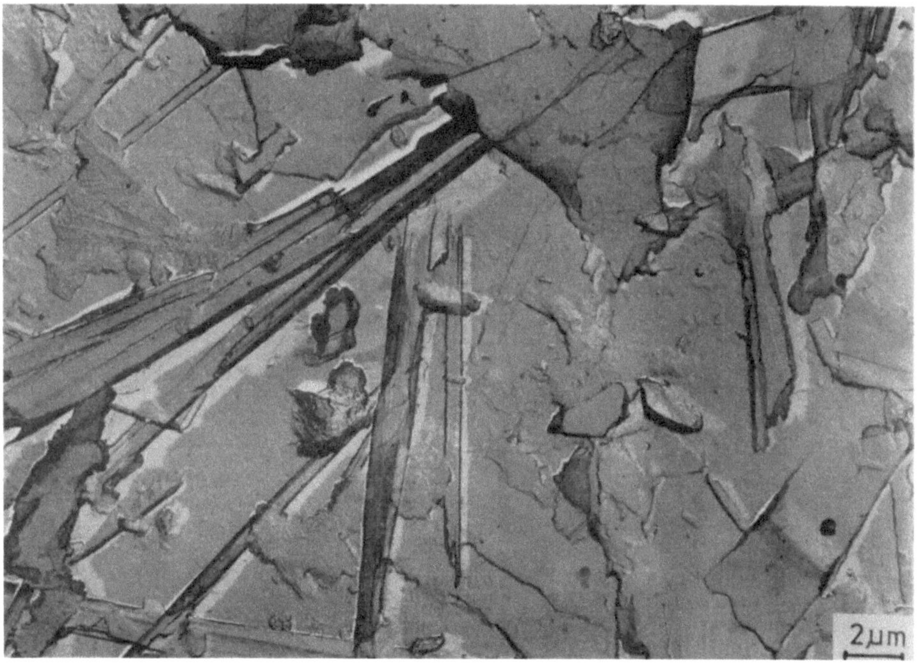

Fig. 155. TEM: *Chrysotile* and *antigorite* (platelets) from weathered serpentinite, Callenberg, GDR

Fig. 156. Replica: *Chrysotile* and *antigorite* (platelets) from weathered serpentinite, Callenberg, GDR

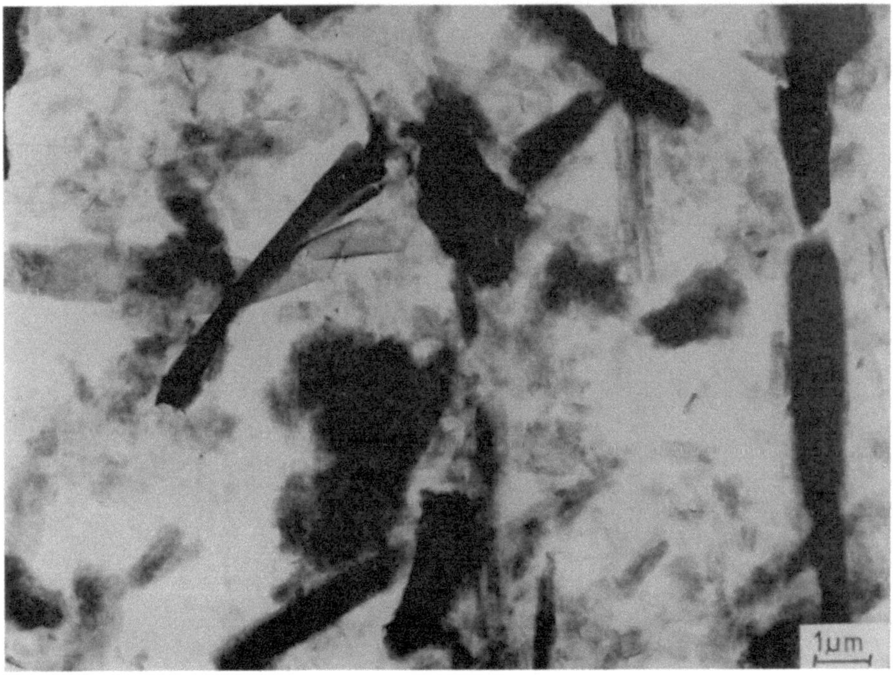

Fig. 157. Replica: *Antigorite* from weathered serpentinite, Callenberg, GDR

Fig. 158. TEM: *Antigorite* from weathered serpentinite, Callenberg, GDR

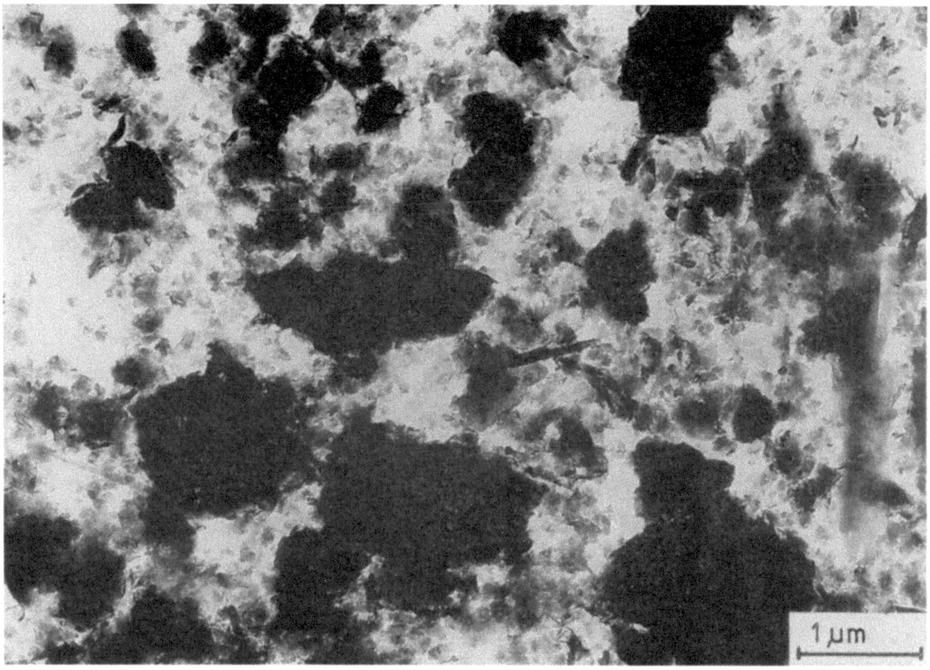

Fig. 159. TEM: *Garnierite* (7 Å) from weathered serpentinite, Noumea, New Caledonia

Fig. 160. Replica: *Garnierite* (7 Å) from weathered serpentinite, Noumea, New Caledonia

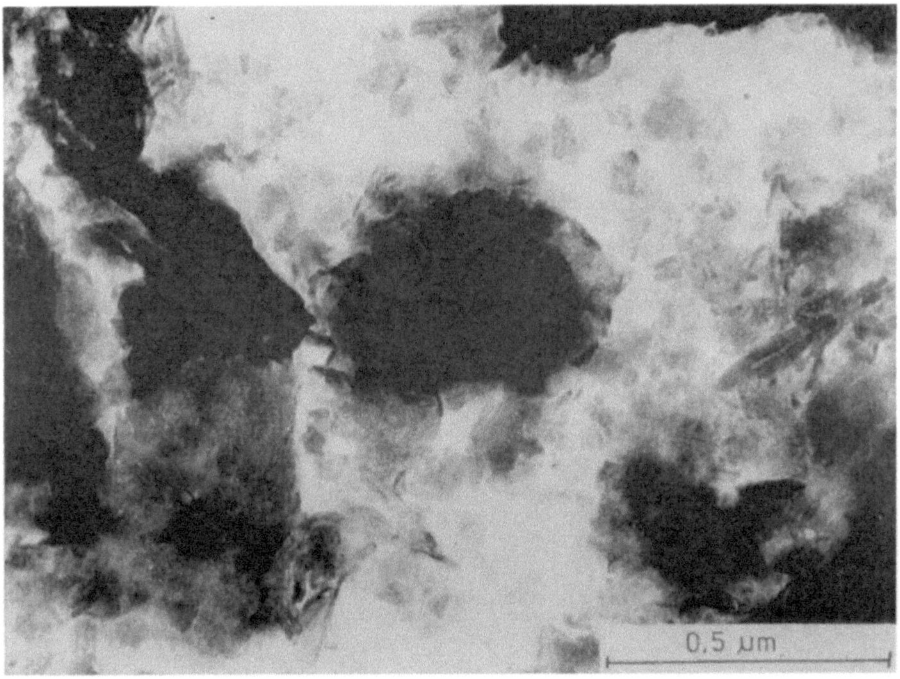

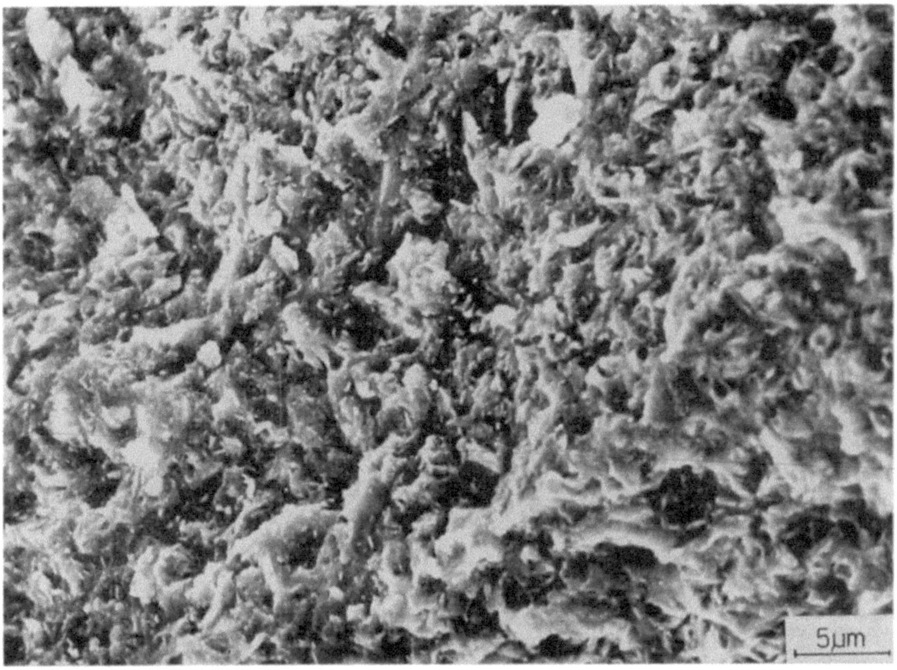

Fig. 161. TEM: *Garnierite* (7 Å) from weathered serpentinite, Noumea, New Caledonia

Fig. 162. SEM: *Garnierite* (7 Å) from weathered serpentinite, Noumea, New Caledonia

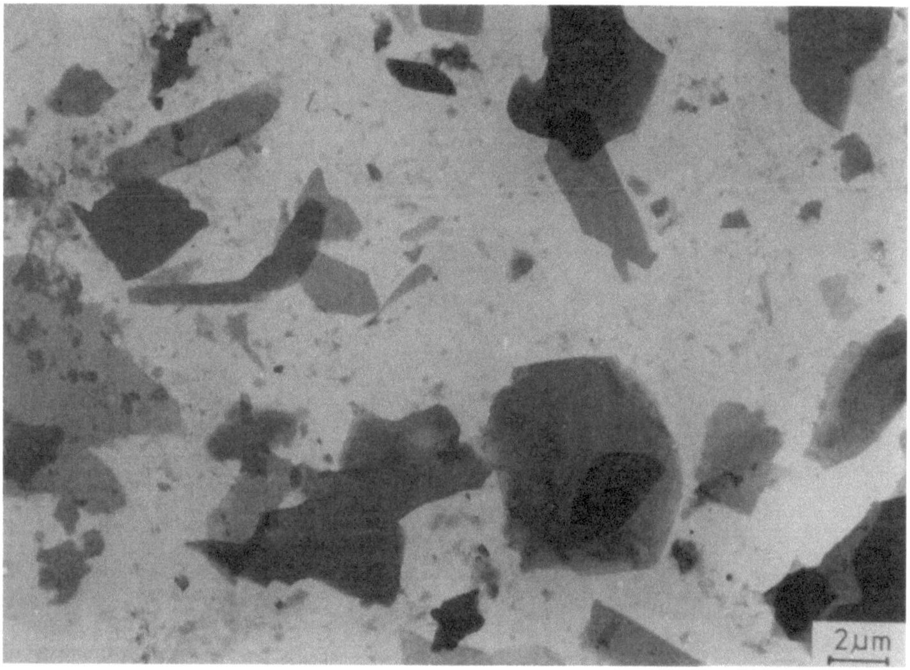

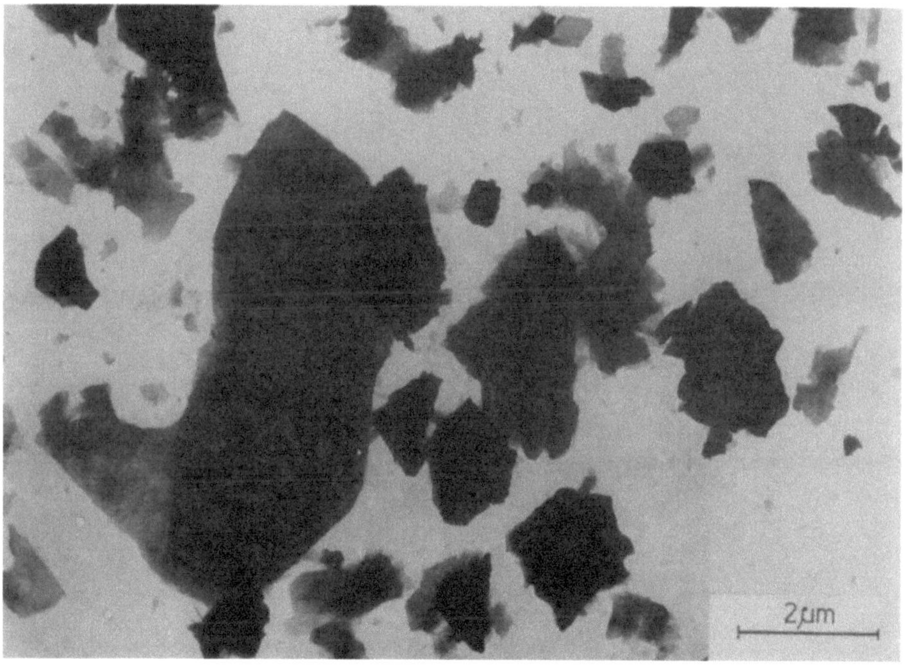

Figs. 163, 164, 165. TEMs: *Pyrophyllite*, Shokozan, Japan

Fig. 164.

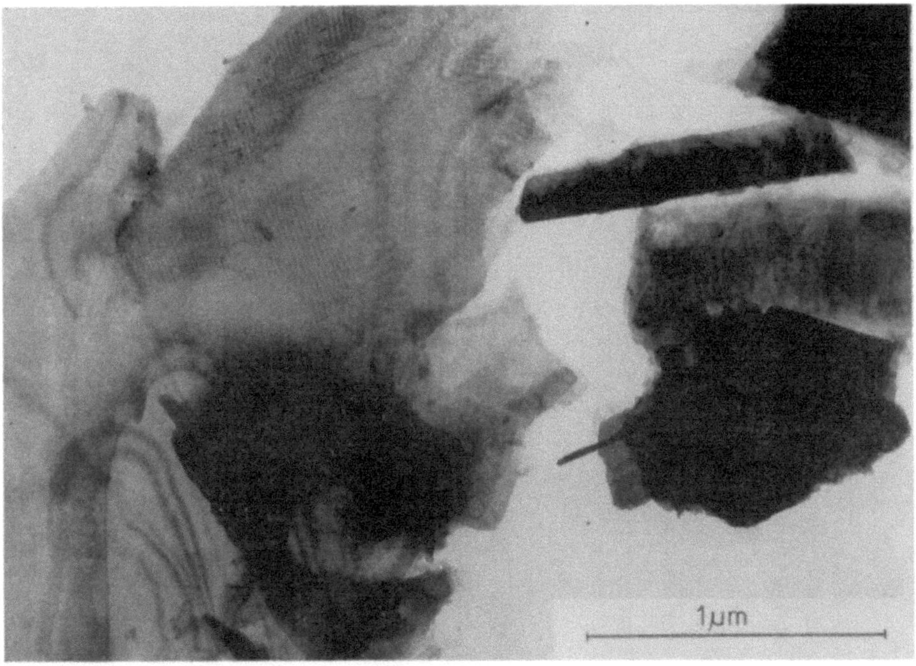

Fig. 165. (*explained at p. 123*)

Fig. 166. Replica: *Pyrophyllite*, Shokozan, Japan

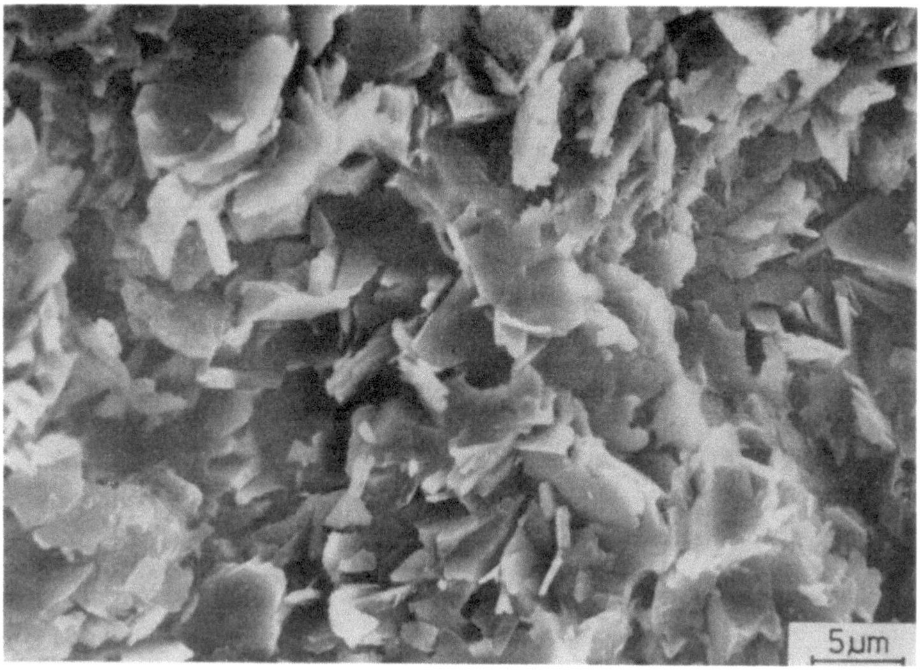

Figs. 167, 168. SEMs: *Pyrophyllite*, Shokozan, Japan

Fig. 168.

Fig. 169. TEM: *Pyrophyllite*, API No. 49, Robbins, North Carolina, USA

Fig. 170. Replica: *Pyrophyllite*, API No. 49, Robbins, North Carolina, USA

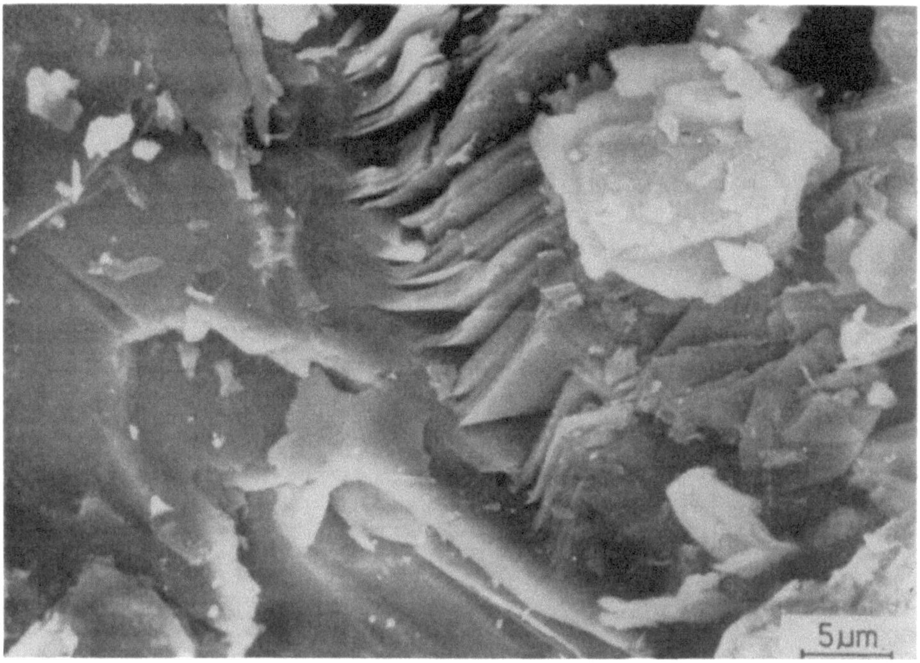

Figs. 171, 172. SEMs: *Pyrophyllite*, API No. 49, Robbins, North Carolina, USA

Fig. 172.

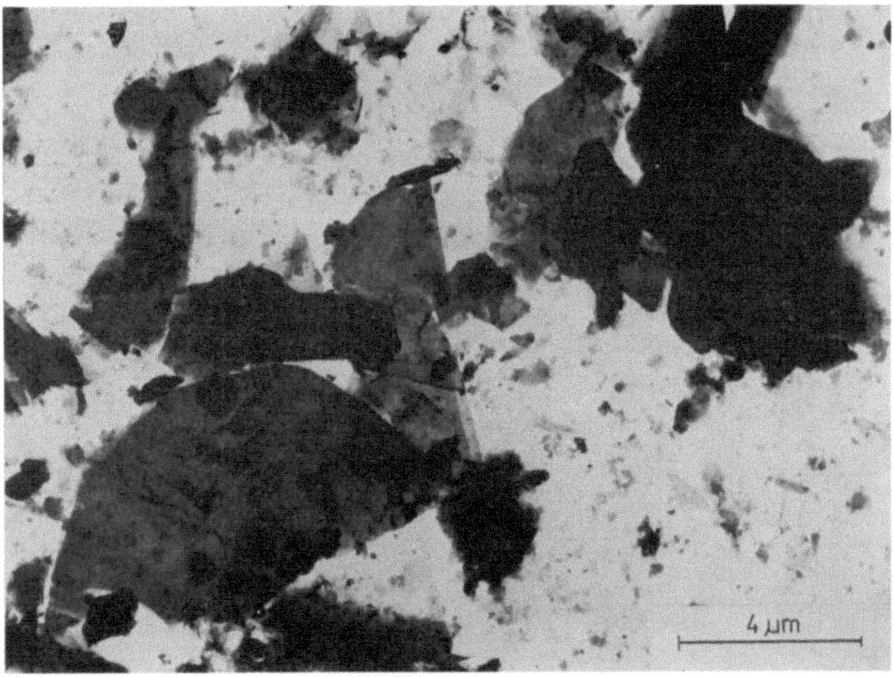

Fig. 173. TEM: *Talc*, Lobsdorf, GDR

Fig. 174. TEM: *Talc*, Callenberg, GDR

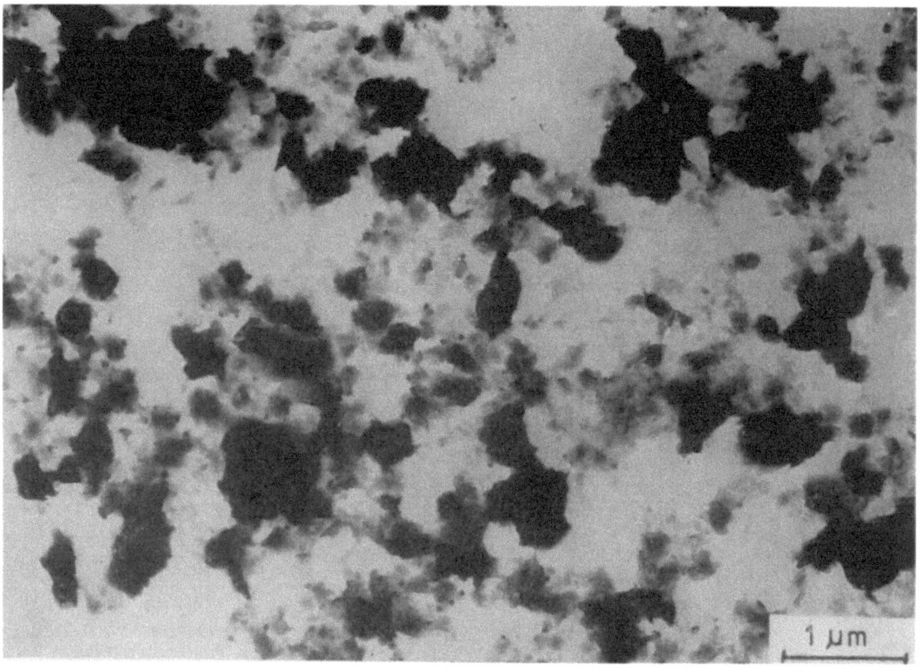

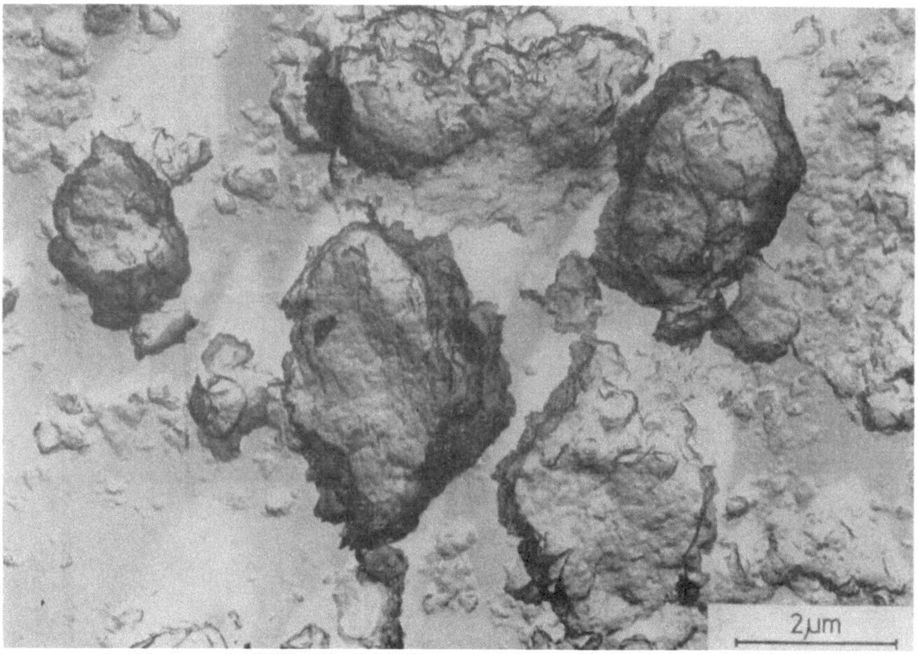

Fig. 175. TEM: *Garnierite* (10 Å) from weathered serpentinite, Callenberg, GDR

Fig. 176. Replica: *Garnierite* (10 Å) from weathered serpentinite, Callenberg, GDR

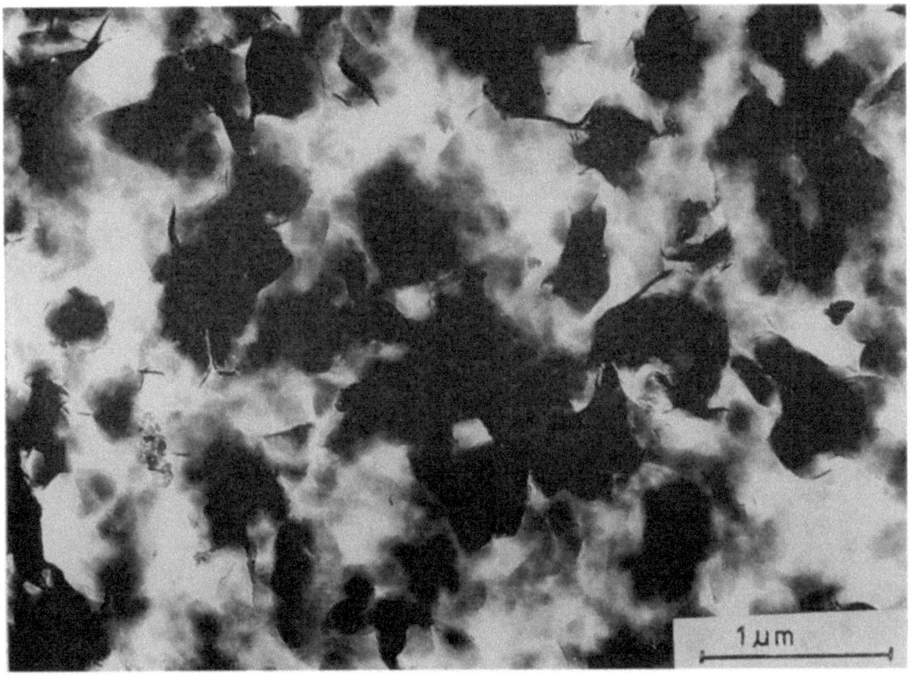

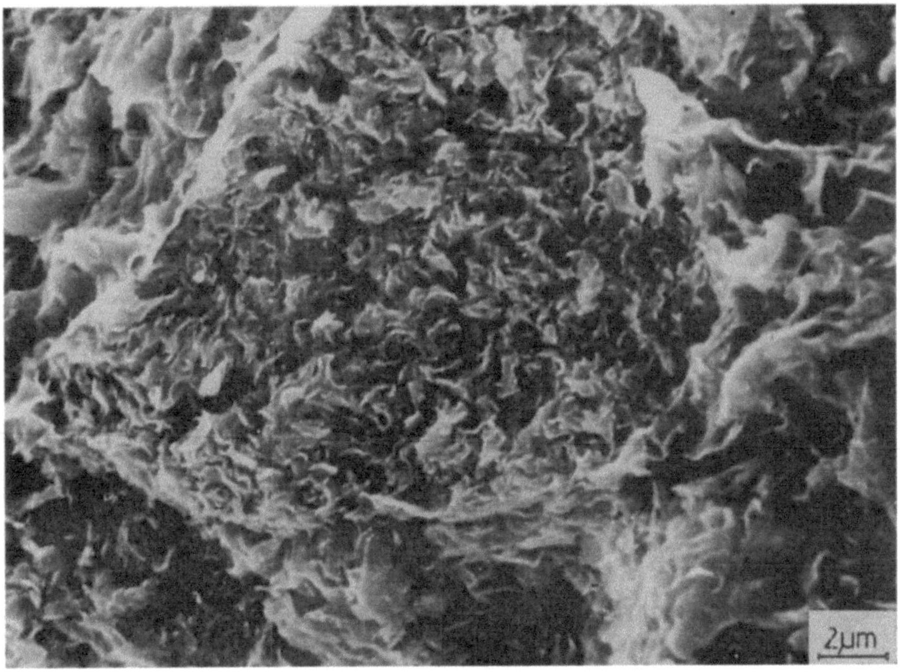

Fig. 177. TEM: *Montmorillonite*, API No. 30a, Bayard, New Mexico, USA

Fig. 178. SEM: *Montmorillonite*, API No. 30a, Bayard, New Mexico, USA

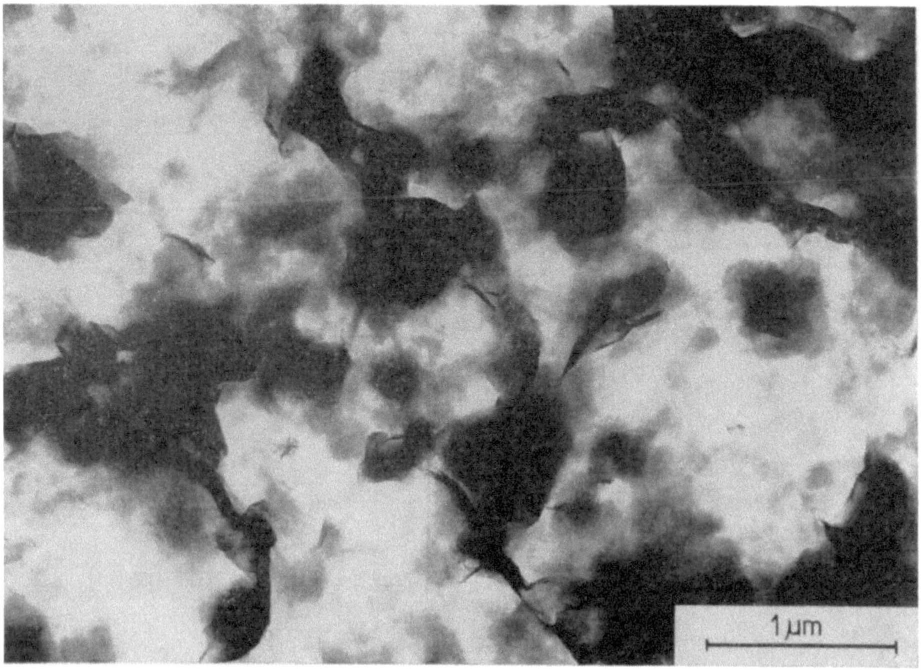

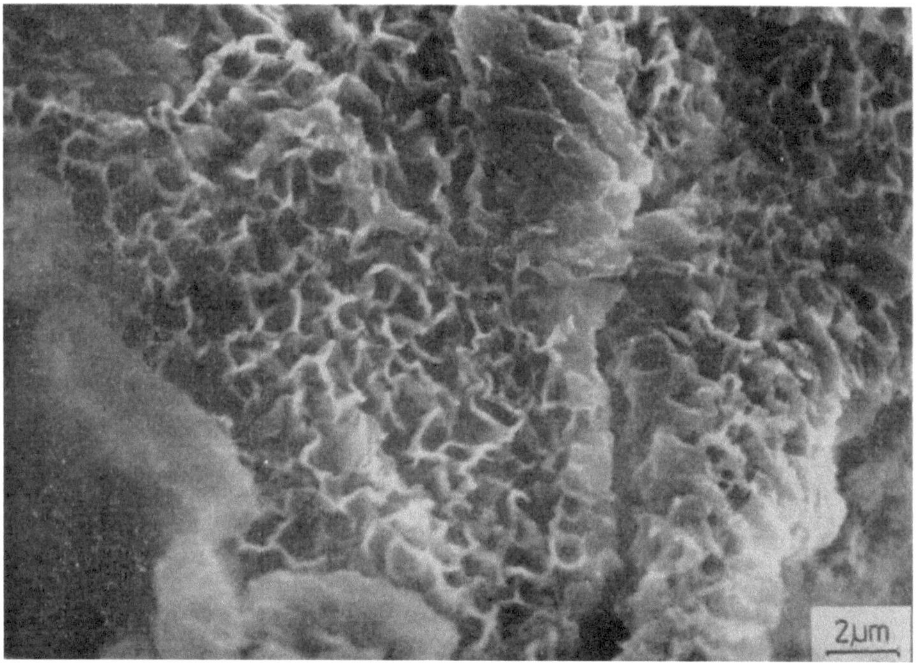

Fig. 179. TEM: *Montmorillonite,* API No. 20, Husband Mine, Polkville, Mississippi, USA

Fig. 180. SEM: *Montmorillonite,* API No. 20, Husband Mine, Polkville, Mississippi, USA

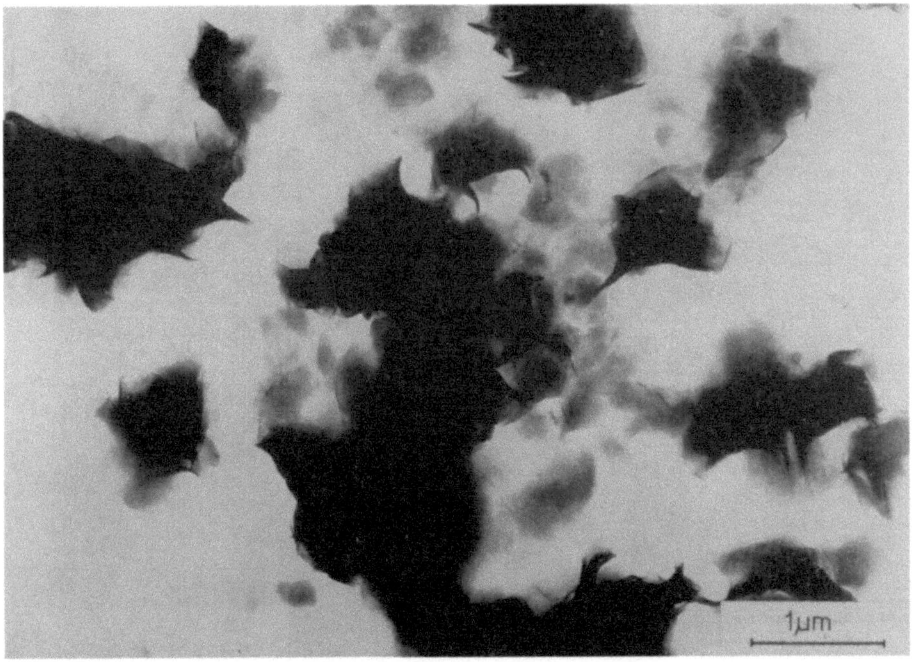

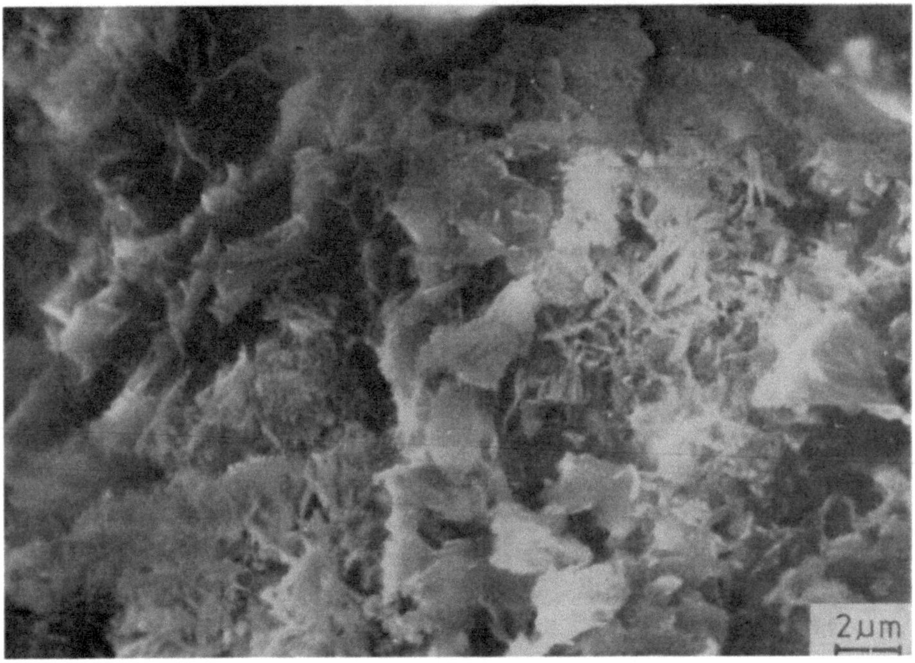

Fig. 181. TEM: *Montmorillonite,* API No. 30, Santa Rita, New Mexico, USA

Fig. 182. SEM: *Montmorillonite,* API No. 30, Santa Rita, New Mexico, USA

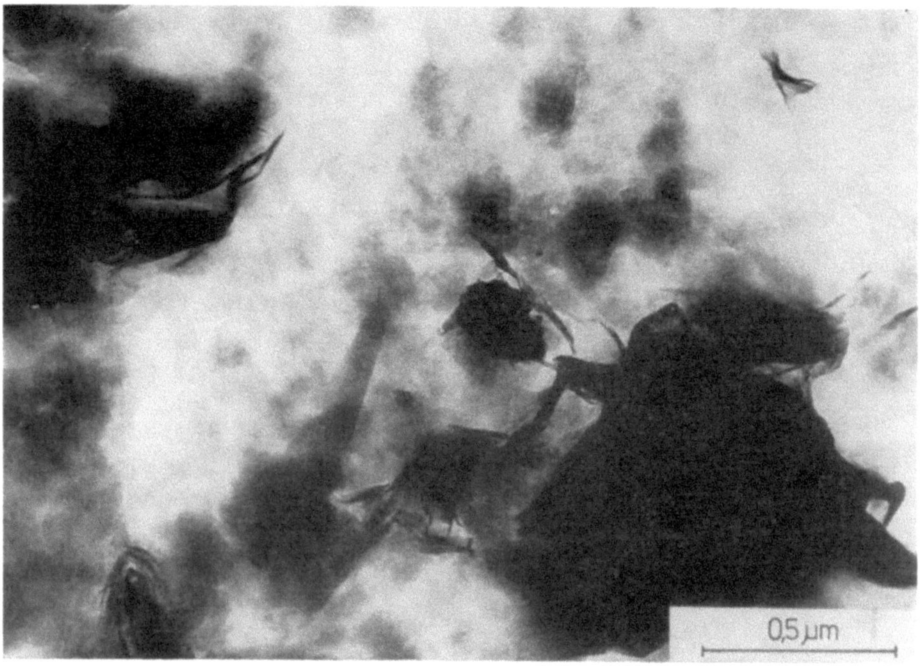

Fig. 183. TEM: *Montmorillonite* (cream-coloured), API No. 22b, Amory, Mississippi, USA

Fig. 184. SEM: *Montmorillonite* (cream-coloured), API No. 22b, Amory, Mississippi, USA

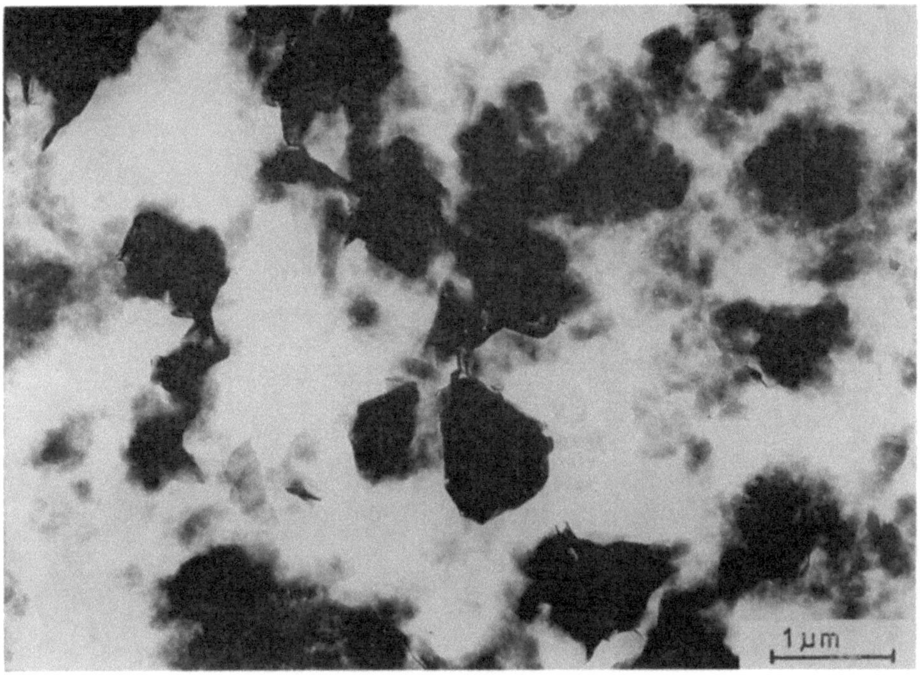

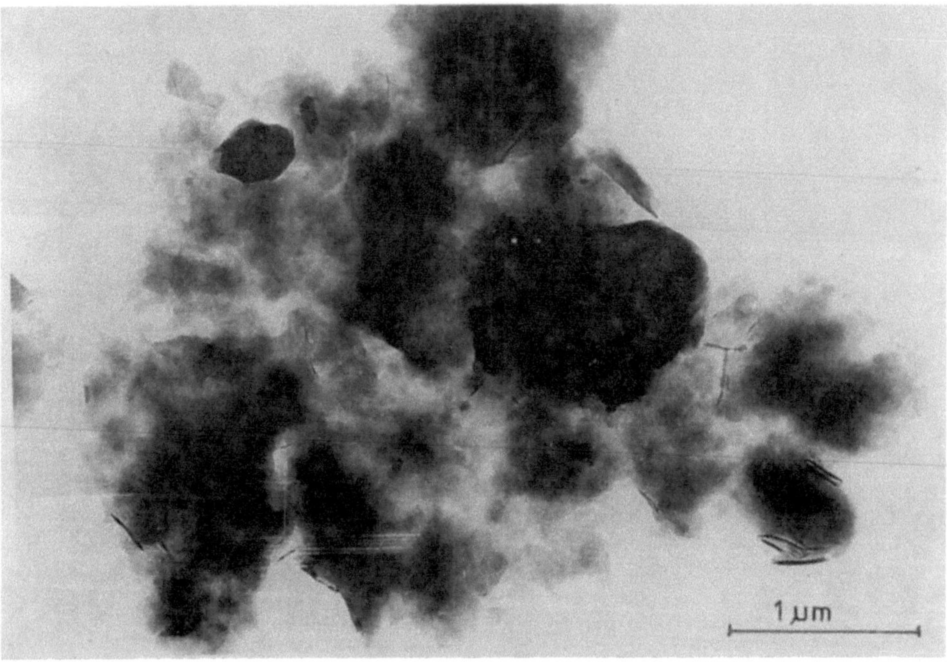

Fig. 185. TEM: *Montmorillonite*, API No. 24, Otay, California, USA

Fig. 186. TEM: *Montmorillonite*, API No. 31, Cameron, Arizona, USA

Fig. 187. TEM: *Montmorillonite*, API No. 27, Belle Fourche, South Dakota, USA

Fig. 188. TEM: *Montmorillonite*, API No. 25, Upton, Wyoming, USA

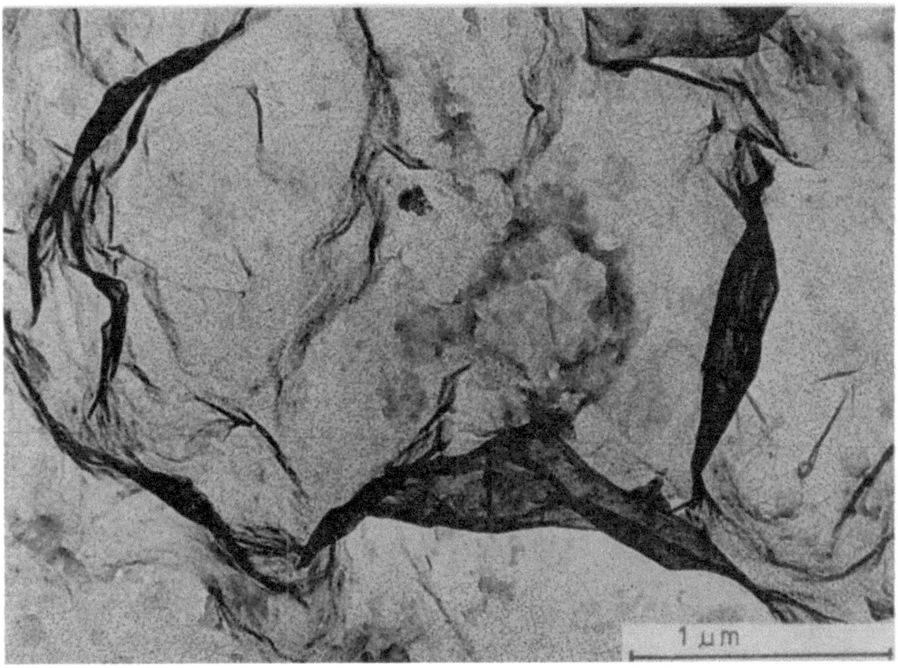

Fig. 189. Replica: *Montmorillonite,* API No. 25, Upton, Wyoming, USA

Fig. 190. Replica: *Montmorillonite,* gold decorated, API No. 25, Upton, Wyoming, USA

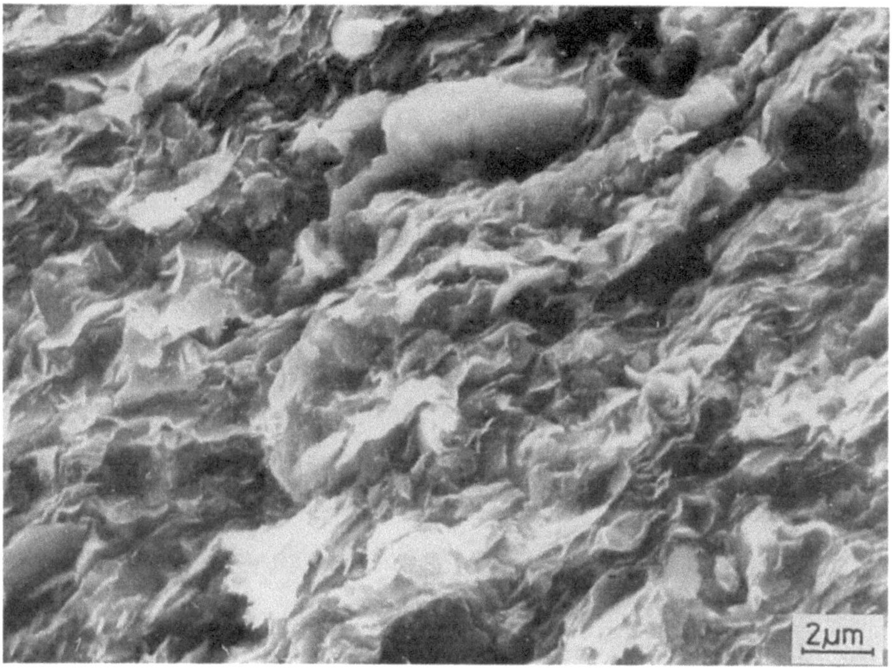

Fig. 191. SEM: *Montmorillonite,* Clay Spur Bentonite, fracture vertical to bedding, Belle Pit, Montana, USA

Fig. 192. SEM: *Montmorillonite,* Clay Spur Bentonite, fracture parallel to bedding, Belle Pit Montana, USA

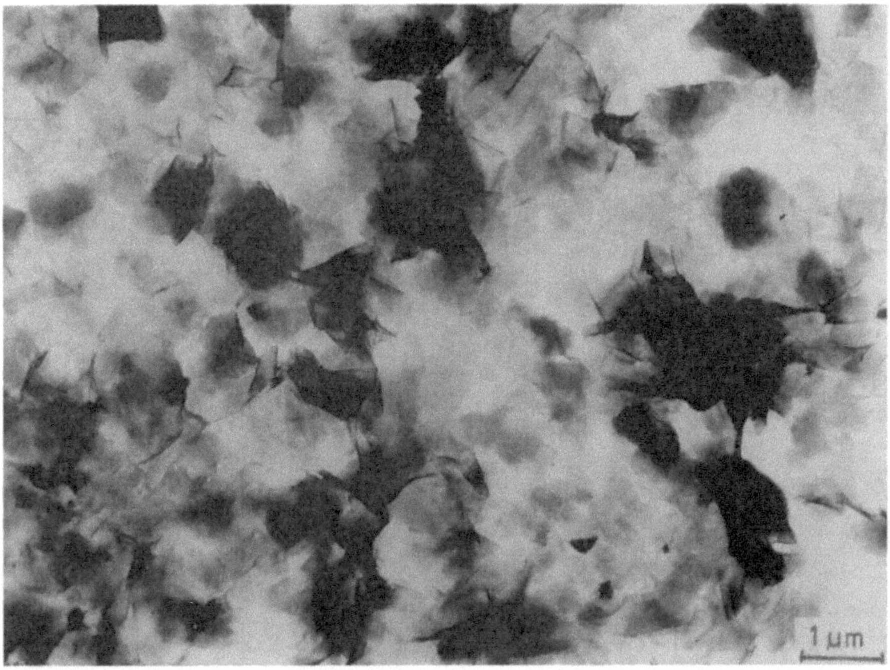

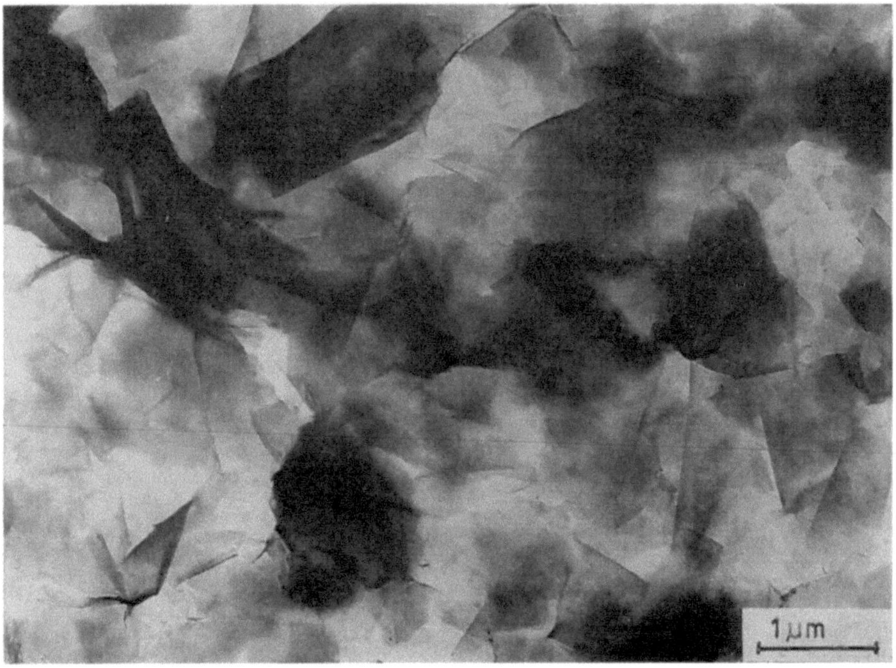

Figs. 193, 194. TEMs: *Montmorillonite* (greenish), not identical with API No. 23, Chambers, Arizona, USA

Fig. 194.

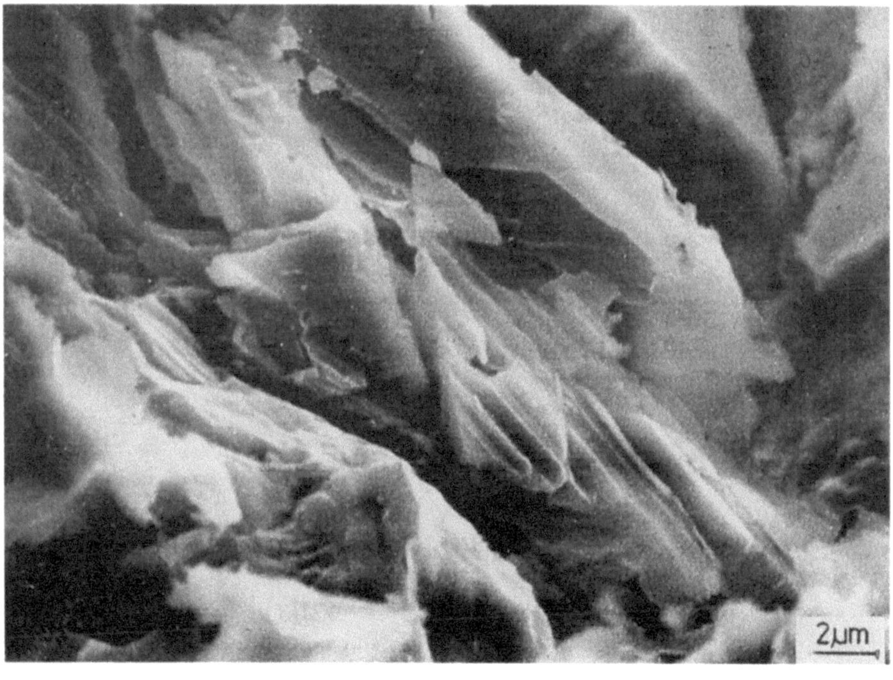

Fig. 195. SEM: *Montmorillonite* (greenish), not identical with API No. 23, Chambers, Arizona, USA

Fig. 196. SEM: *Montmorillonite*, Tazewell, Virginia, USA

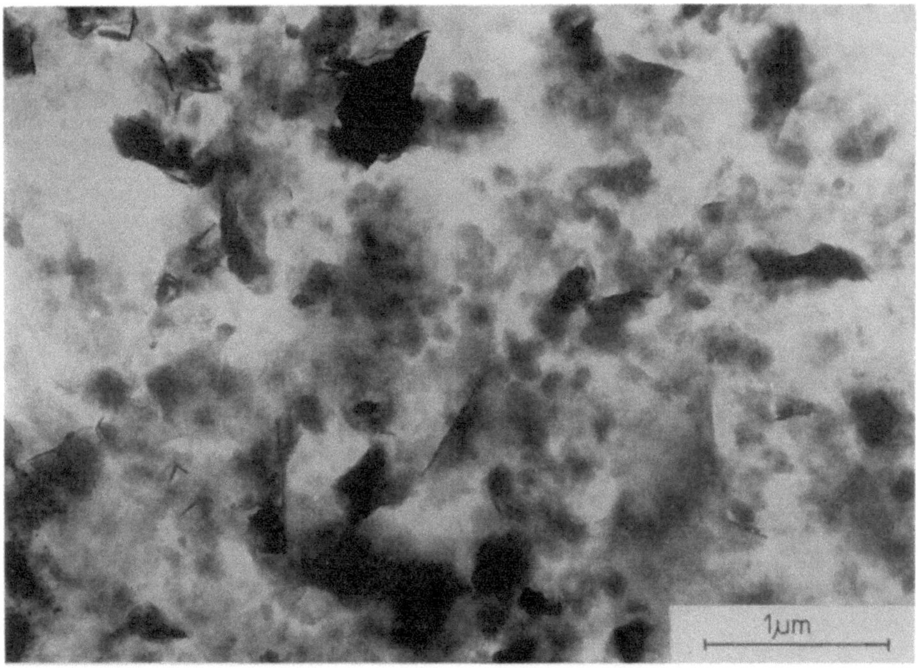

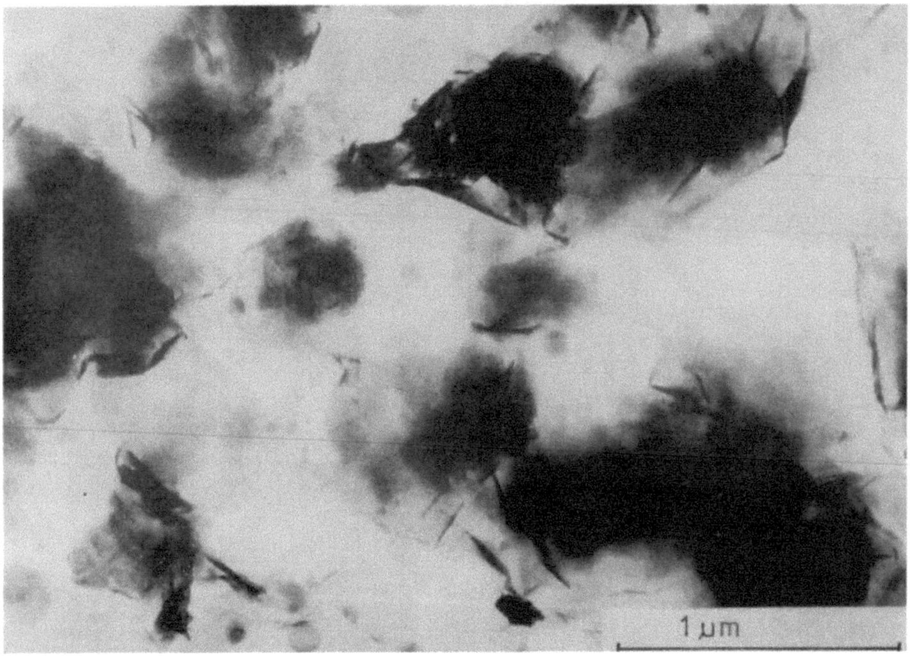

Fig. 197. TEM: *Montmorillonite*, Branany, Czechoslovakia

Fig. 198. TEM: *Montmorillonite*, Jelšovy Potok, Czechoslovakia

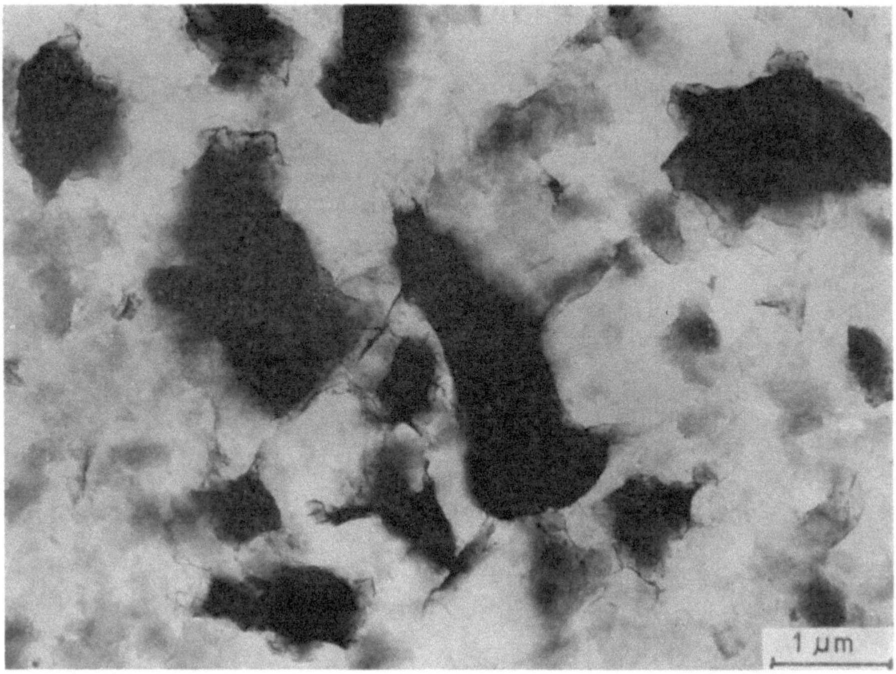

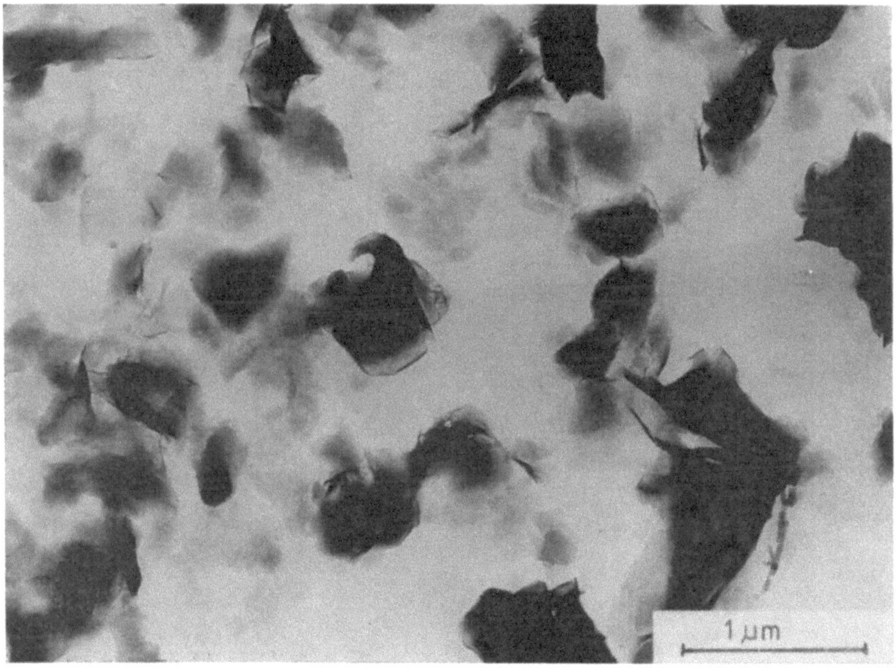

Fig. 199. TEM: *Montmorillonite*, Stare Kremnice, Czechoslovakia

Fig. 200. TEM: *Montmorillonite*, Großer Dolmar, Thuringia, GDR

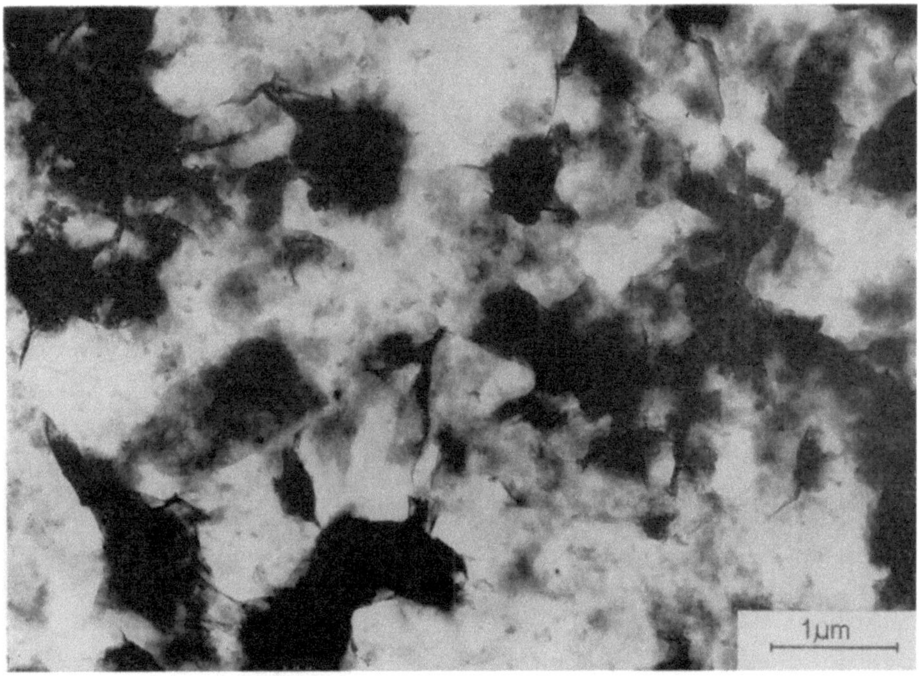

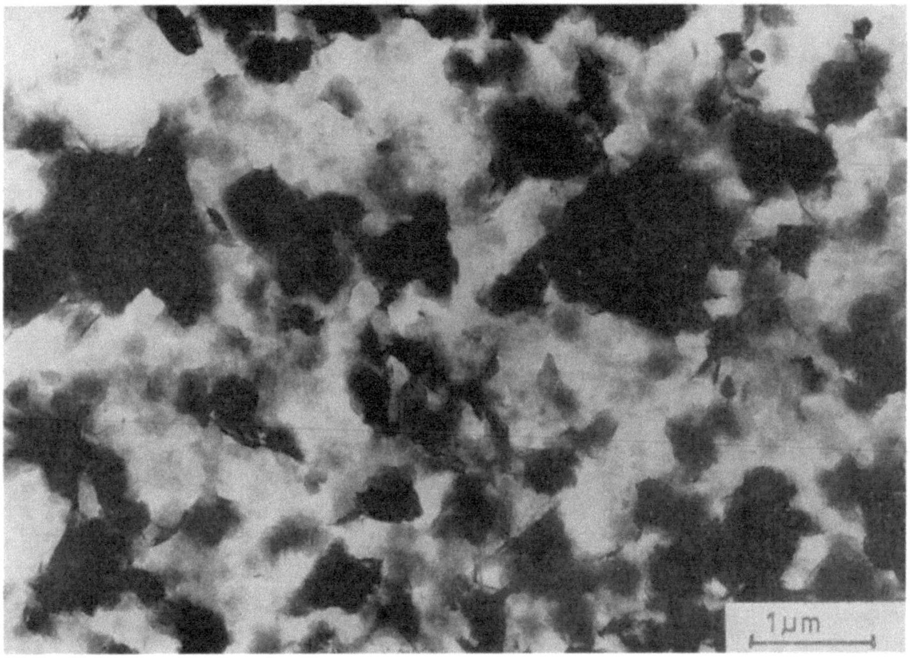

Fig. 201. TEM: *Montmorillonite*, Istenmezeje, Hungary

Fig. 202. TEM: *Montmorillonite*, Krimskiy Kit, Ukrainian SSR, USSR

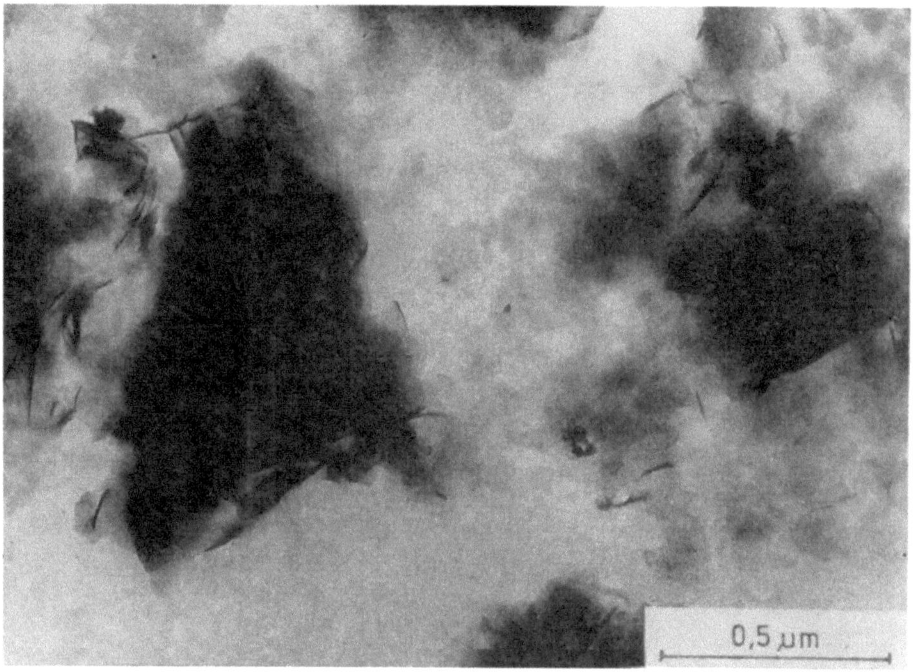

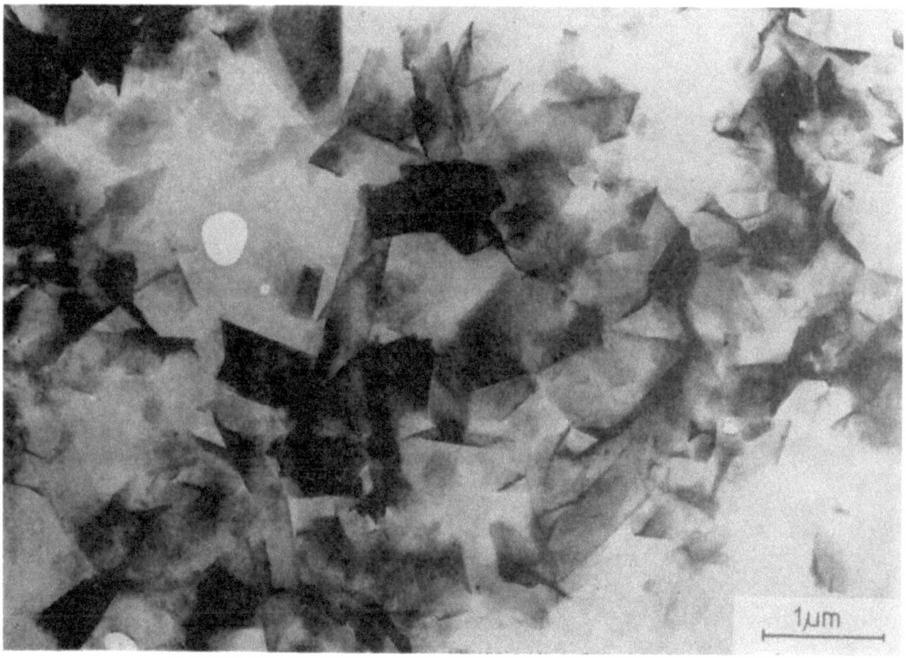

Fig. 203. TEM: *Montmorillonite*, Geisenheim, FRG

Fig. 204. TEM: *Montmorillonite*, hydrothermal, Michalkovo, Bulgaria

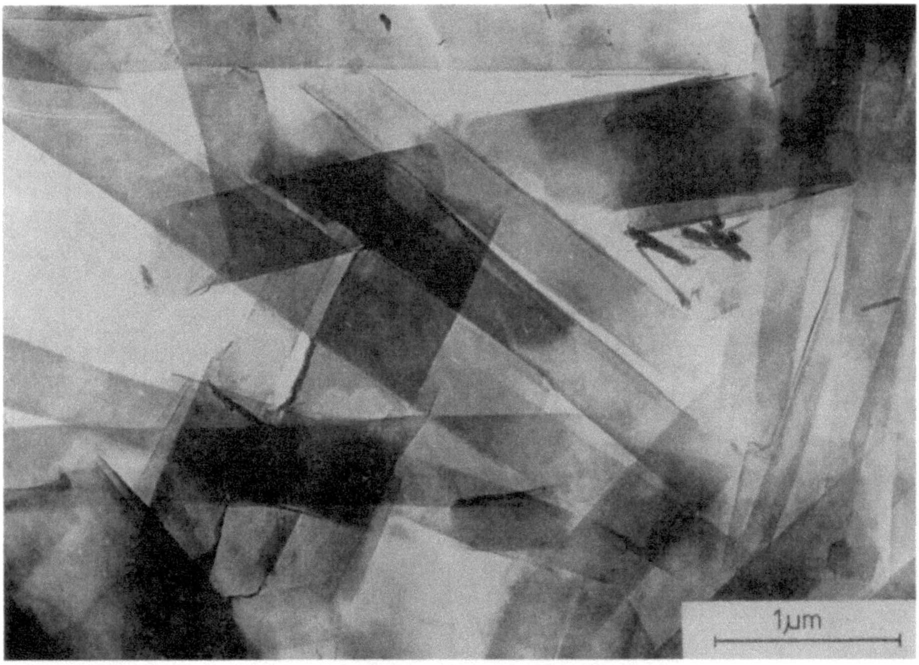

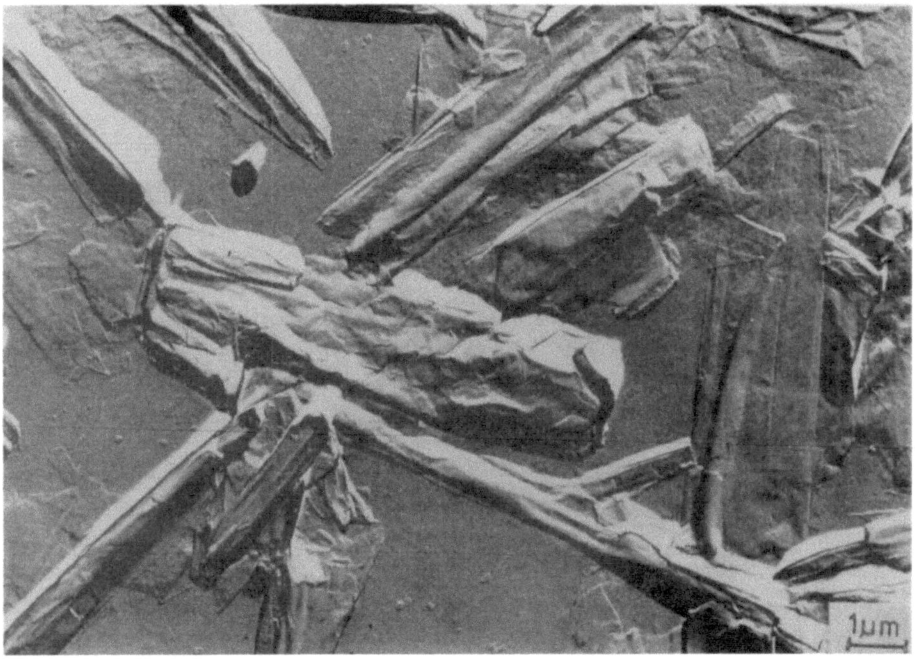

Fig. 205. TEM: *Montmorillonite*, hydrothermal, Michalkovo, Bulgaria

Fig. 206. Replica: *Montmorillonite*, hydrothermal, Michalkovo, Bulgaria

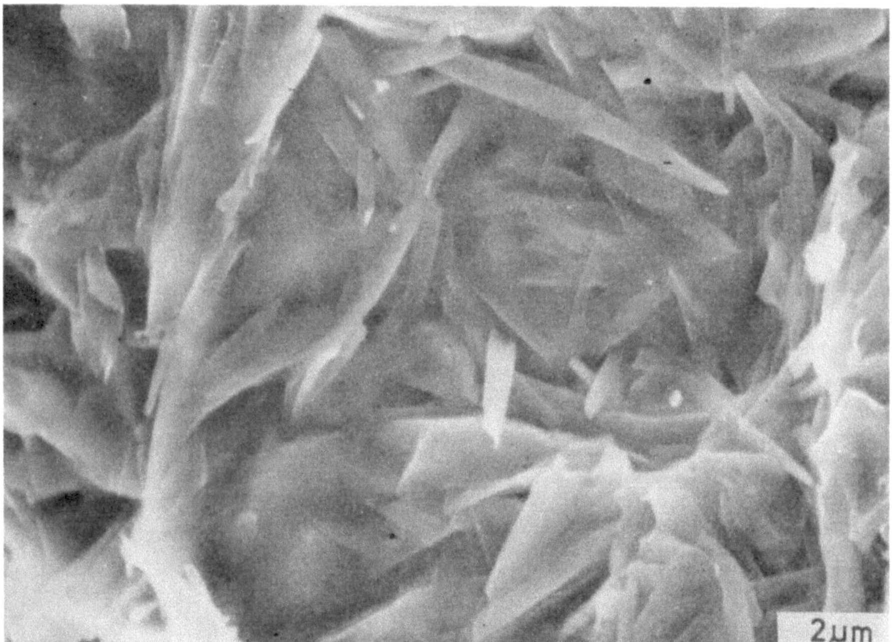

Figs: 207, 208, 209, 210. SEMs: *Montmorillonite*, hydrothermal, Michalkovo, Bulgaria

Fig. 208

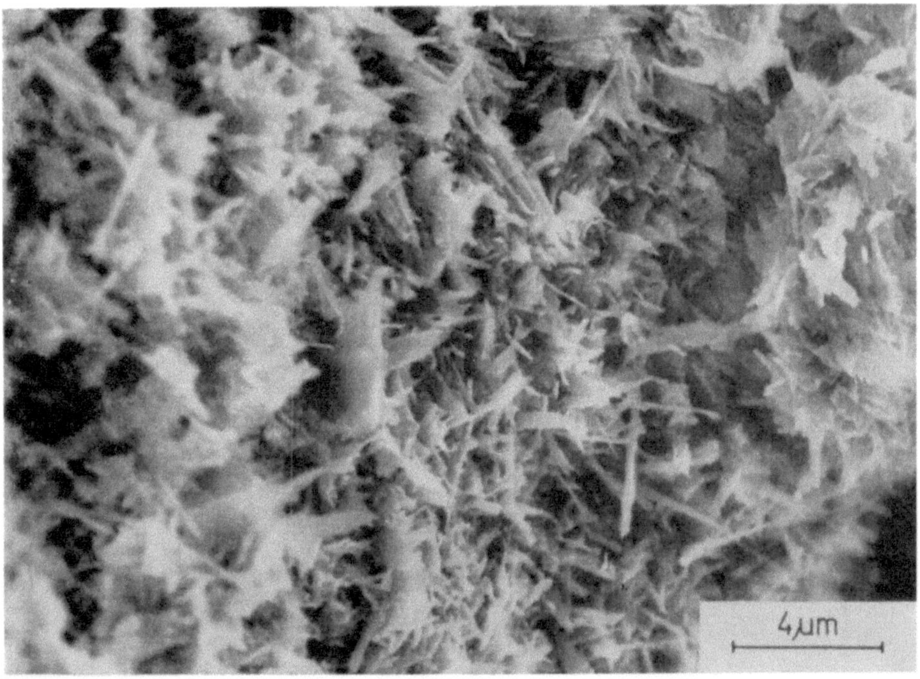

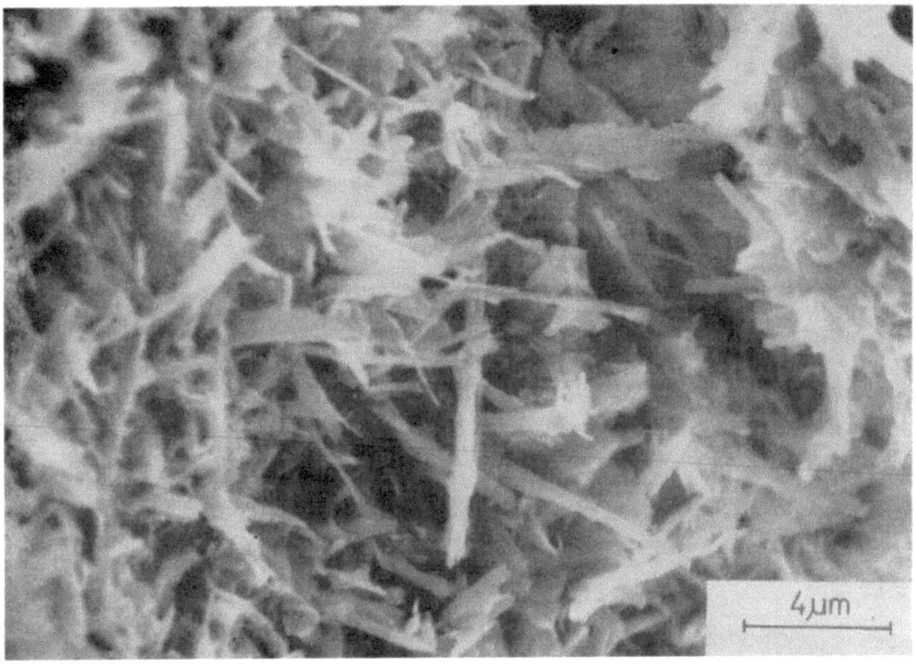

Fig. 209.

(*explained at p. 145*)

Fig. 210.

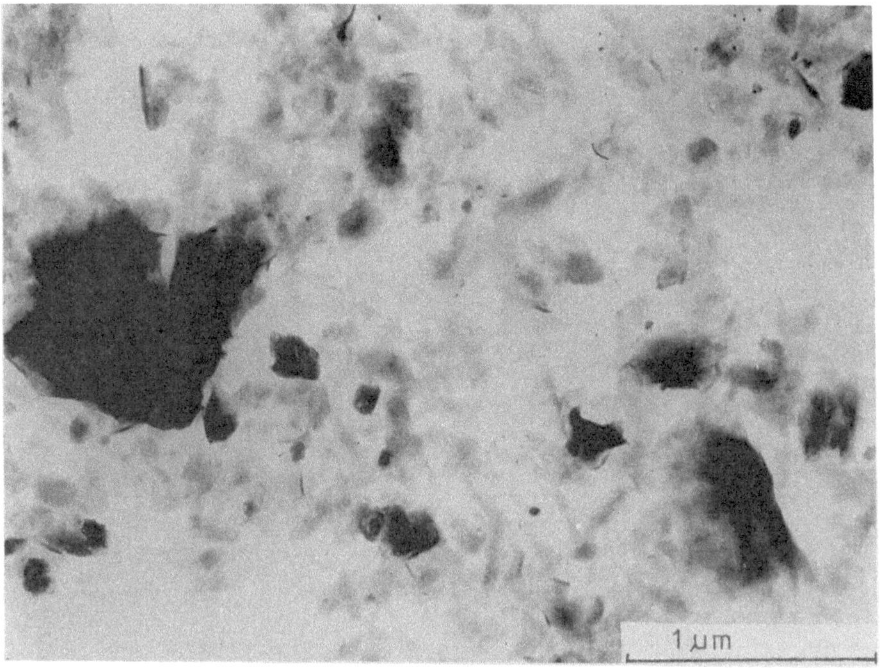

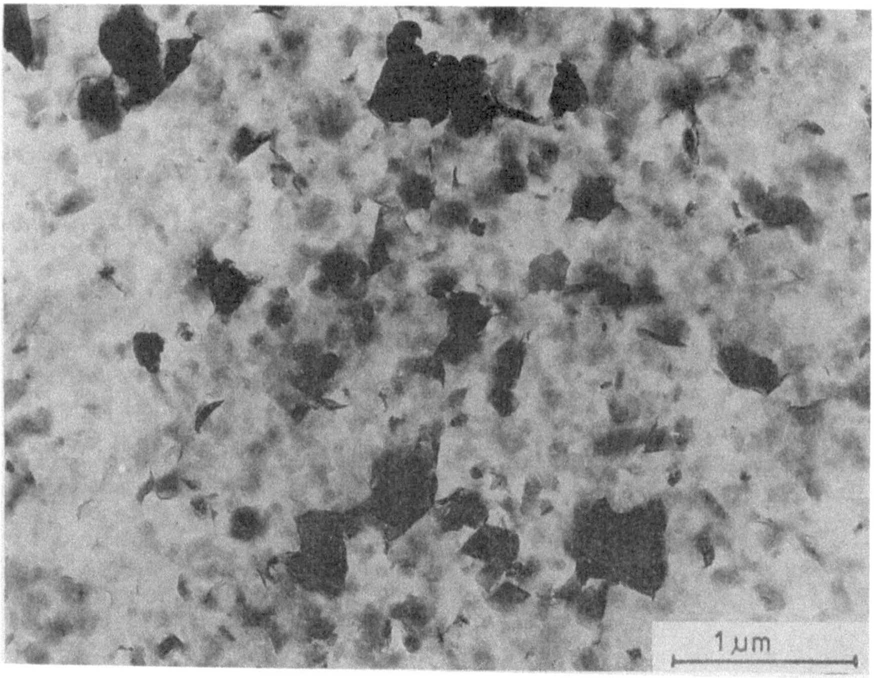

Fig. 211. TEM: *Beidellite* (bentonite), Ostritz, Lusatia, GDR

Fig. 212. TEM: *Beidellite* from bentonized Tertiary basaltic tuff, Mittelherwigsdorf, Lusatia, GDR

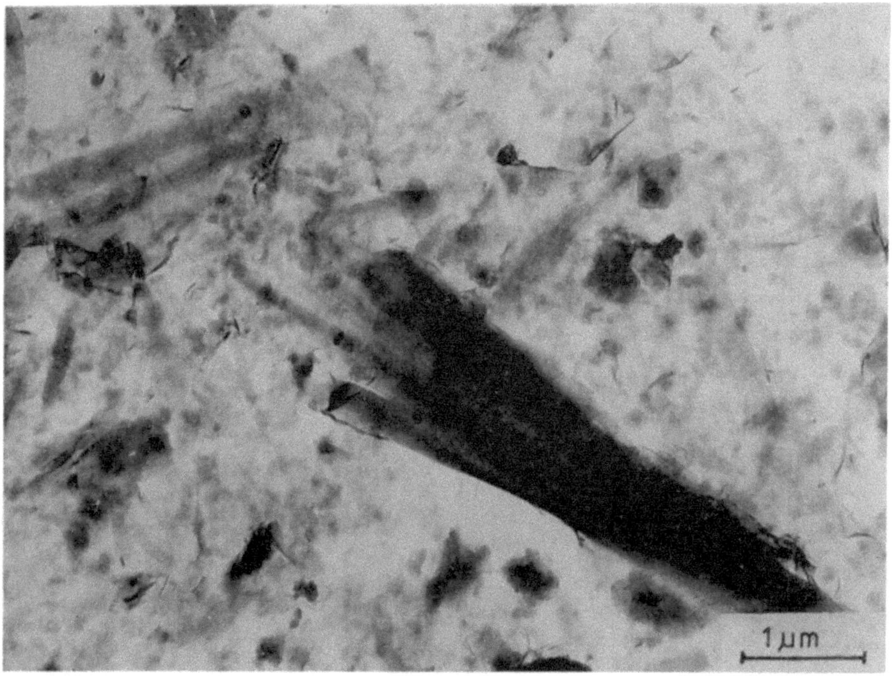

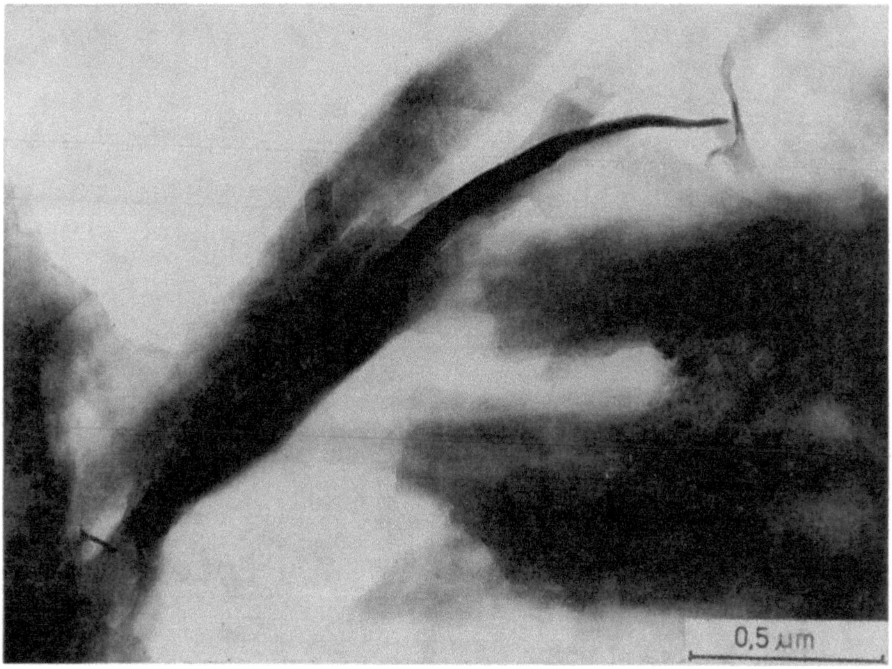

Figs. 213, 214. TEMs: *Nontronite*, API No. 33a, Garfield, Washington, USA

Fig. 214.

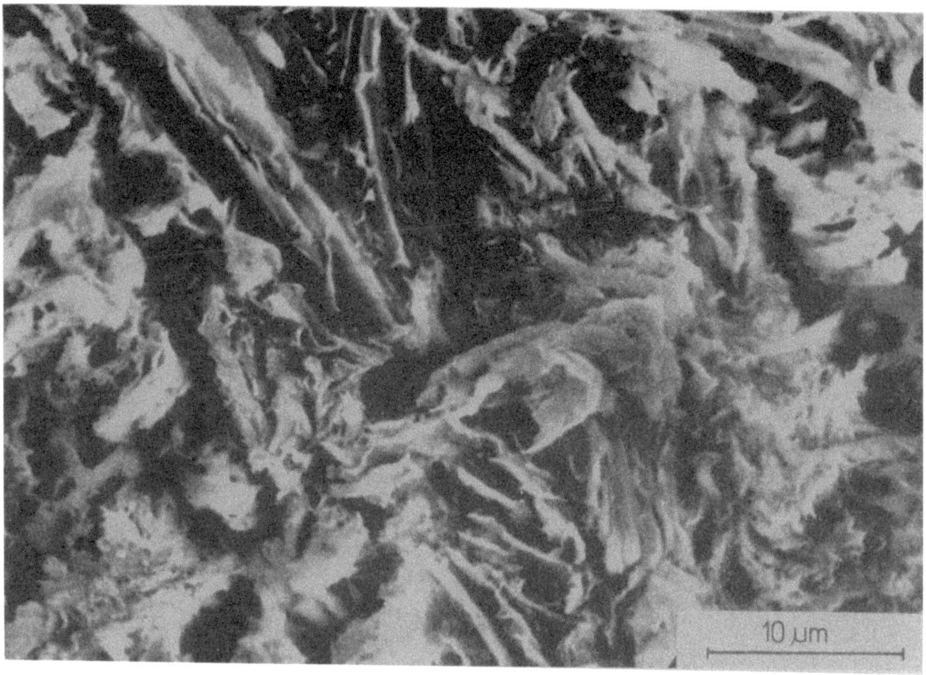

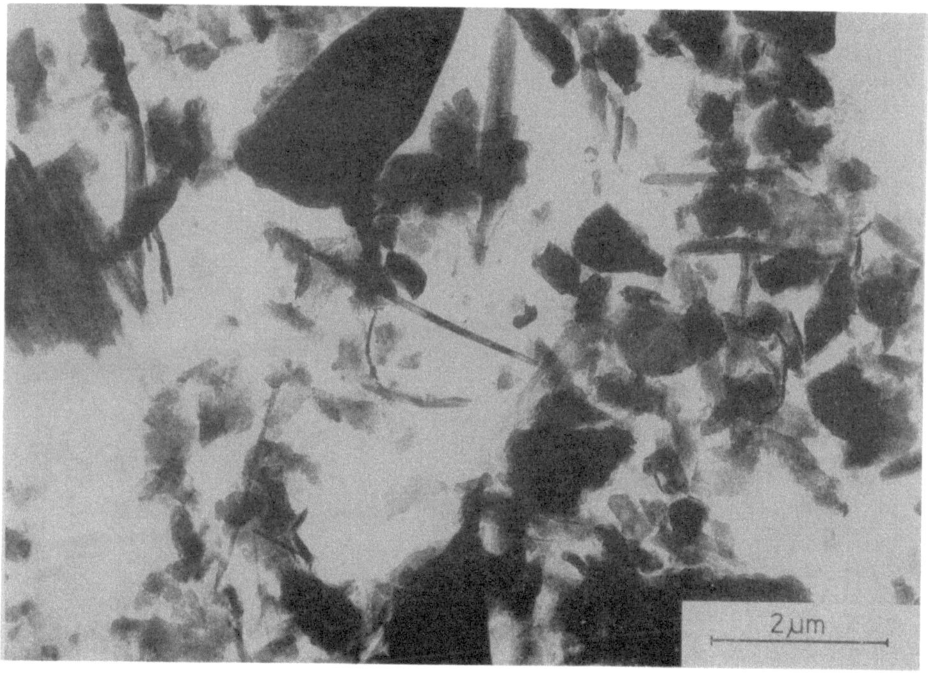

Fig. 215. SEM: *Nontronite*, API No. 33a, Garfield, Washington, USA

Fig. 216. TEM: *Nontronite*, Nontron, Dordogne, France

Fig. 217. Replica: *Nontronite* from pegmatite, Dolny Bory, Czechoslovakia

Fig. 218. SEM: *Nontronite* from pegmatite, Dolny Bory, Czechoslovakia

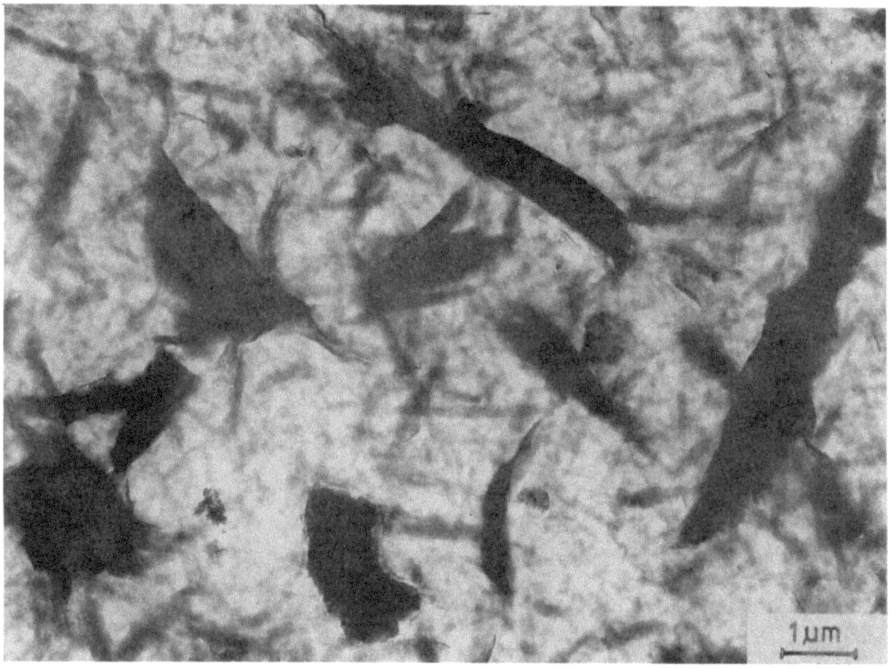

Fig. 219. TEM: *Nontronite*, Pfaffenreuth, FRG

Fig. 220. SEM: *Nontronite*, Rajasthan, India

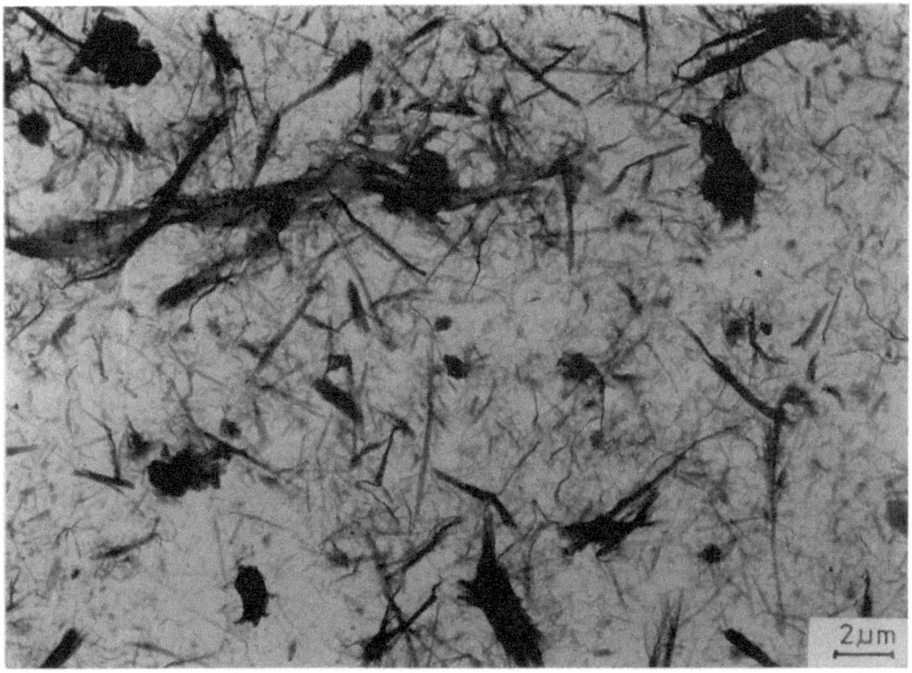

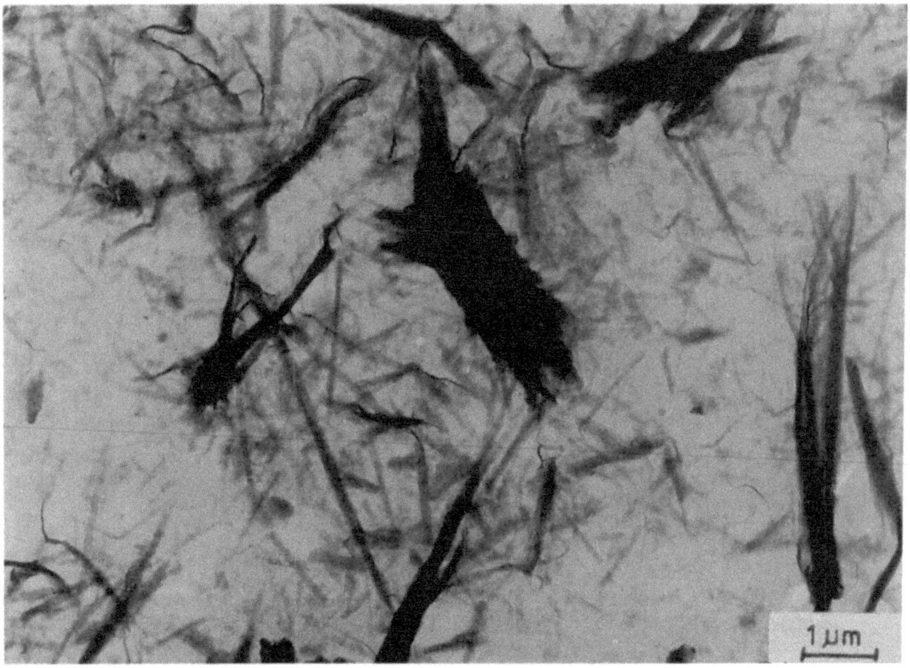

Figs. 221, 222. TEMs: *Nontronite*, Seidau near Bautzen, GDR

Fig. 222.

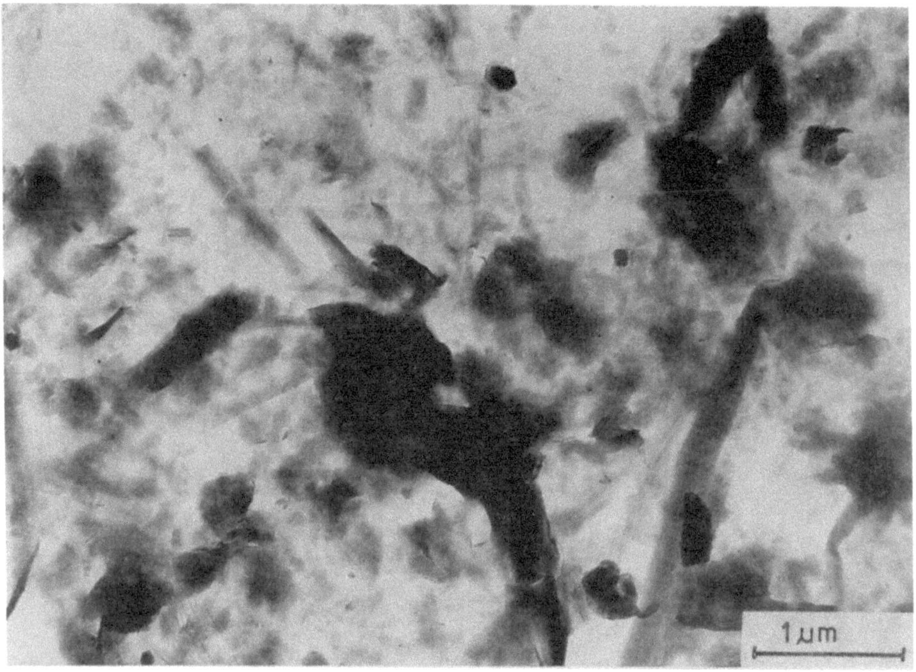

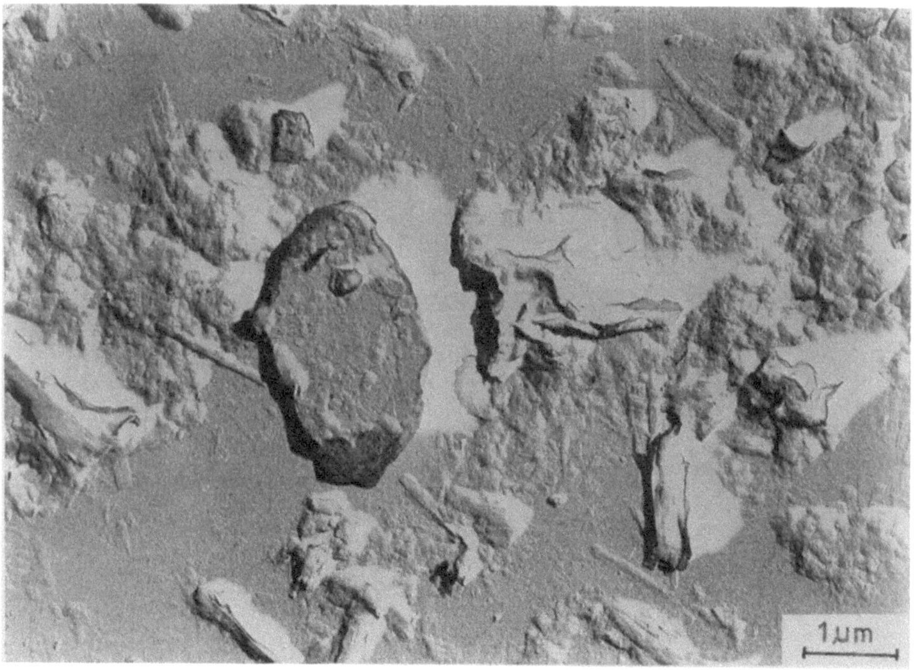

Fig. 223. TEM: *Nontronite*, Meensen, FRG

Fig. 224. Replica: *Nontronite*, Meensen, FRG

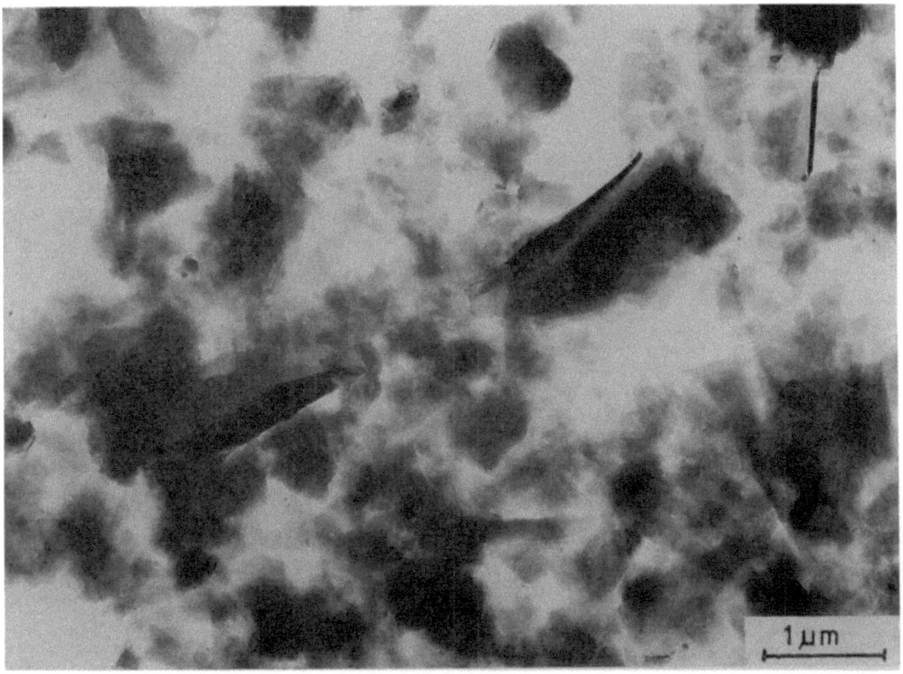

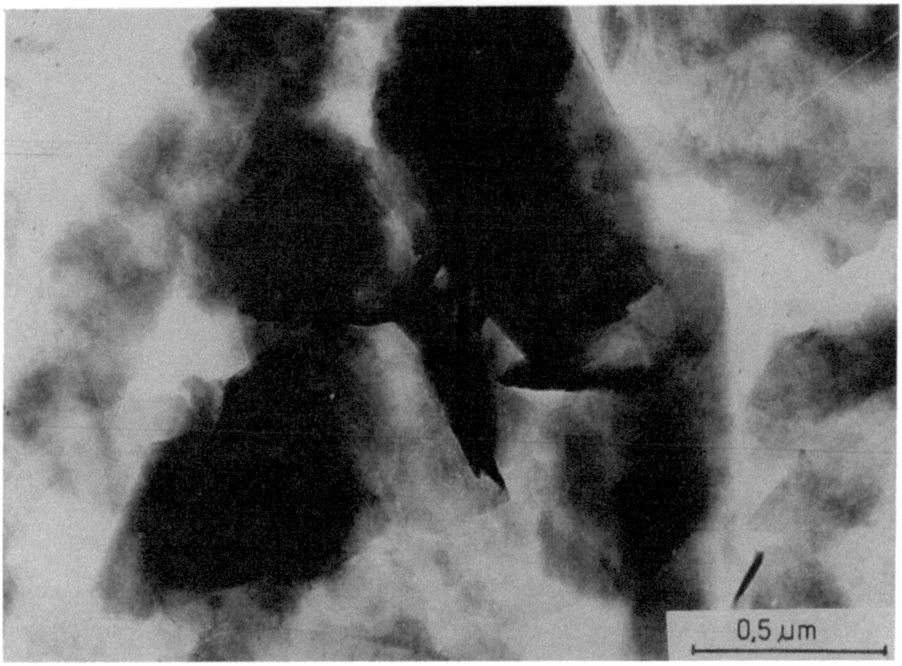

Figs. 225, 226. TEMs: *Saponite*, Cap Lizard, Cornwall, UK

Fig. 226.

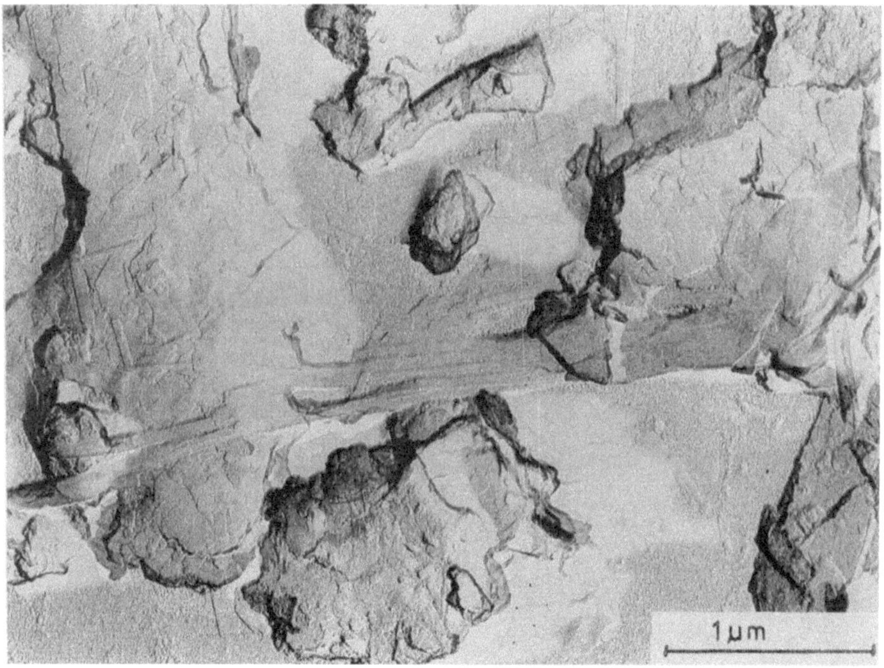

Fig. 227, 228. Replicas: *Saponite*, Cap Lizard, Cornwall, UK

Fig. 228.

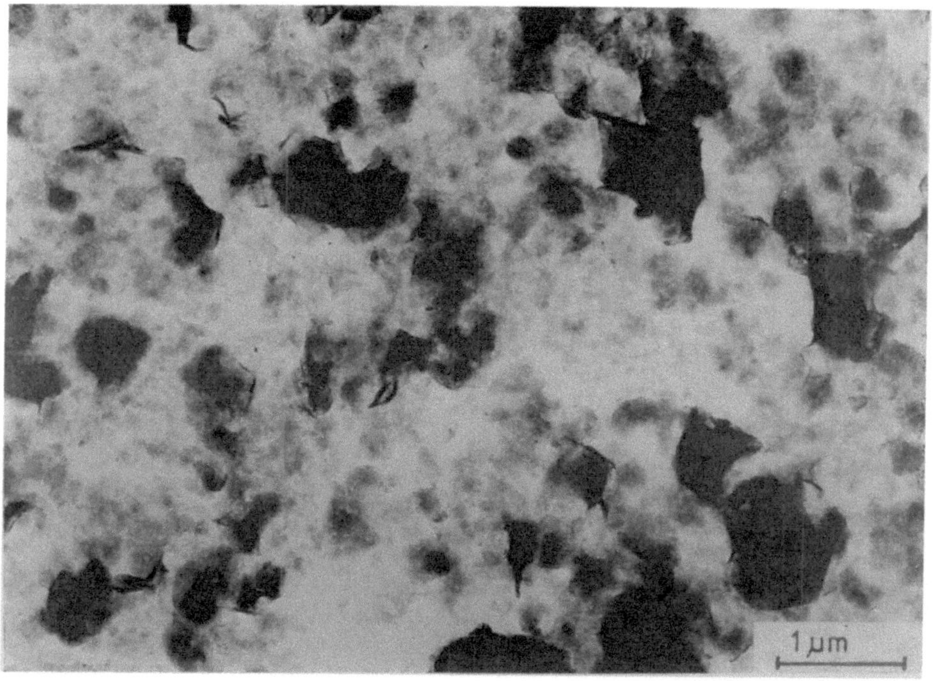

Fig. 229. TEM: *Saponite*, Szklary (Wiry), Poland

Fig. 230. Replica: *Saponite*, Szklary (Wiry), Poland

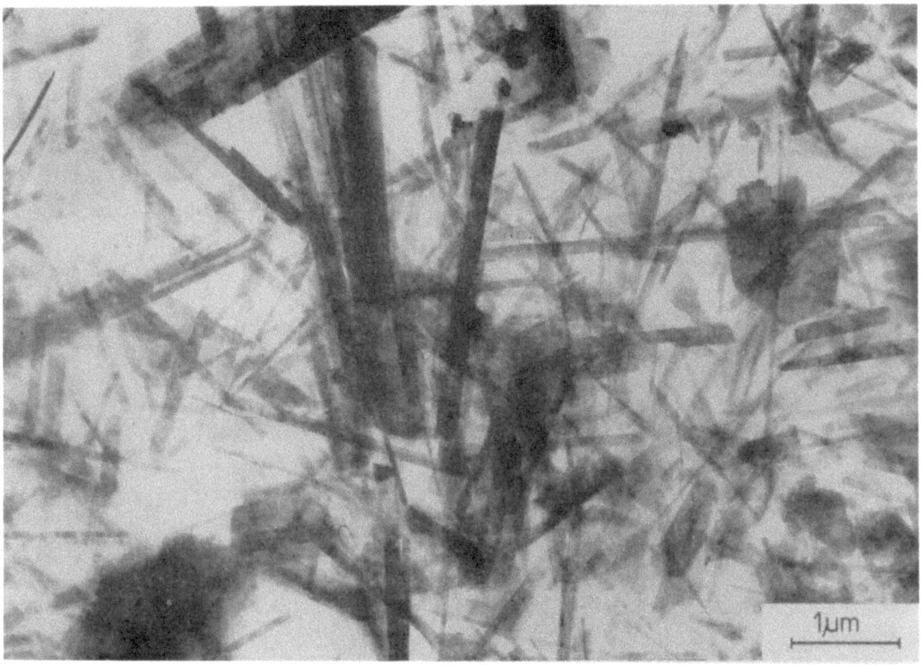

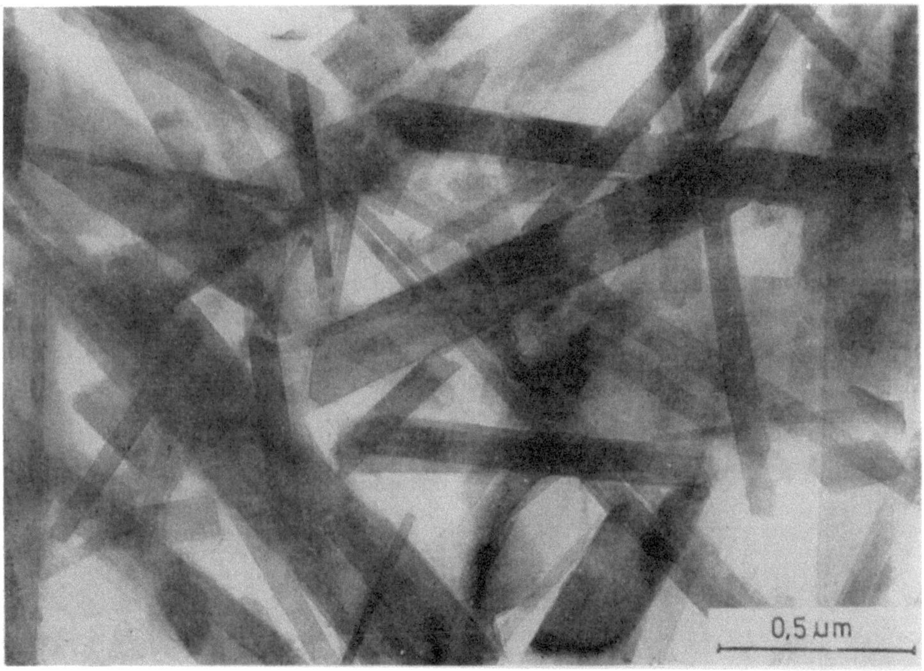

Figs. 231, 232. TEMs: *Saponite*, Pöhla, GDR

Fig. 232.

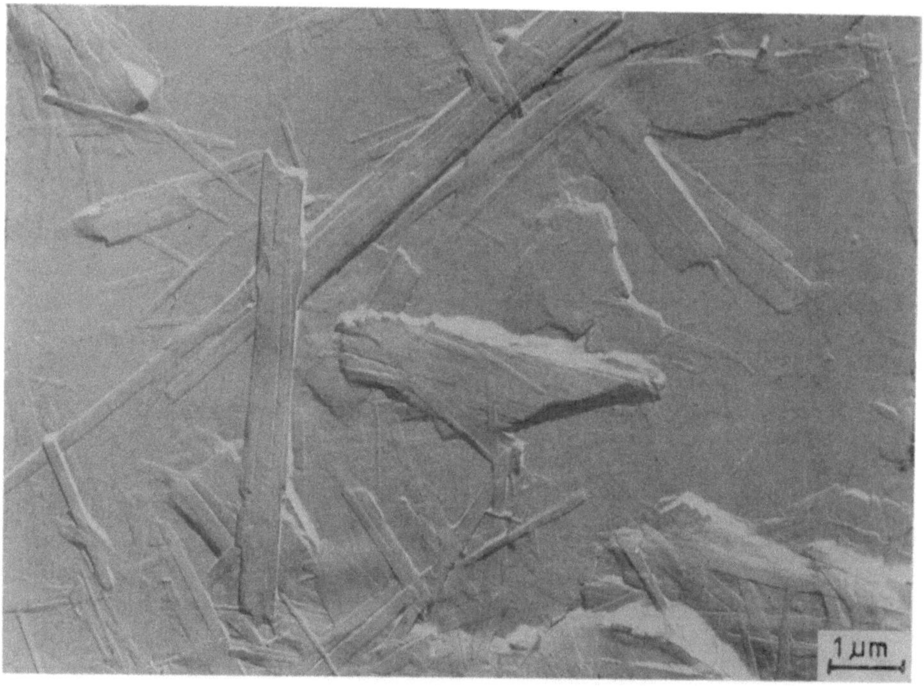

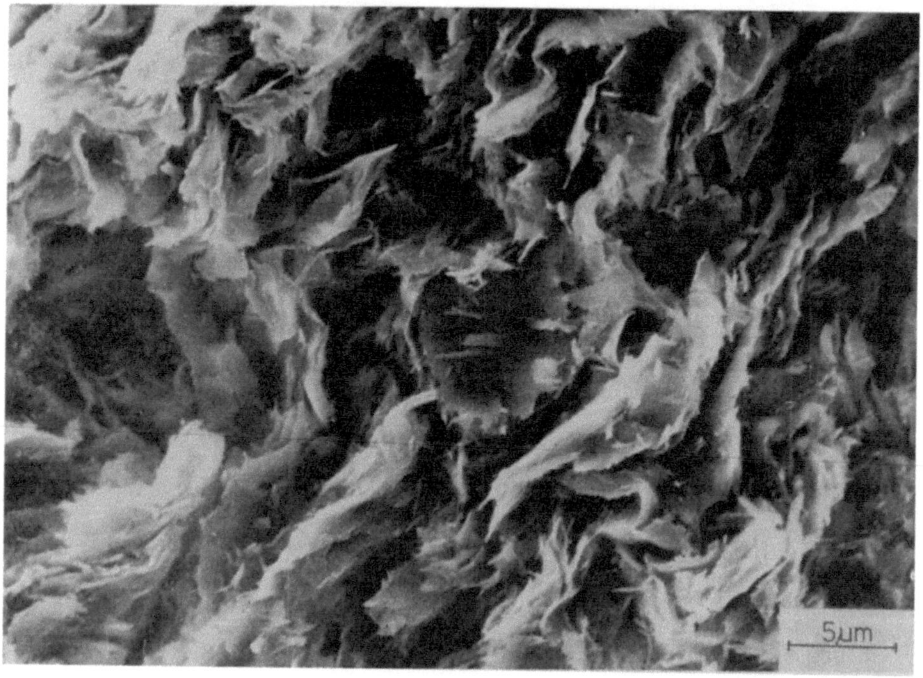

Fig. 233. Replica: *Saponite*, Pöhla, GDR

Fig. 234. SEM: *Saponite*, Pöhla, GDR

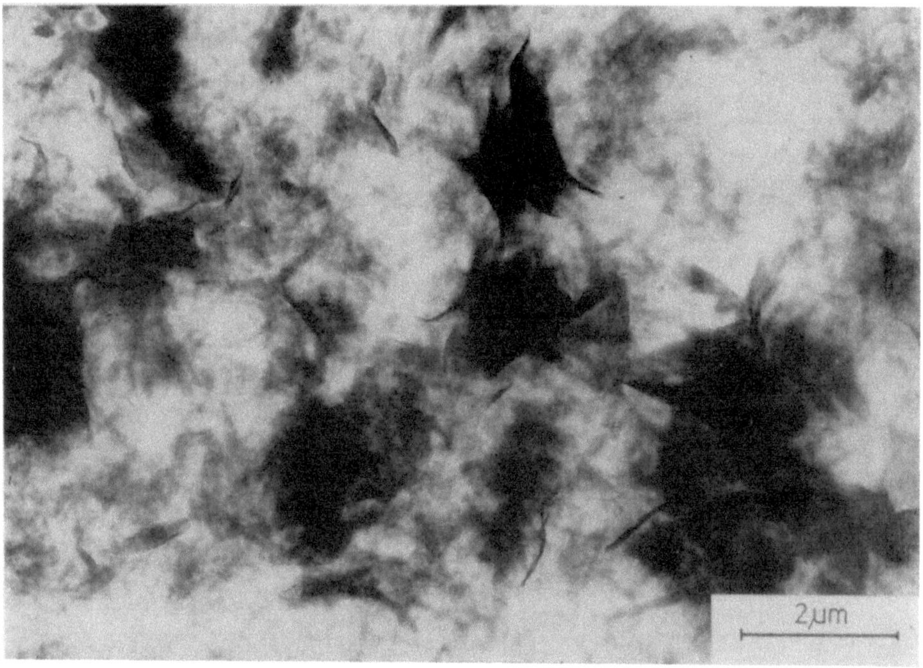

Fig. 235. TEM: *Hectorite*, Hector, California, USA

Fig. 236. Replica: *Hectorite*, Hector, California, USA

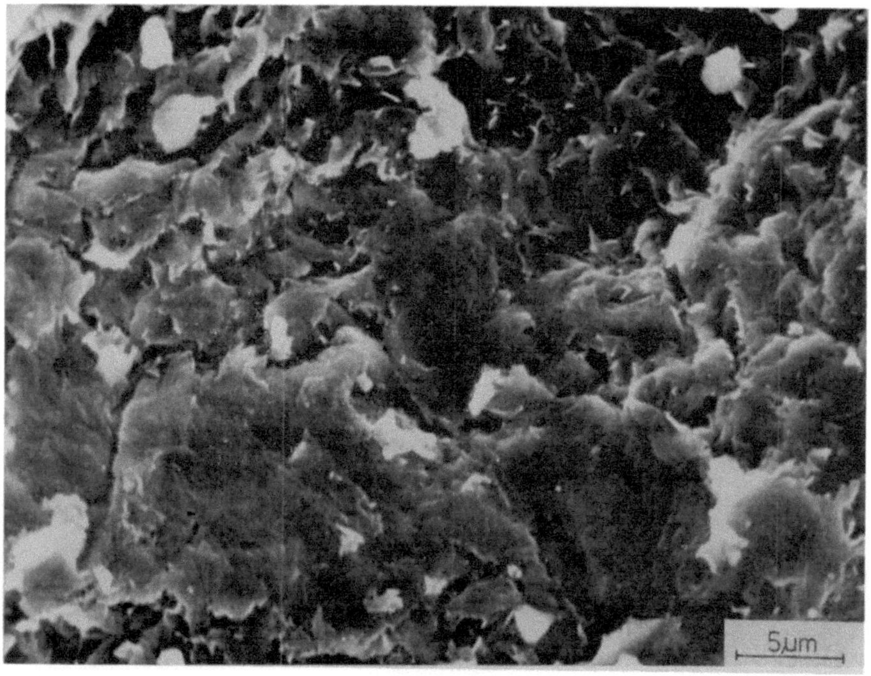

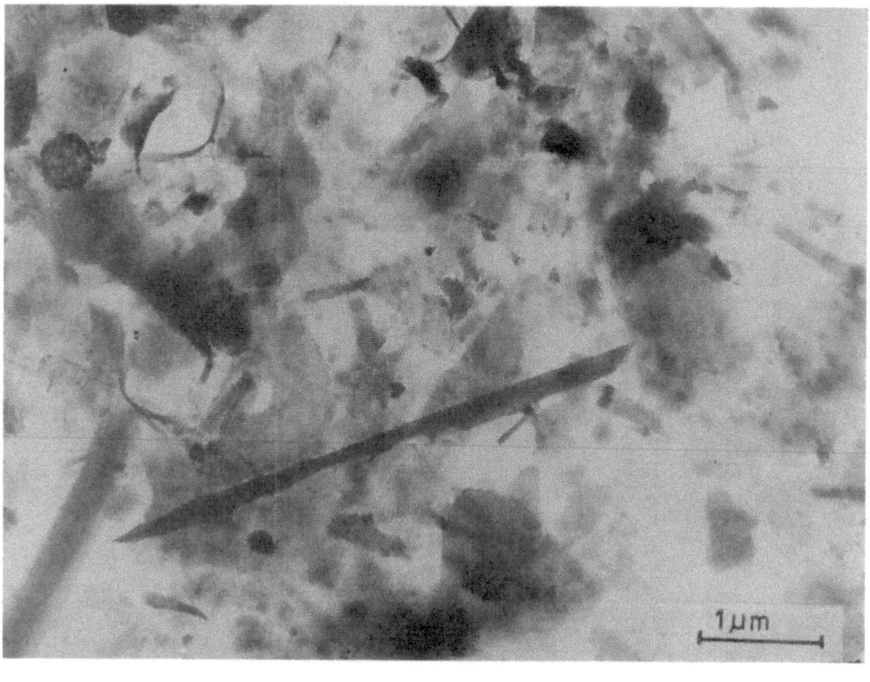

Fig. 237. SEM: *Hectorite*, Hector, California, USA

Figs. 238, 239. TEMs: *Vermiculite* ("zonolite"), Libby, Montana, USA

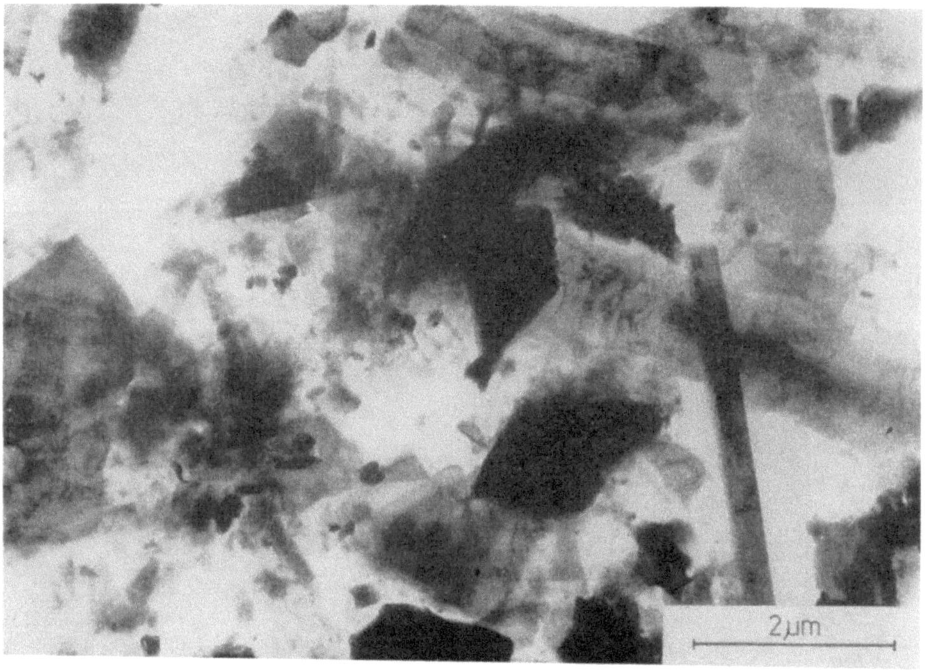

Fig. 239.

Fig. 240. Replica: *Vermiculite* ("zonolite"), Libby, Montana, USA

Figs. 241, 242. SEMs: *Vermiculite* ("zonolite"), Libby, Montana, USA

Fig. 242.

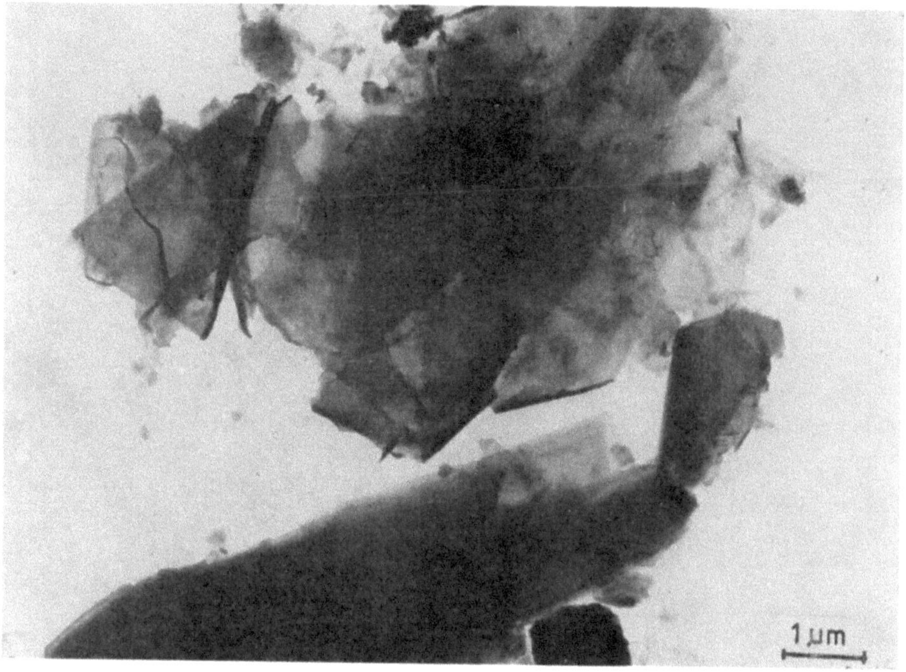

Fig. 243. TEM: *Vermiculite*, Piker Peach, Colorado, USA

Fig. 244. Replica: *Vermiculite*, Piker Peach, Colorado, USA

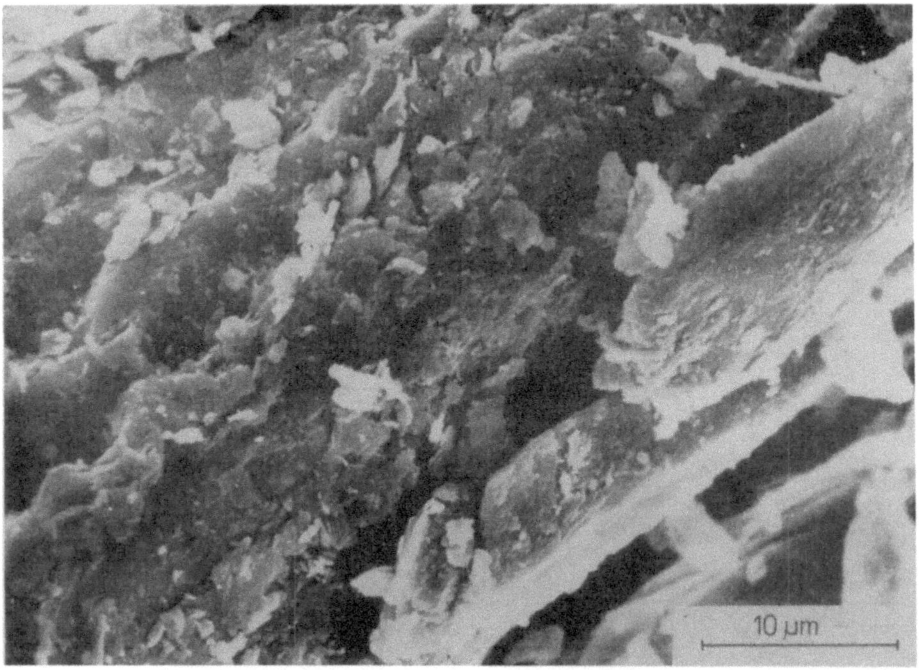

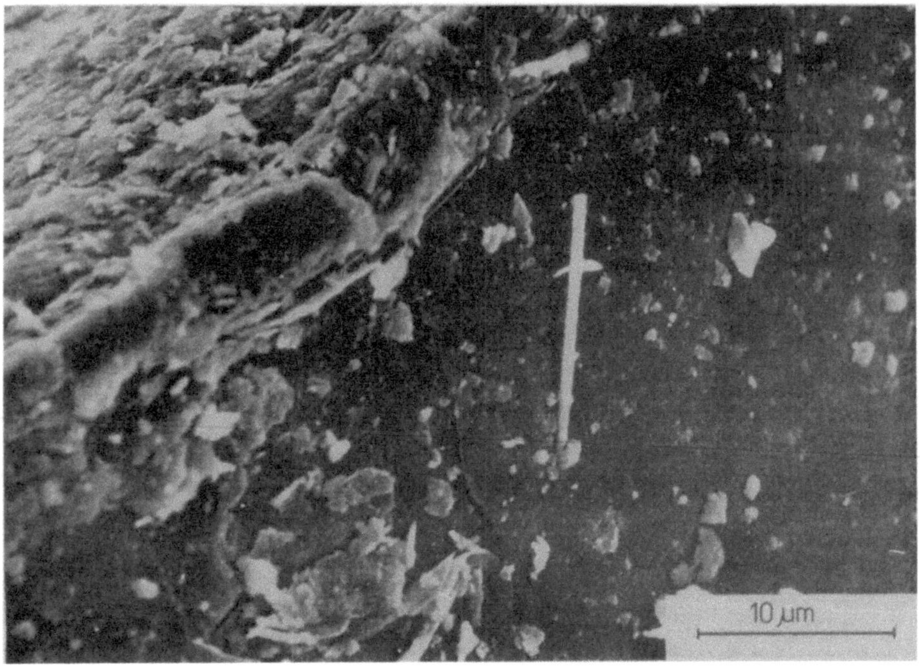

Figs. 245, 246. SEMs: *Vermiculite*, Piker Peach, Colorado, USA

Fig. 246.

Figs. 247, 248. Replicas: *Vermiculite*, Transvaal, South Africa

Fig. 248.

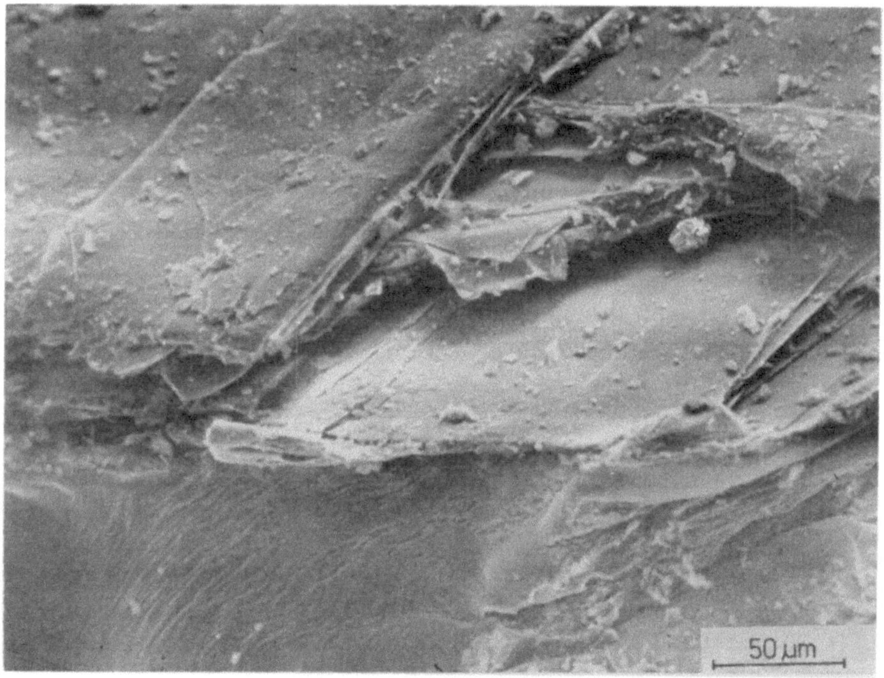

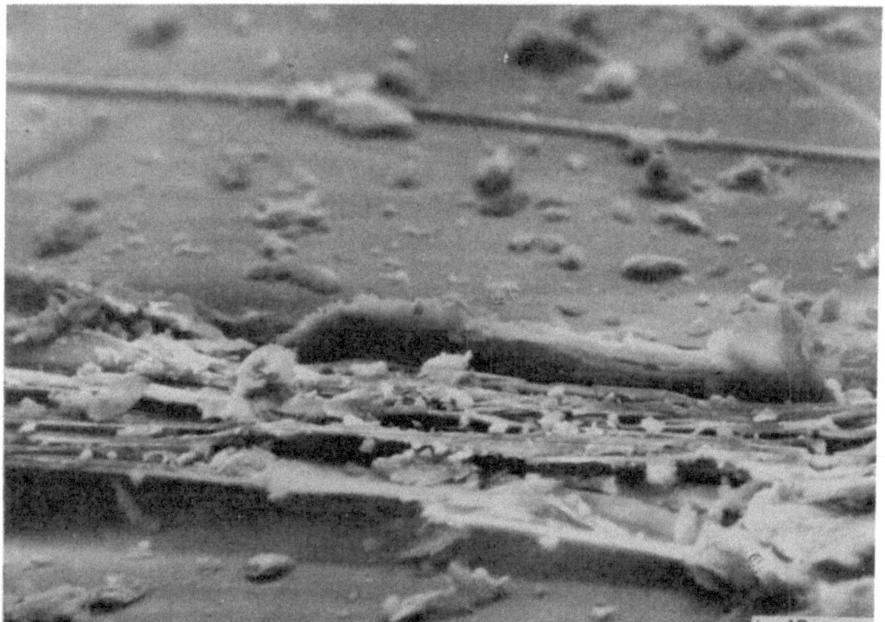

Figs. 249, 250. SEMs: *Vermiculite*, Transvaal, South Africa

Fig. 250.

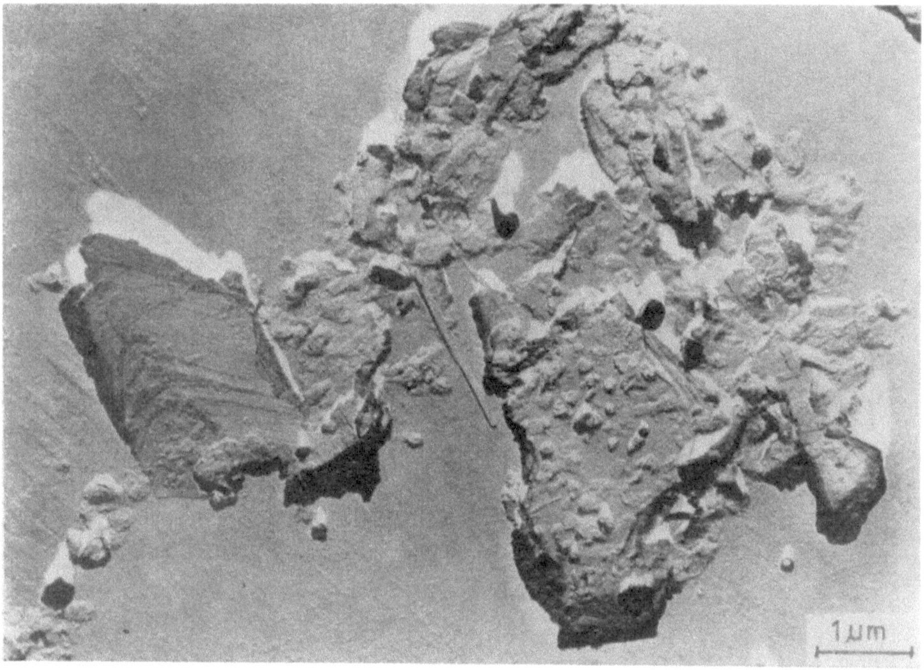

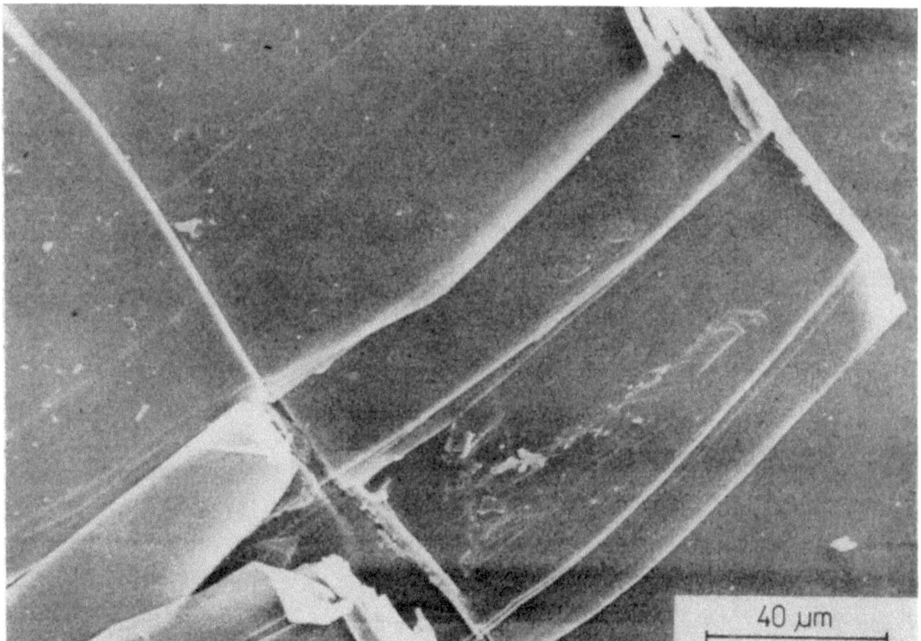

Fig. 251. Replica: *Vermiculite*, Murmansk, RSFSR, USSR

Fig. 252. SEM: *Vermiculite*, Murmansk, RSFSR, USSR

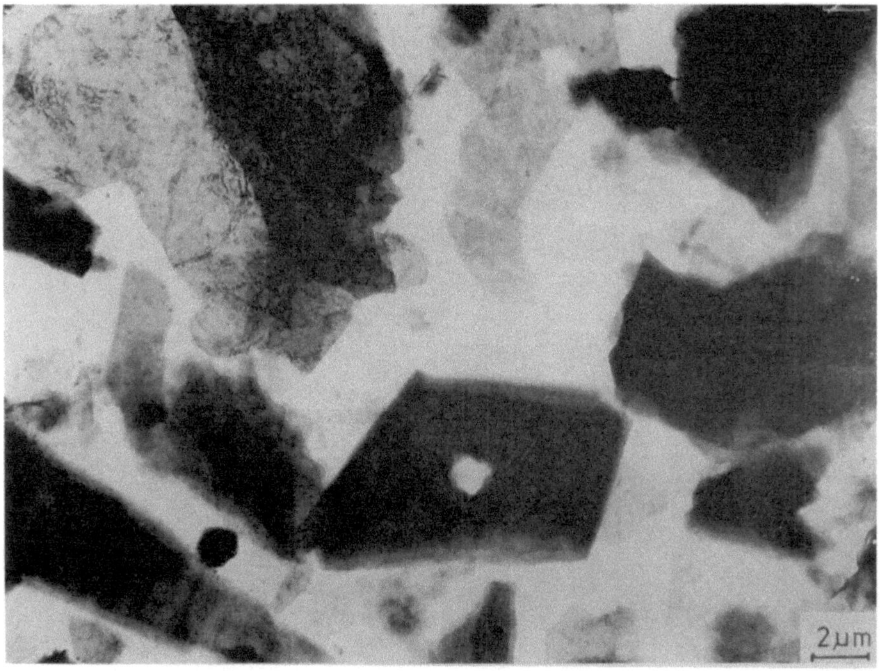

Figs. 253, 254. TEMs: *Vermiculite* from weathered serpentinite, Callenberg, GDR

Fig. 254.

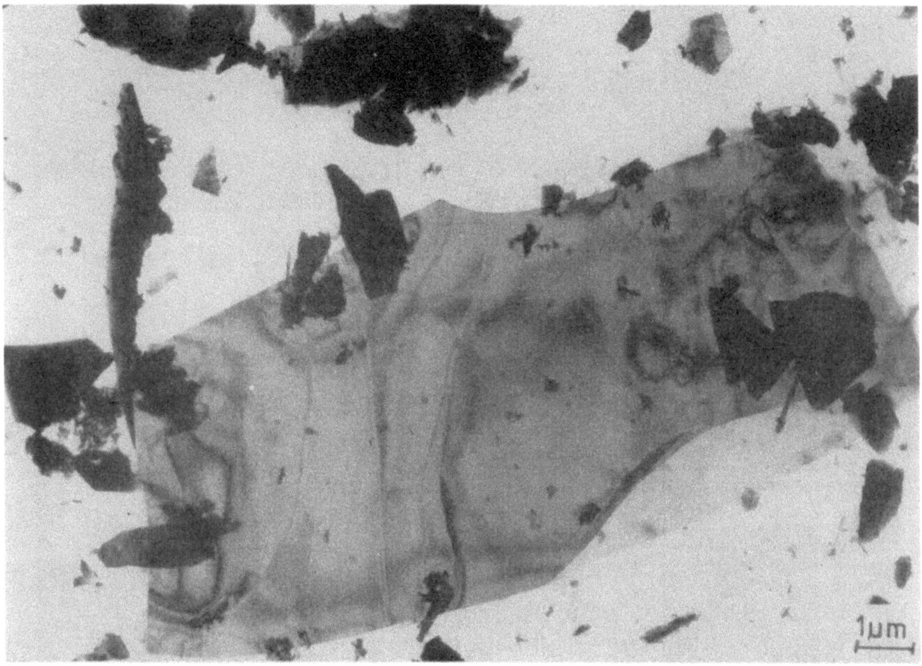

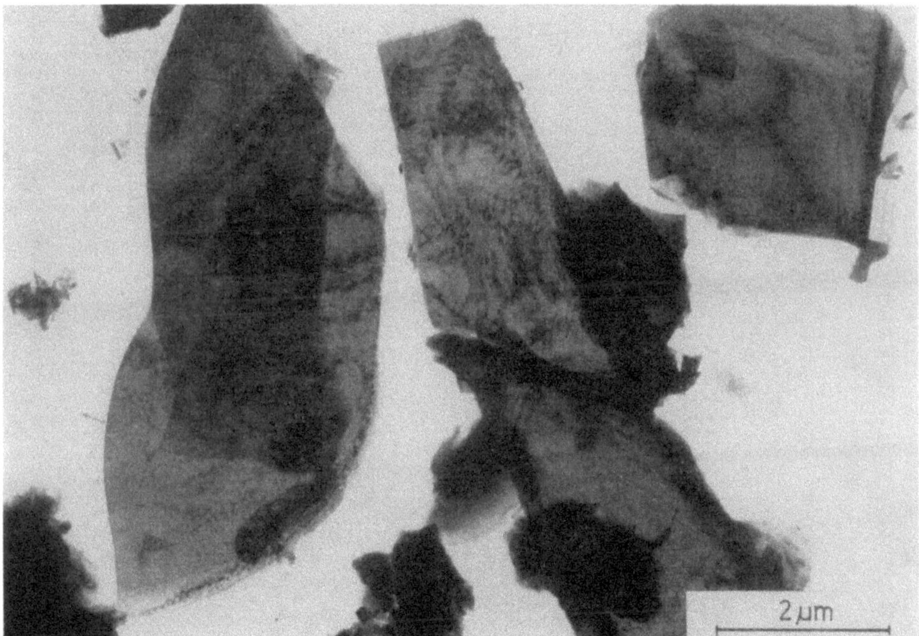

Figs. 255, 256. TEMs: *Muscovite*, Zermatt, Switzerland

Fig. 256.

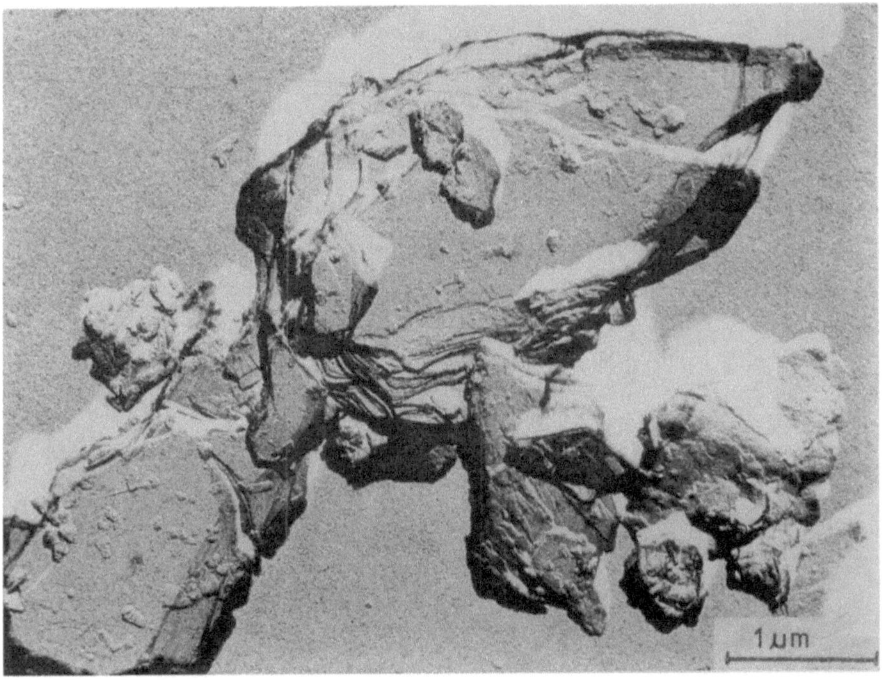

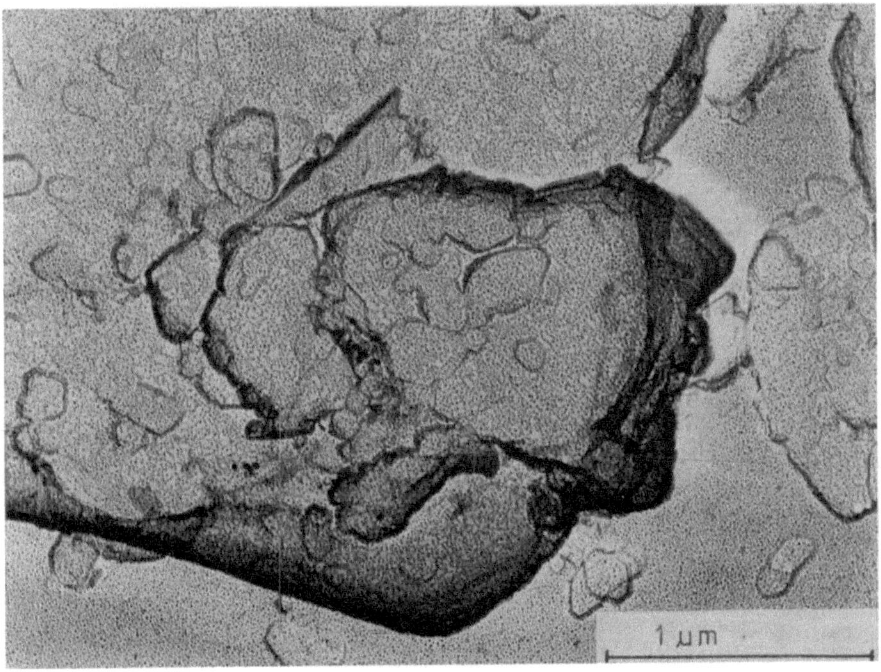

Fig. 257. Replica: *Muscovite* (gilbertite) from pneumatolytic vein, Ehrenfriedersdorf, GDR

Fig. 258. Replica: *Muscovite*, gold decorated, Nova Več, Czechoslovakia

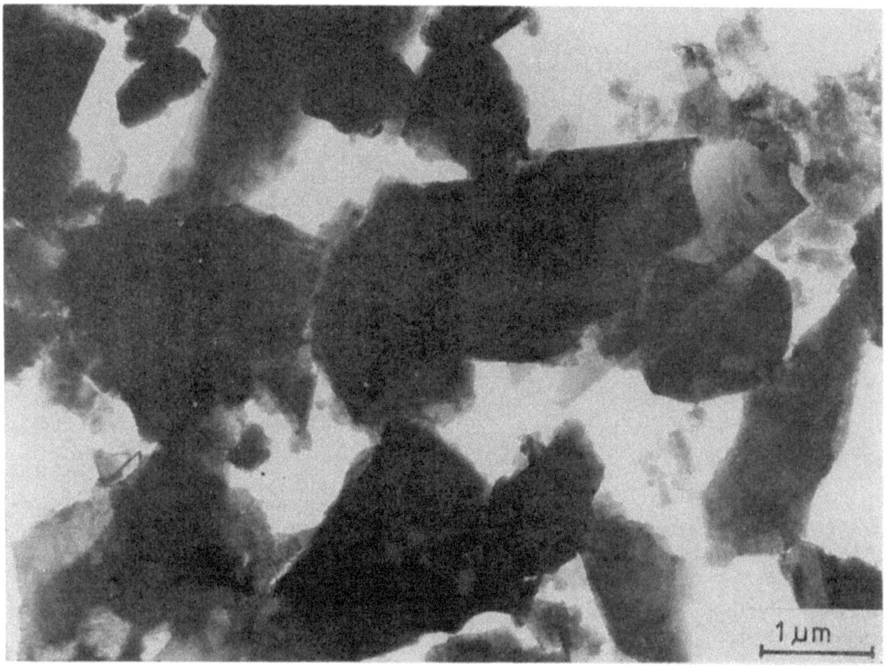

Fig. 259. TEM: *Paragonite,* Bockau, GDR

Figs. 260, 261. Replicas: *Paragonite,* Bockau, GDR

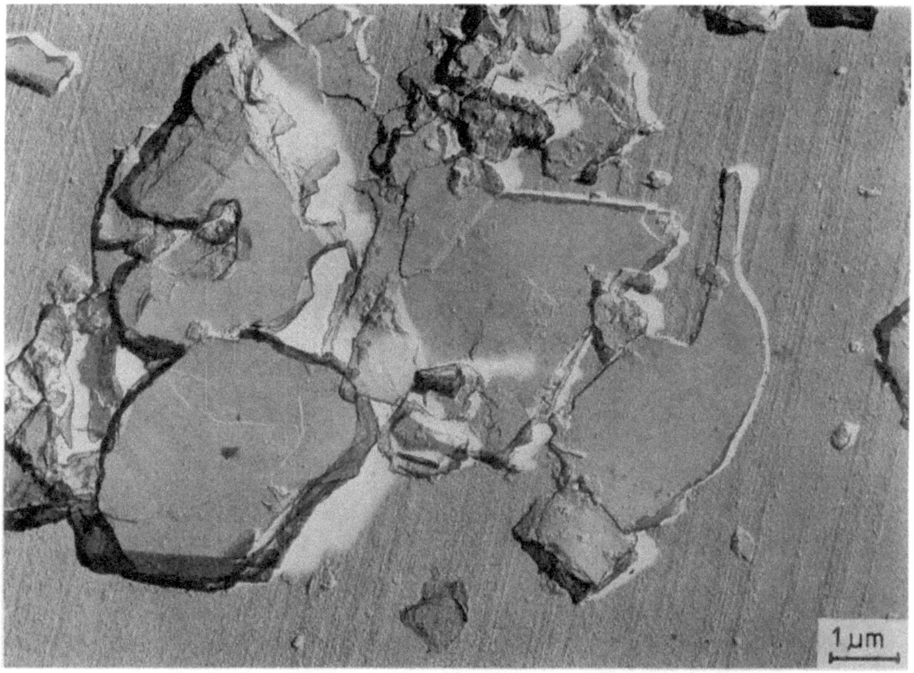

Fig. 261. *(explained at p. 171)*

Fig. 262. Replica: ***Paragonitic schist***, surface of fracture, Bockau, GDR

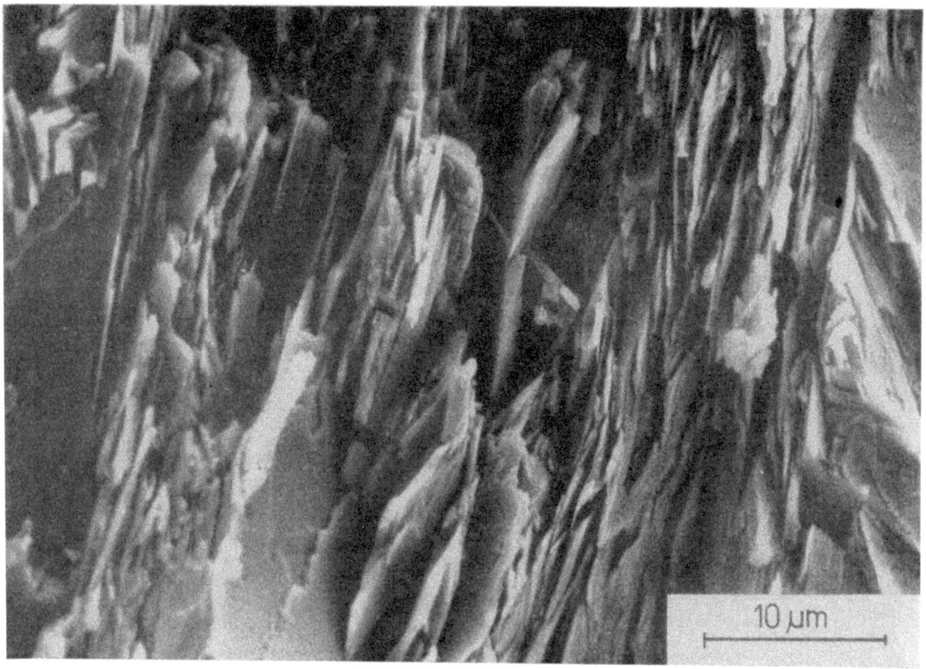

Fig. 263. SEM: *Paragonitic schist*, fracture vertical to bedding, Bockau, GDR

Fig. 264. SEM: *Paragonitic schist*, fracture parallel to bedding, Bockau, GDR

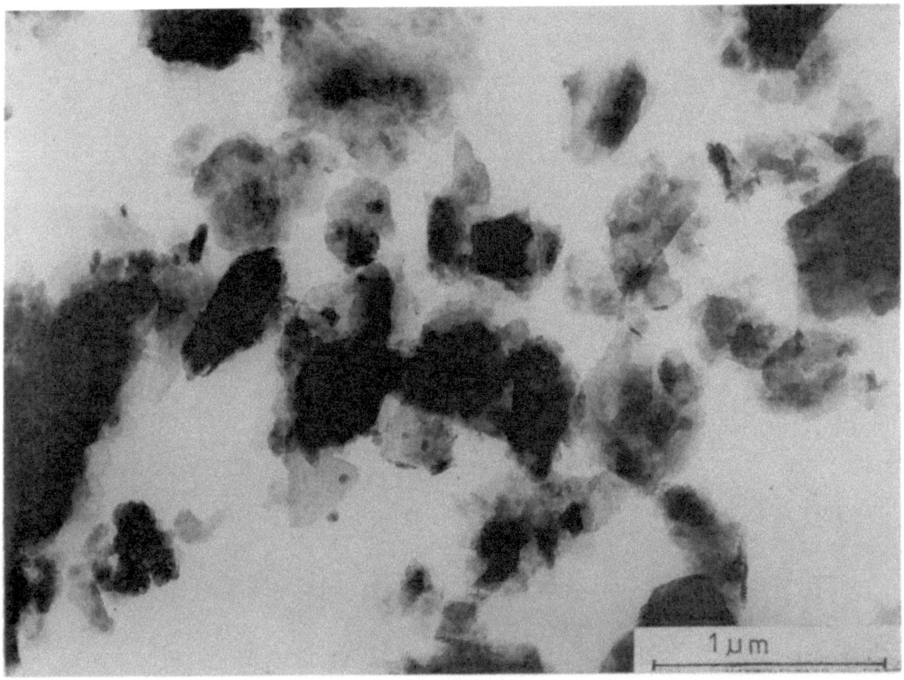

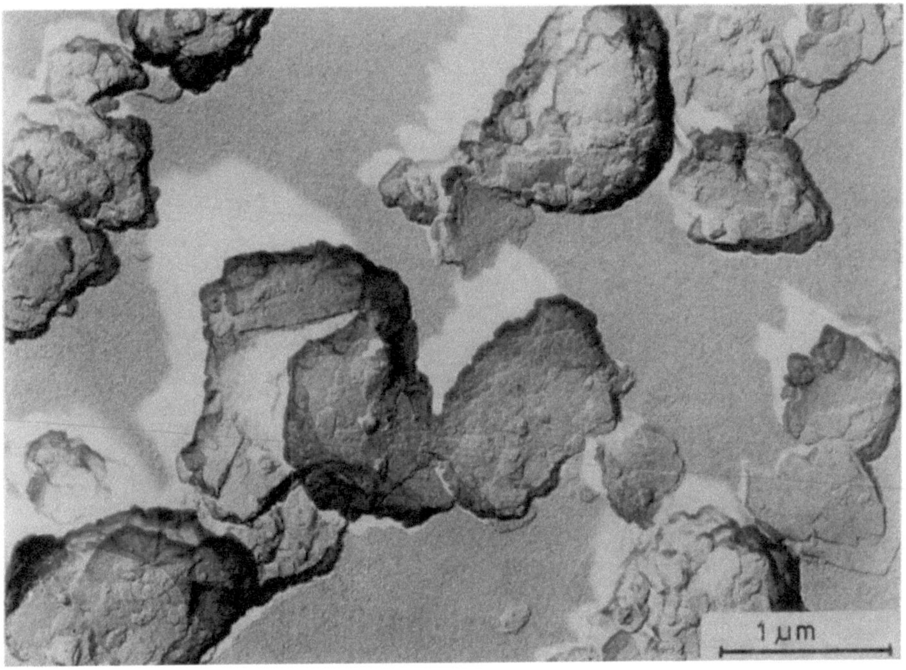

Fig. 265. TEM: *Illite,* Rochester, New York, USA

Fig. 266. Replica: *Illite,* Rochester, New York, USA

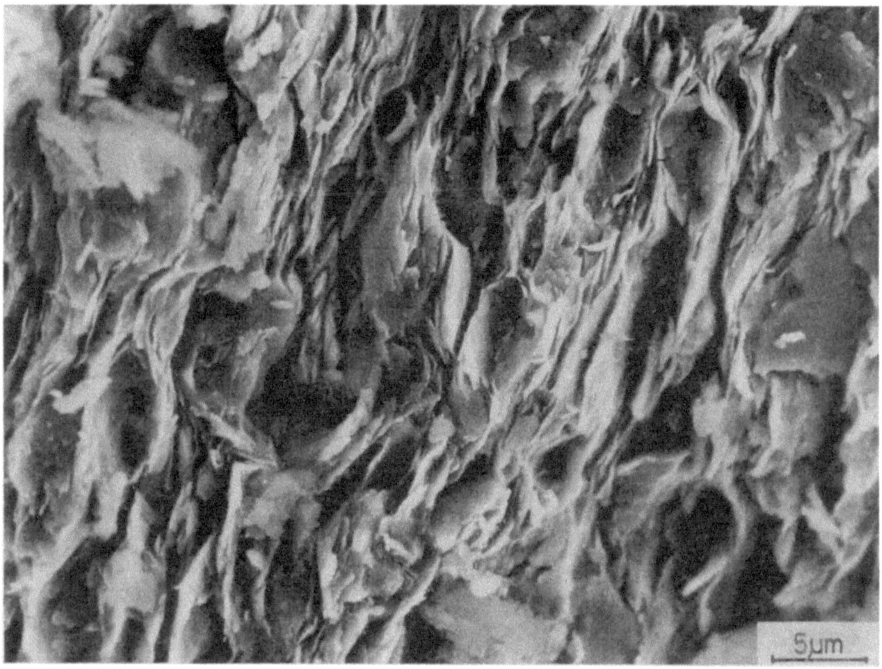

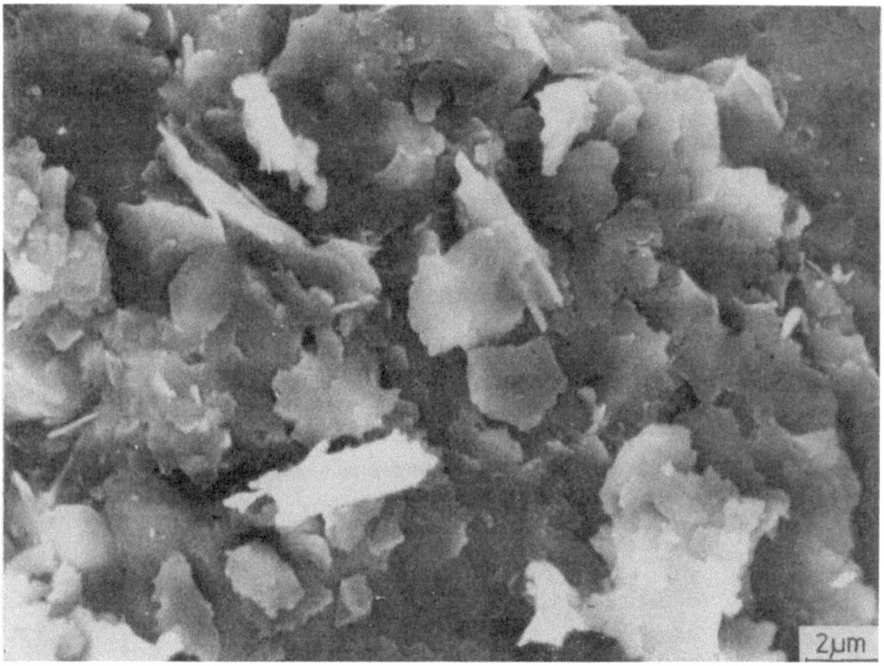

Fig. 267. SEM: *Illite-bearing shale*, fracture vertical to bedding, Rochester, New York, USA

Fig. 268. SEM: *Illite-bearing shale*, fracture parallel to bedding, Rochester, New York, USA

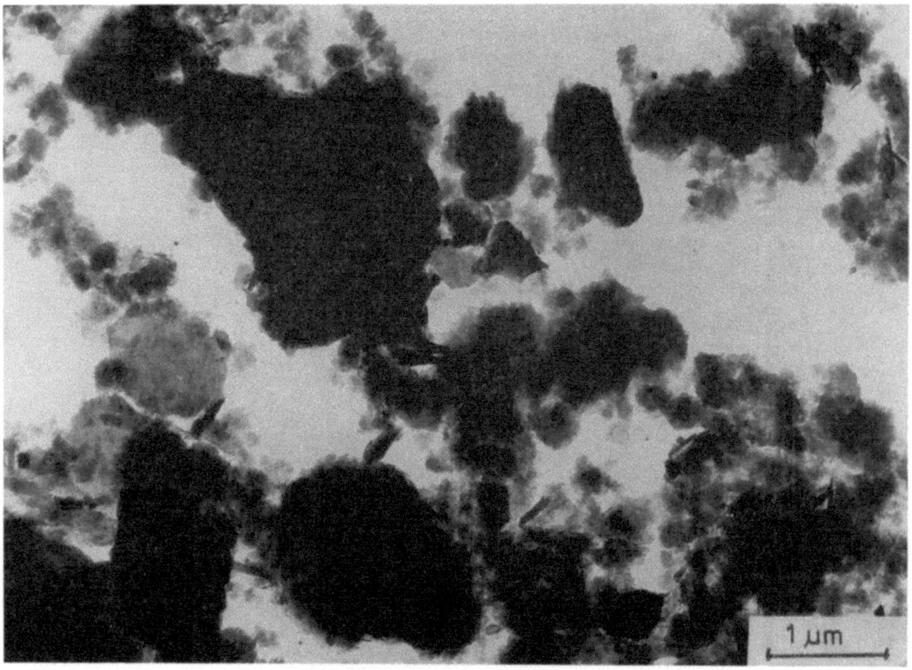

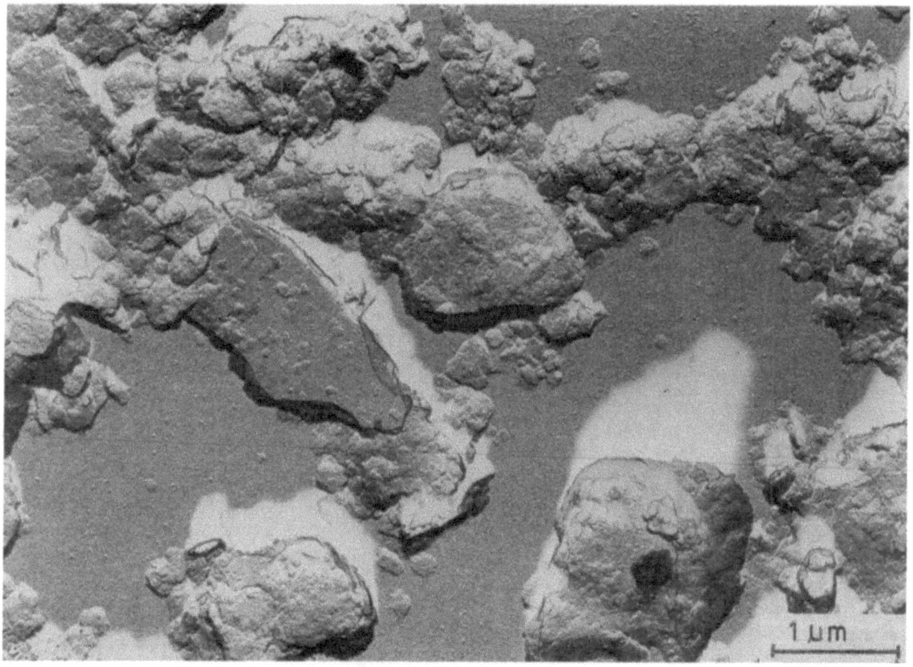

Fig. 269. TEM: *Illite*, API No. 35, Fithian, Illinois, USA

Fig. 270. Replica: *Illite*, API No. 35, Fithian, Illinois, USA

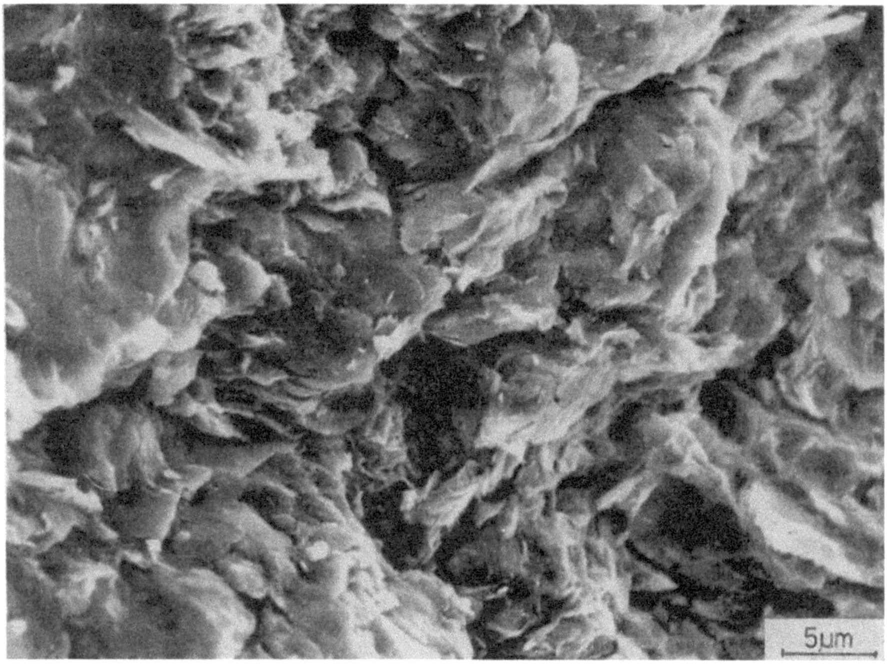

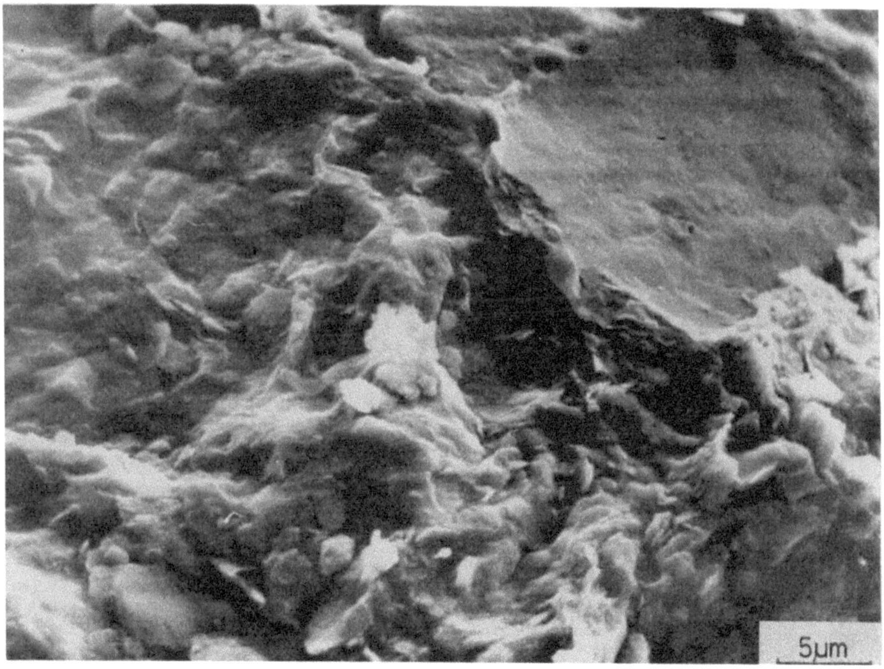

Figs. 271, 272. SEMs: *Illite*, API No. 35, Fithian, Illlinois, USA

Fig. 272.

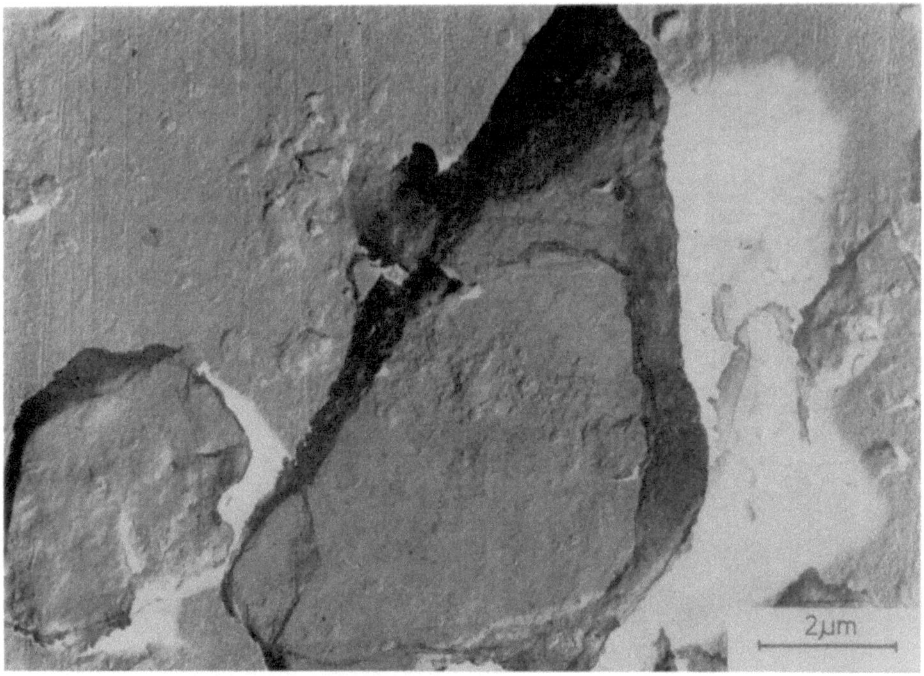

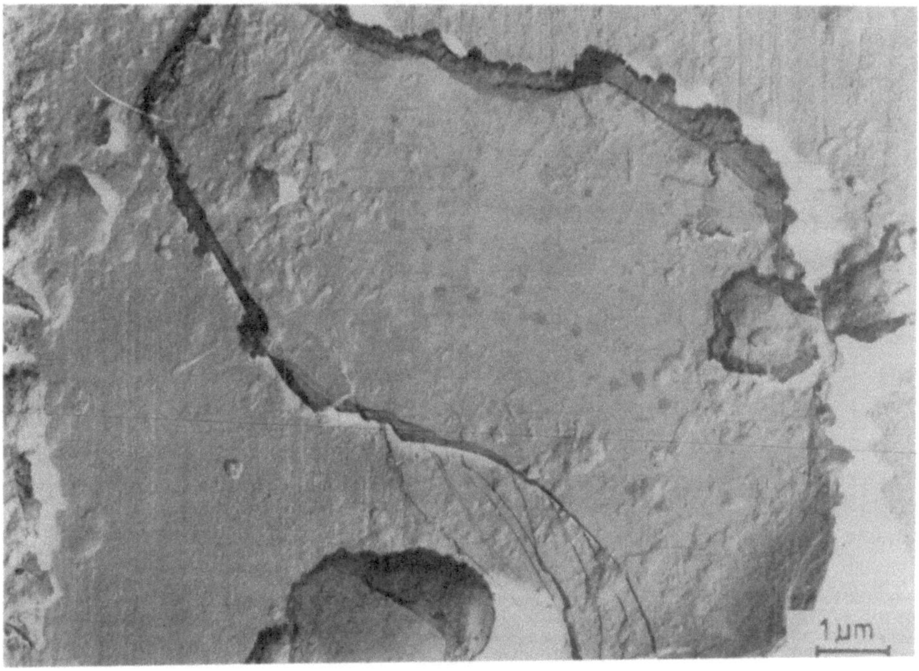

Figs. 273, 274. Replicas: *Illite*, Bad Dürrenberg, GDR

Fig. 274.

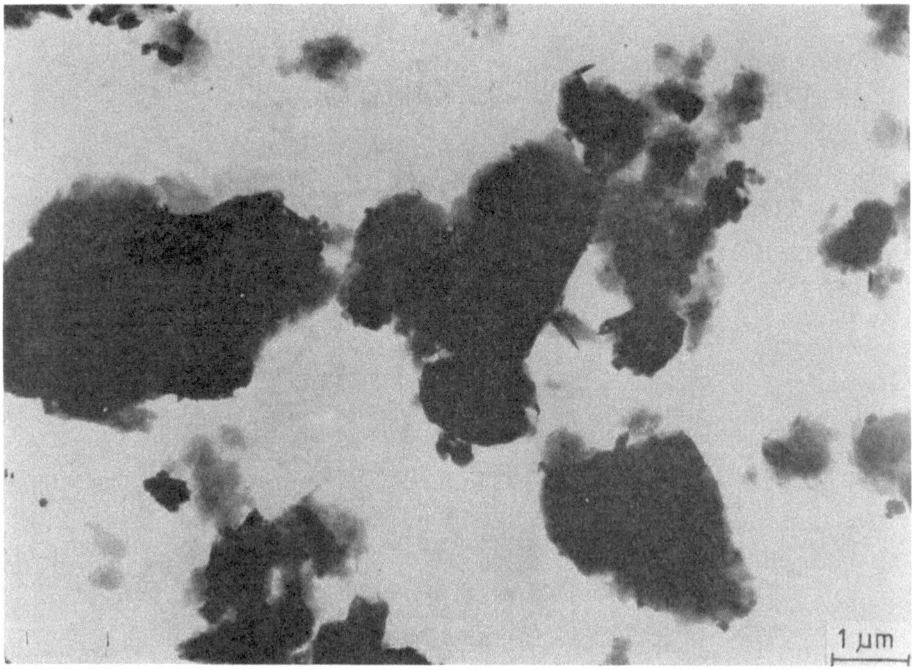

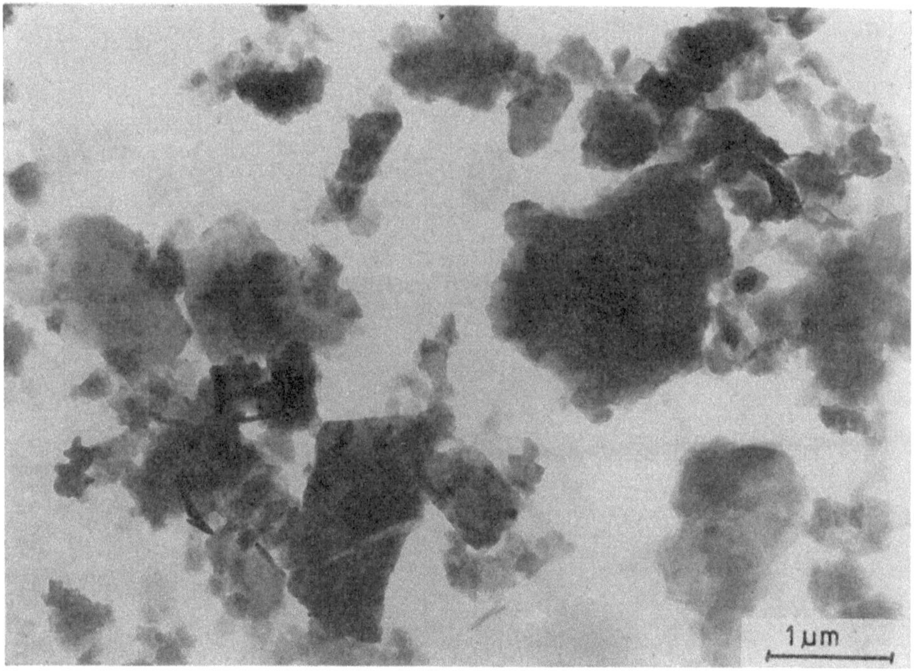

Figs. 275, 276. TEMs: *Illite*, API No. 36, Morris, Illinois, USA

Fig. 276.

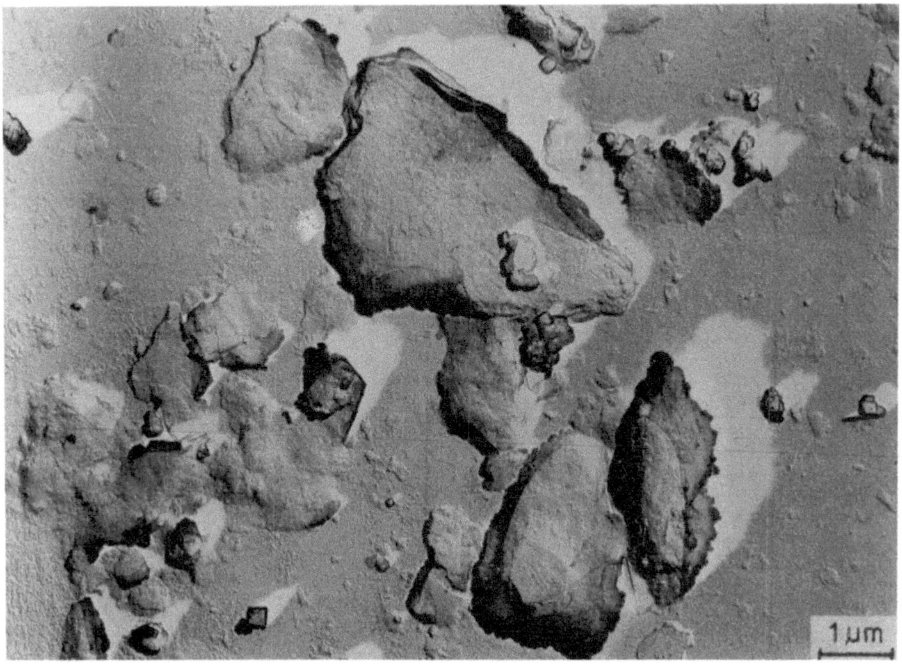

Figs. 277, 278. Replicas: *Illite*, API No. 36, Morris, Illinois, USA

Fig. 278.

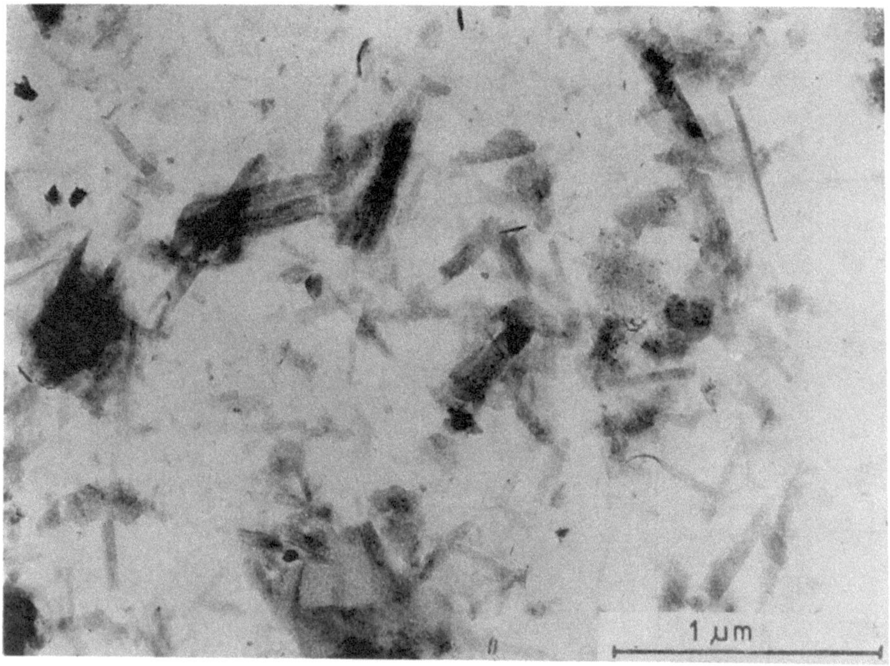

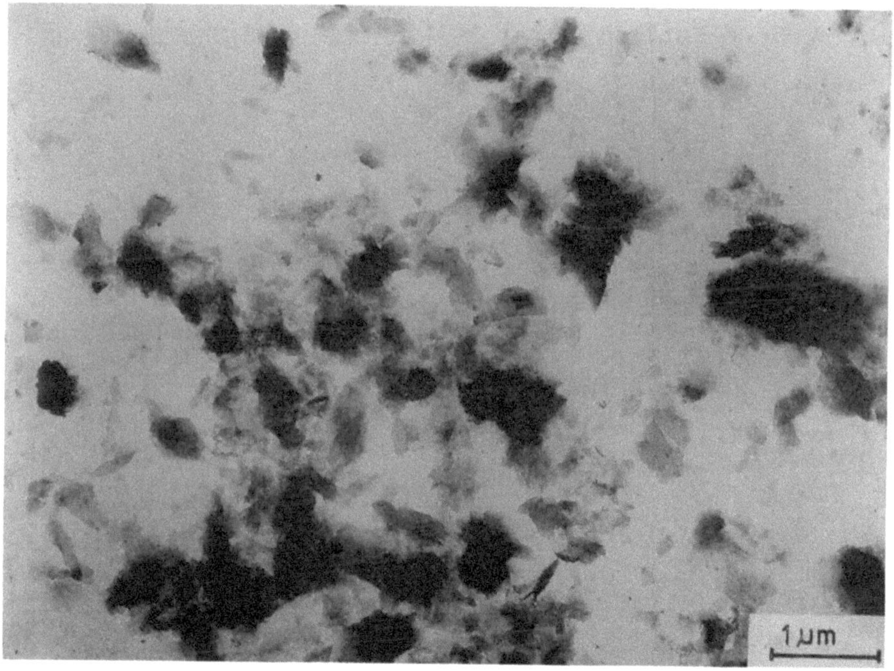

Fig. 279. TEM: *Glauconite* from pellets, drill-core, Lychen, GDR

Fig. 280. TEM: *Glauconite* from drill-core, Lychen, GDR

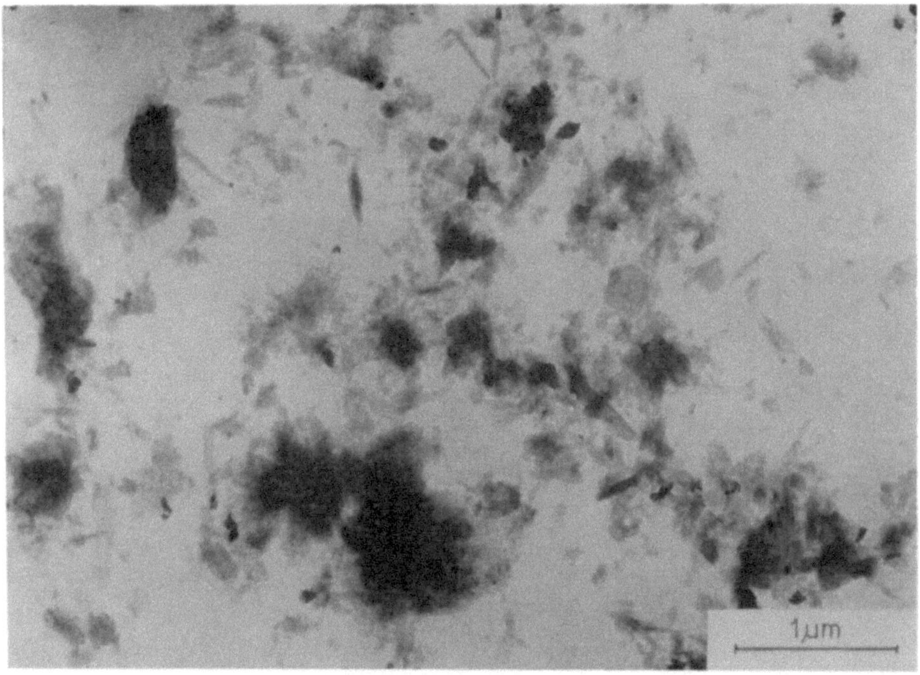

Fig. 281. TEM: *Glauconite* from drill-core, Usedom, GDR

Fig. 282. SEM: *Glauconite* from pellets, Friedland, GDR

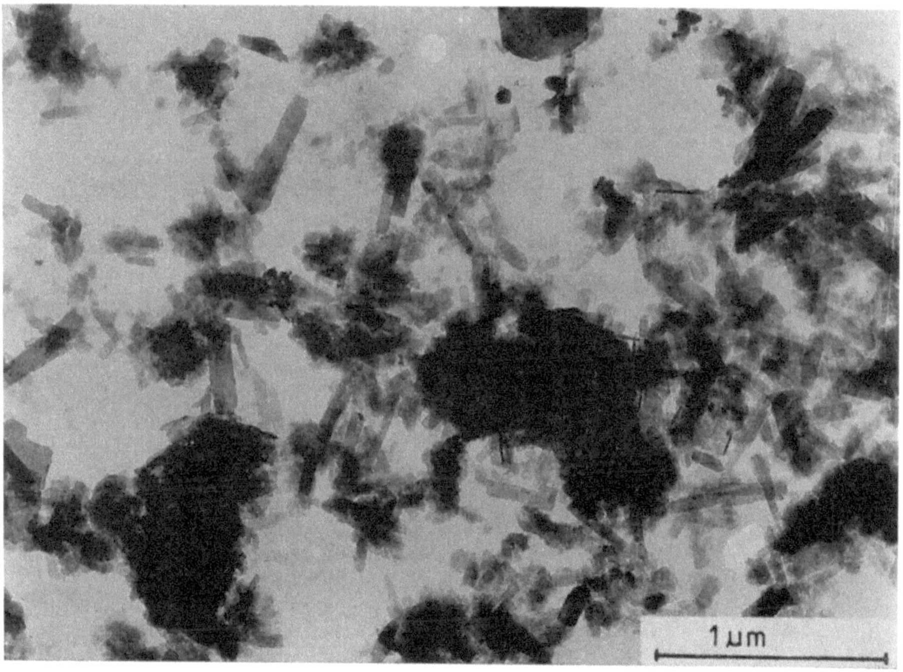

Fig. 283. SEM: Surface of *glauconite* of Fig. 282, Friedland, GDR

Figs. 284, 285. TEMs: *Seladonite*, Kadaň, Czechoslovakia

Fig. 285. (*explained at p. 183*)

Fig. 286. Replica: *Seladonite*, Kadaň, Czechoslovakia

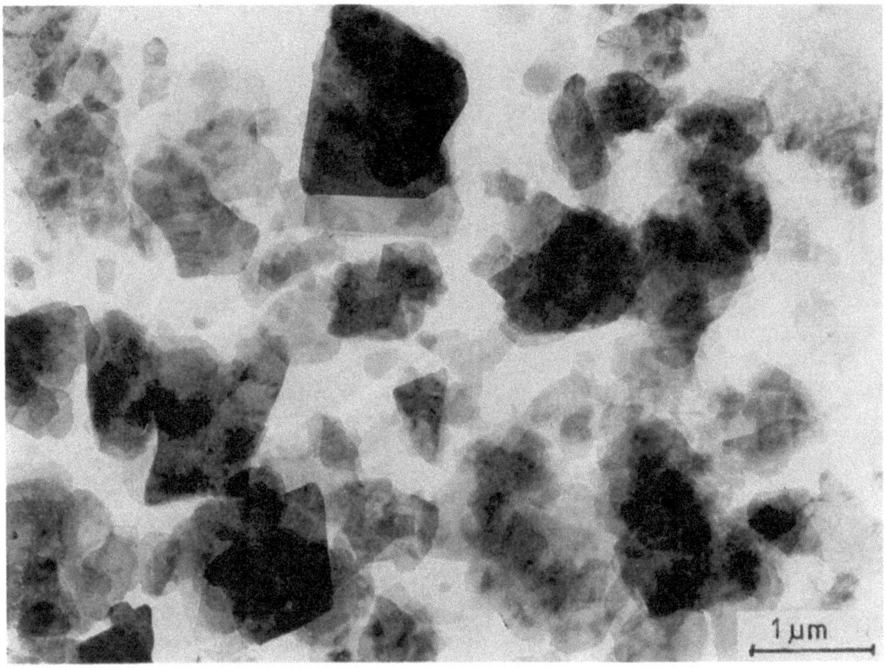

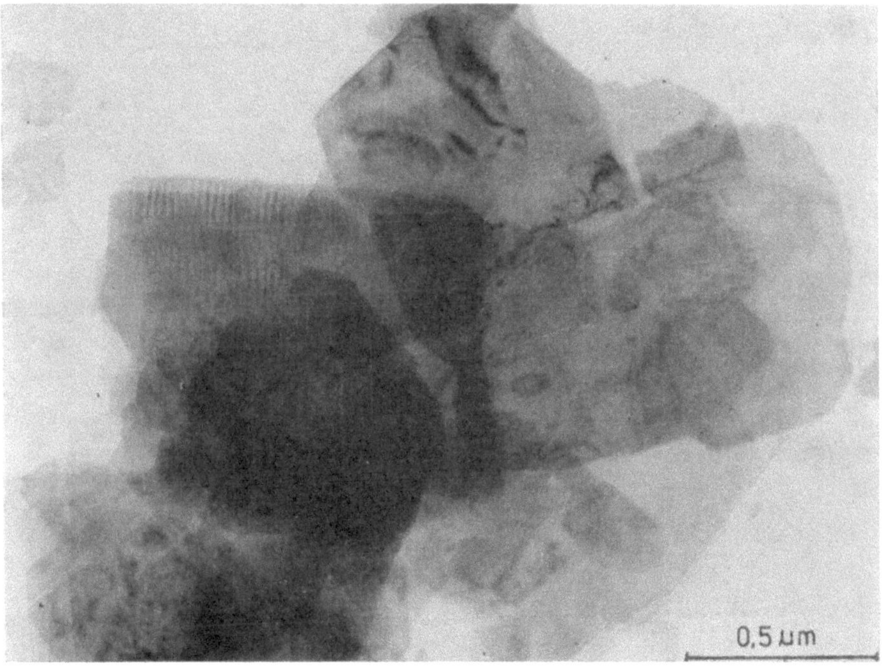

Figs. 287, 288. TEMs: *Sudoite*, Kosaka Mine, Akita Prefecture, Japan

Fig. 288.

Fig. 289. Replica: *Sudoite*, Kosaka Mine, Akita Prefecture, Japan

Fig. 290. SEM: *Sudoite*, Kosaka Mine, Akita Prefecture, Japan

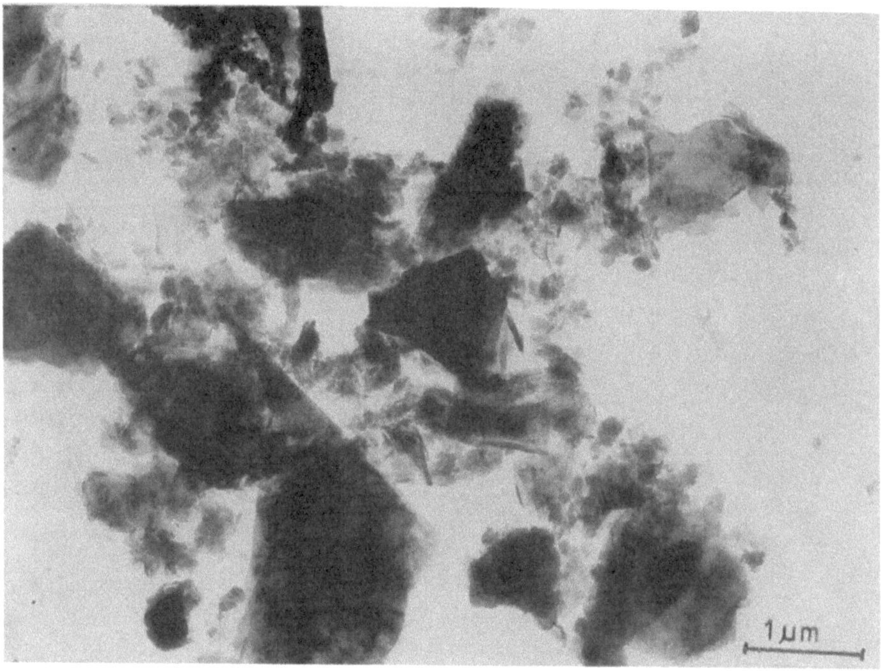

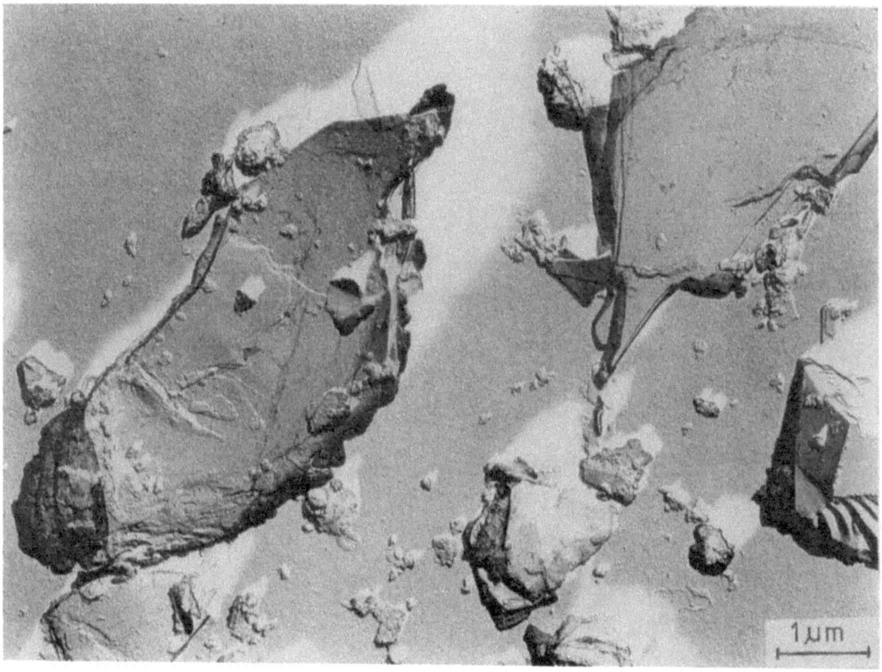

Fig. 291. TEM: *Chlorite*, Pennsylvania, USA

Fig. 292. Replica: *Chlorite*, Pennsylvania, USA

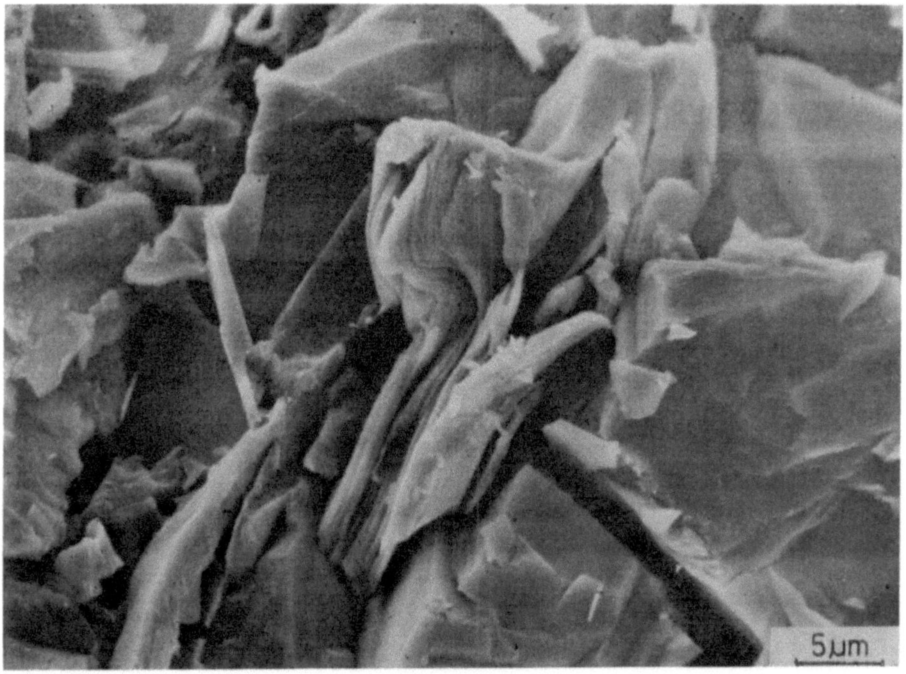

Figs. 293, 294. SEMs: *Chlorite*, Pennsylvania, USA

Fig. 294.

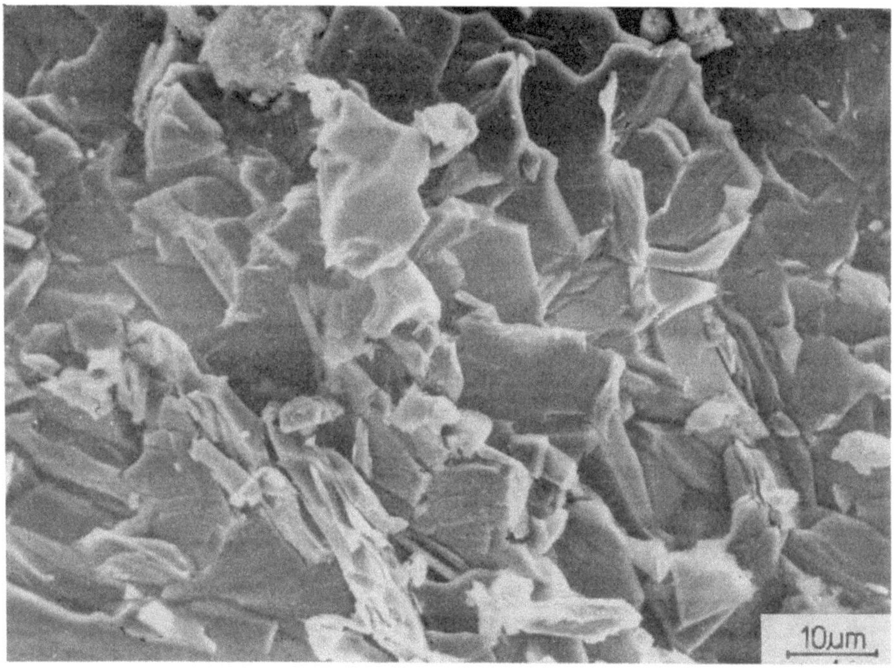

Figs. 295, 296. SEMs: *Chlorite*, Pennsylvania, USA

Fig. 296.

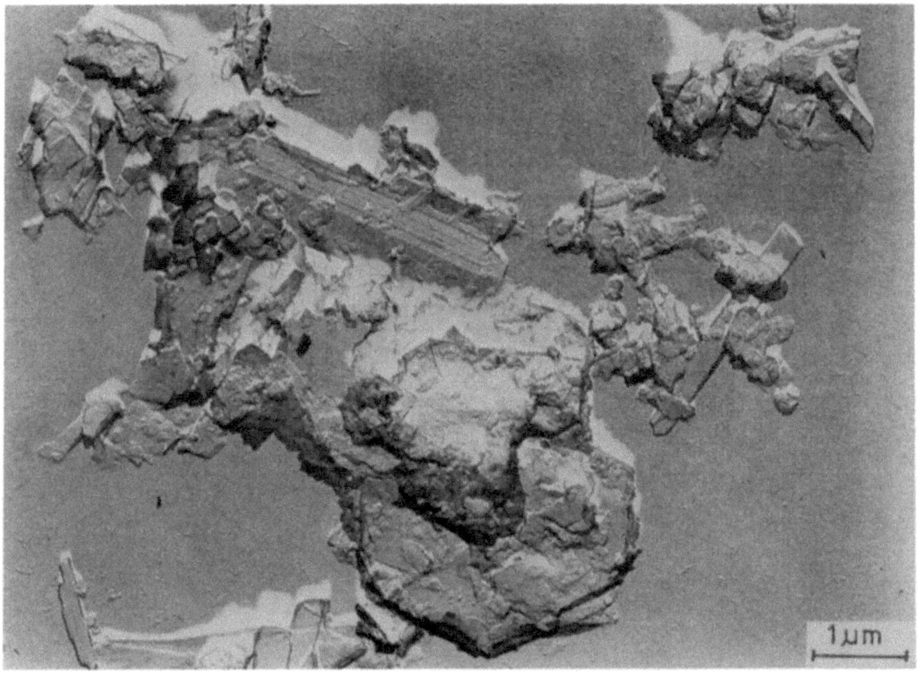

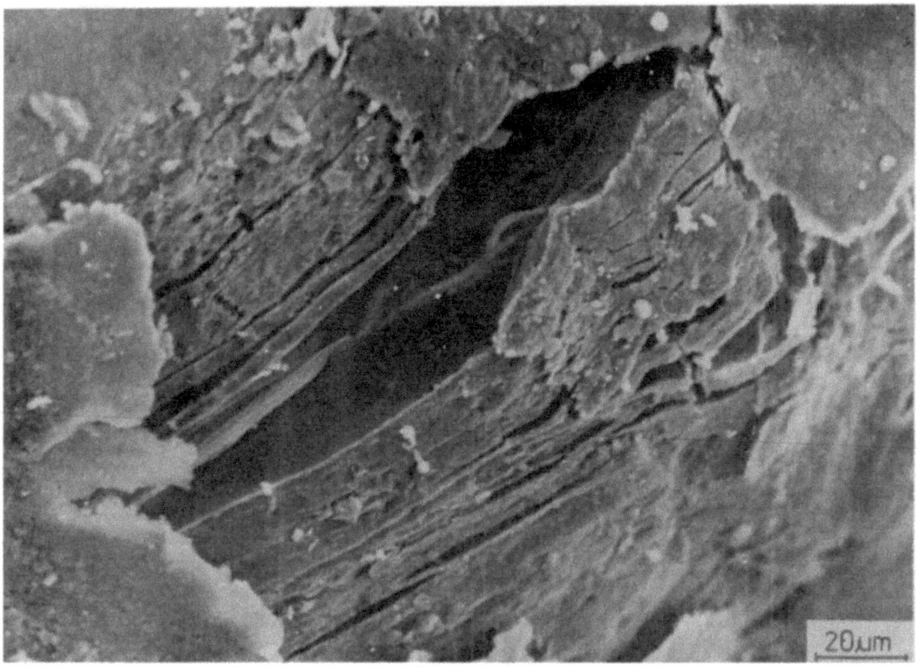

Fig. 297. Replica: *Pennine,* Rimpfischwänge near Zermatt, Switzerland

Figs. 298, 299, 300. SEM: *Pennine,* Rimpfischwänge near Zermatt, Switzerland

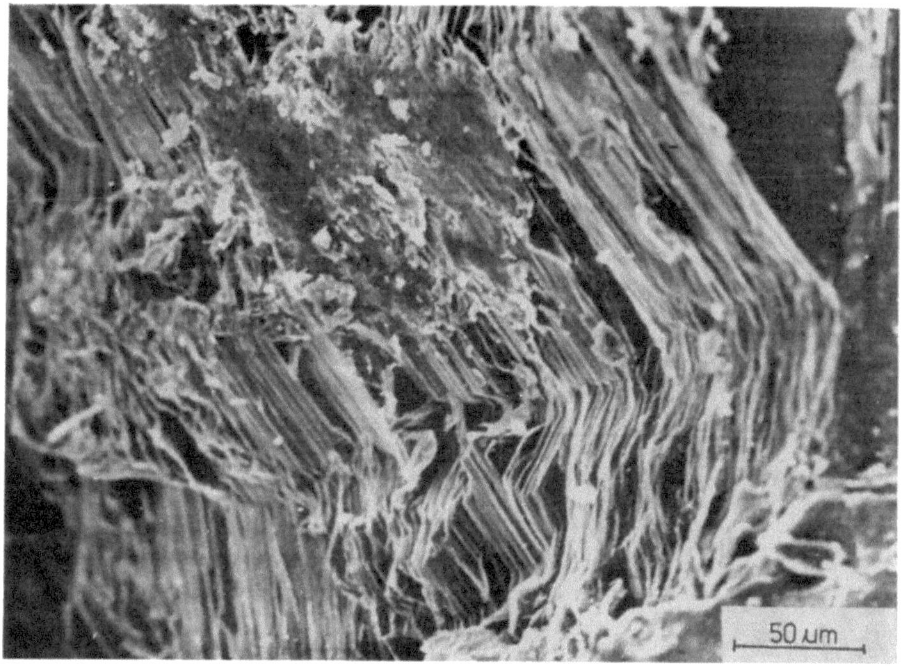

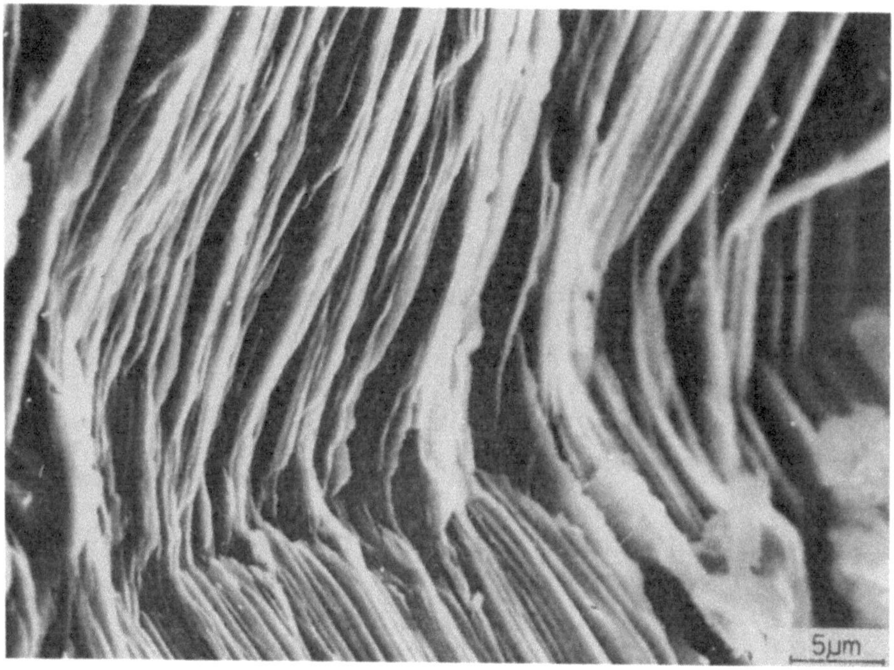

Fig. 299.

Fig. 300.

Figs. 301, 302, 303, 304. SEMs: *Clinochlore*, Piano di Mussa, Italy

Fig. 302.

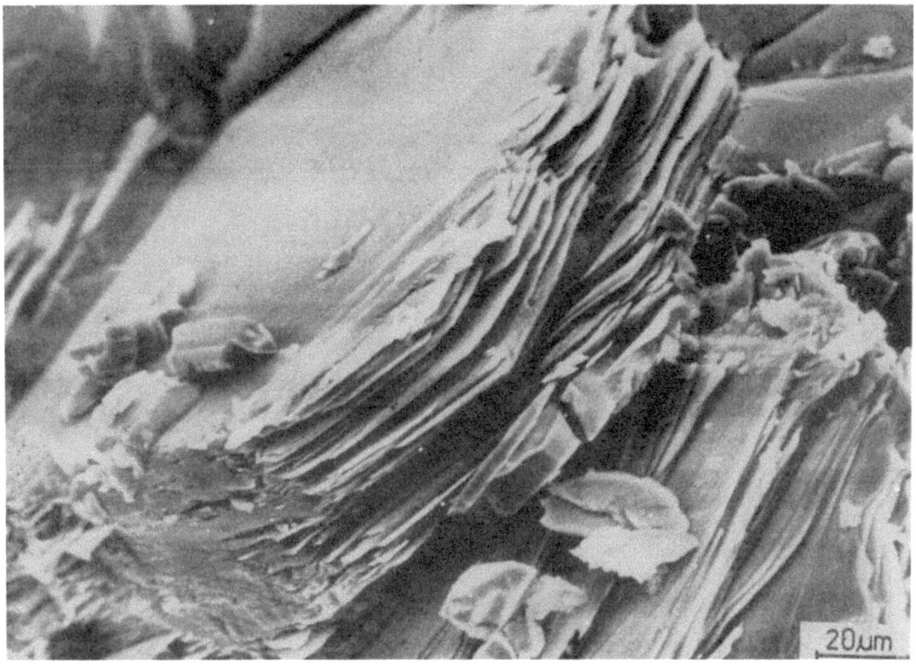

Fig. 303.

Fig. 304.

Fig. 305. TEM: *Prochlorite*, Chester, Vermont, USA

Fig. 306. Replica: *Prochlorite*, Chester, Vermont. USA

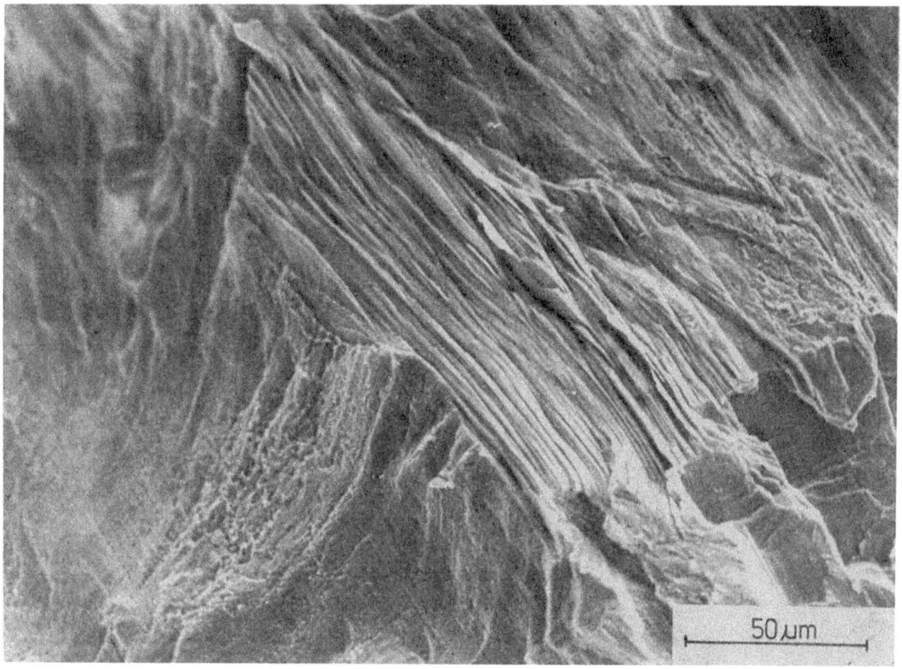

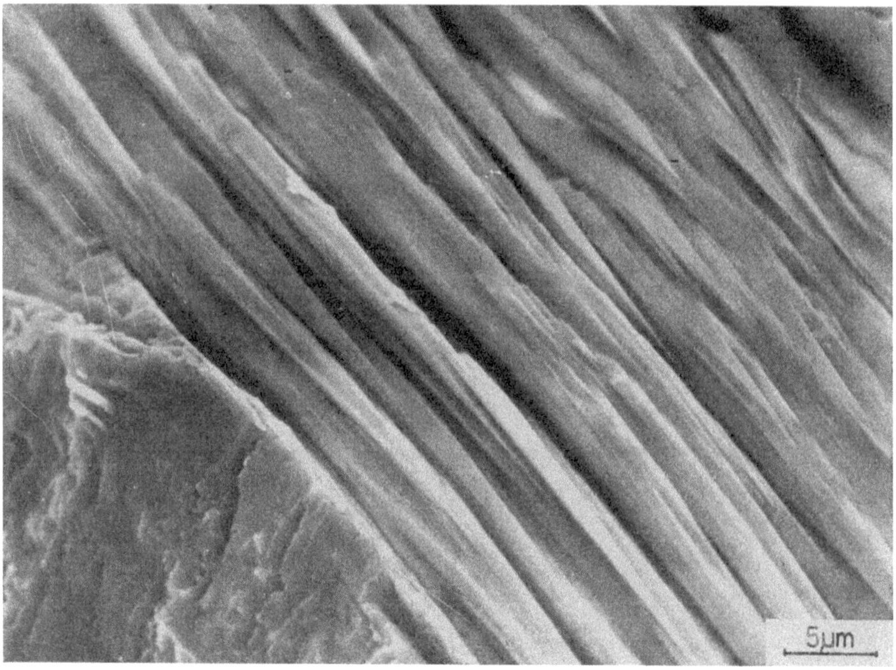

Figs. 307, 308. SEMs: *Prochlorite*, Chester, Vermont, USA

Fig. 308.

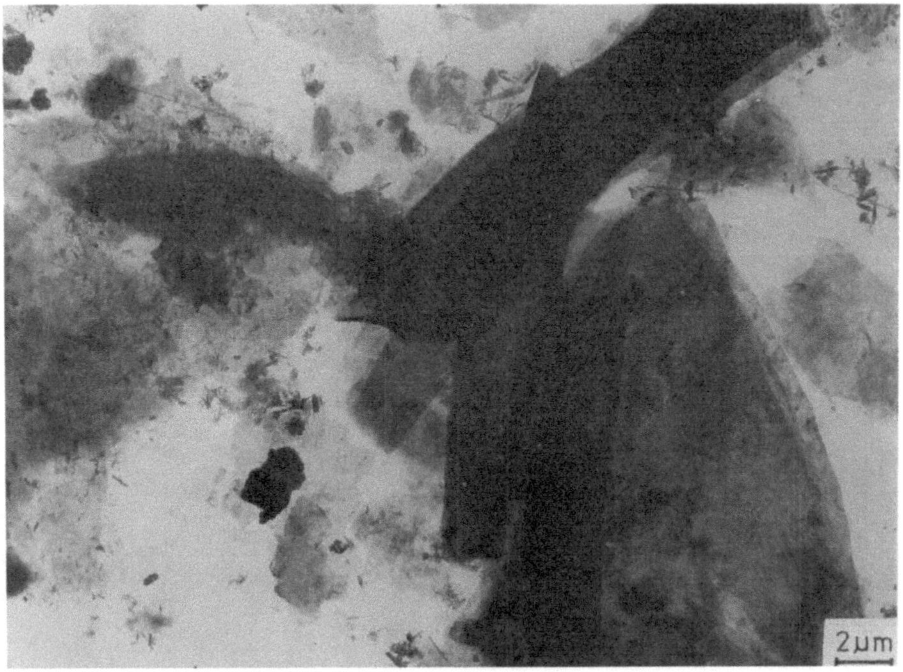

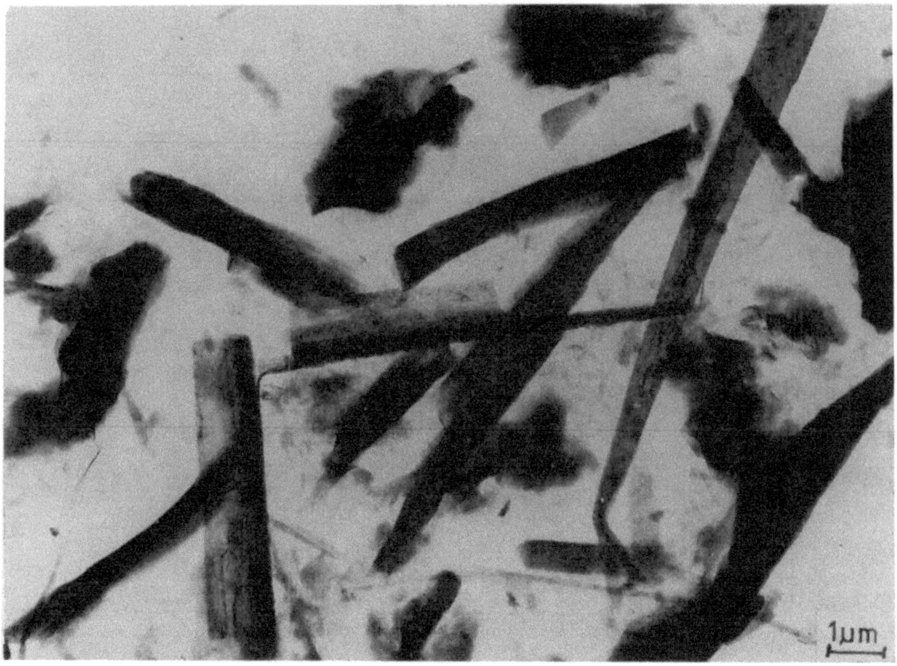

Figs. 309, 310. TEMs: ***Chlorite,*** Callenberg, GDR

Fig. 310.

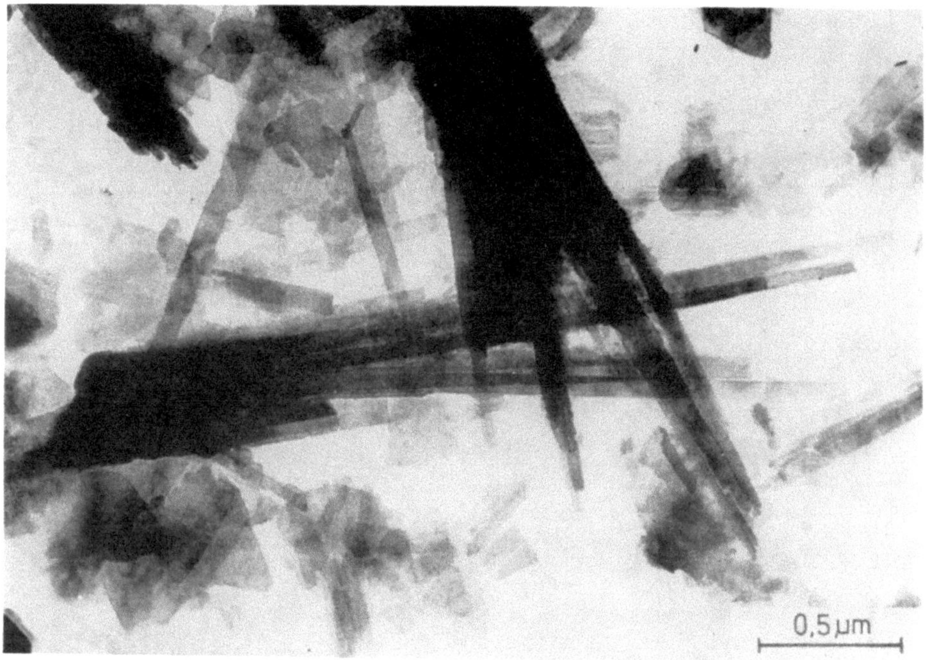

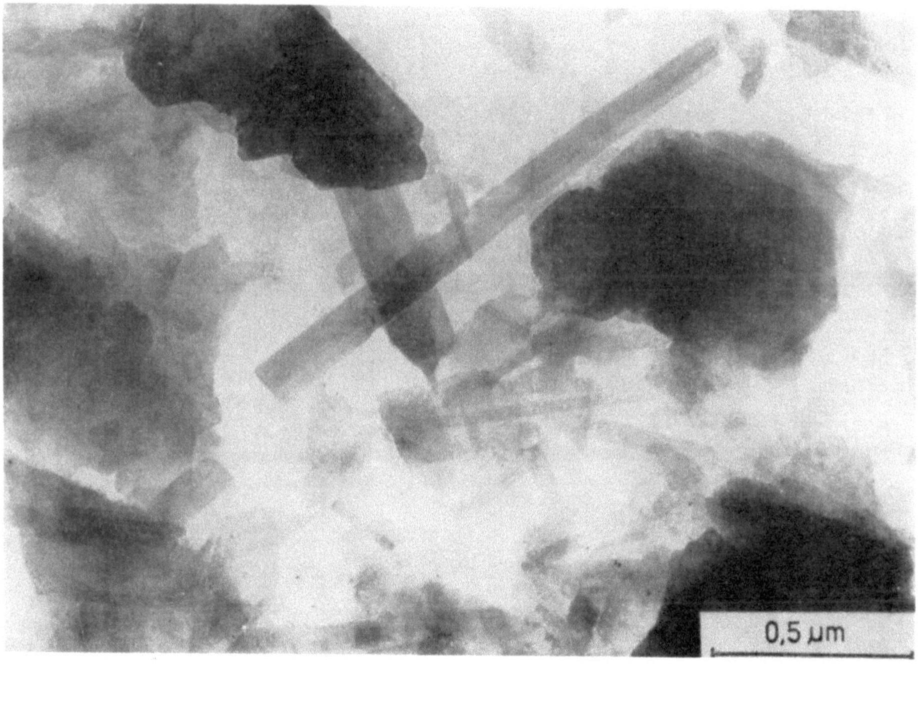

Figs. 311, 312. TEMs: *Muscovite-montmorillonite mixed-layer* (Mu 80%, Mo 20%) and kaolinite from weathered pitchstone, Seilitz near Meißen, GDR

Fig. 312.

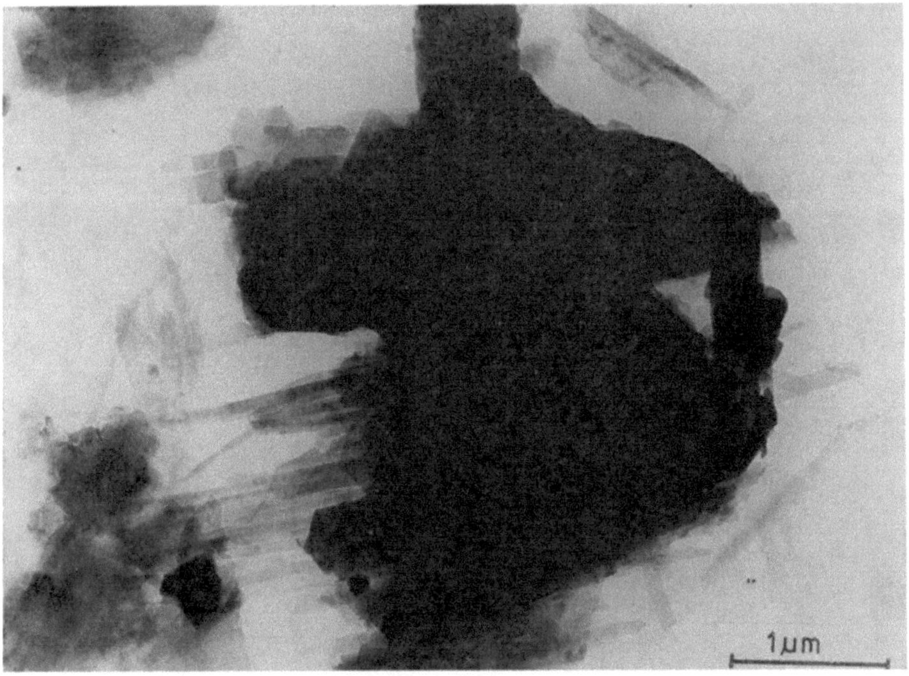

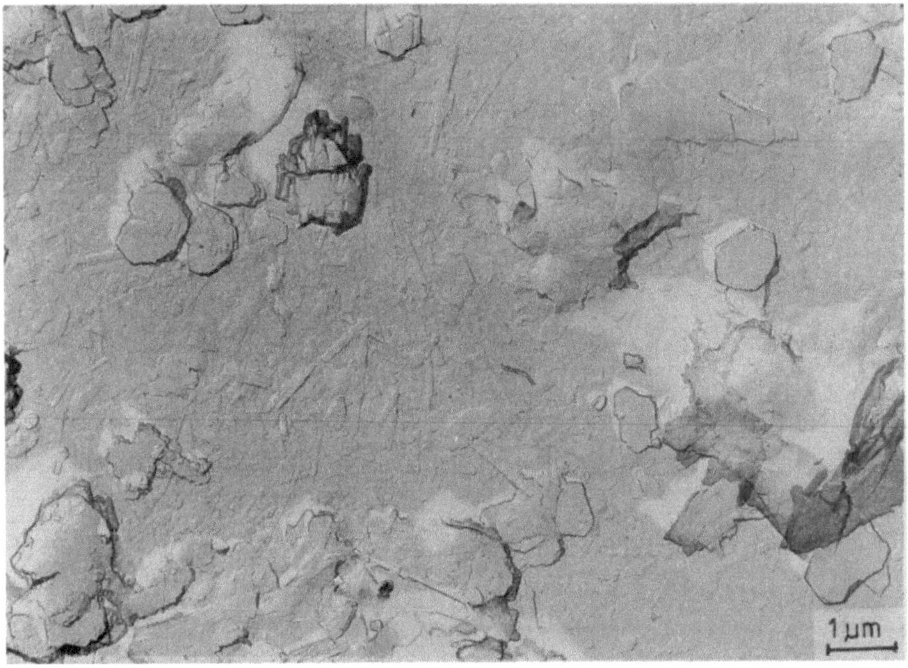

Fig. 313. TEM: *Muscovite-montmorillonite mixed-layer* (Mu 80%, Mo 20%), Seilitz near Meißen, GDR

Fig. 314. Replica: *Muscocvite-montmorillonite mixed-layer* (Mu 80%, Mo 20%) and kaolinite from weathered pitchstone, Seilitz near Meißen, GDR

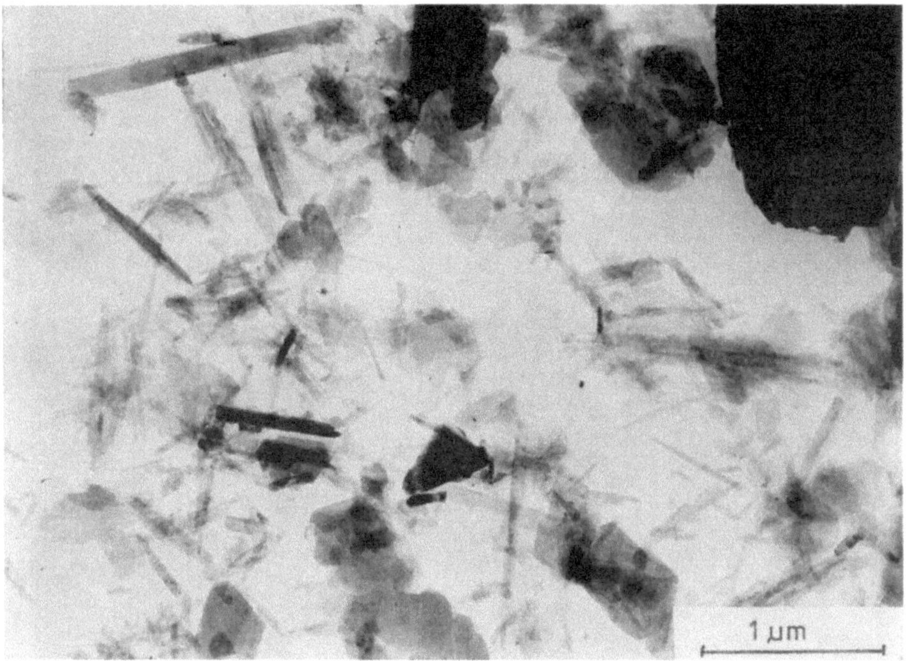

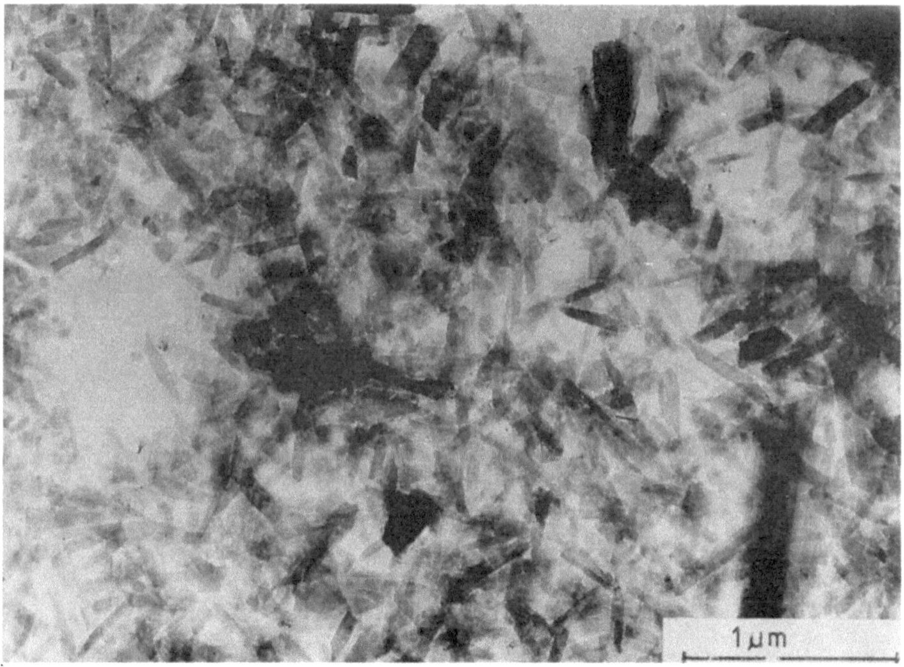

Fig. 315. TEM: *Muscovite-montmorillonite mixed-layer* (Mu 80%, Mo 20%) and kaolinite, Schletta near Meißen, GDR

Fig. 316. TEM: *Muscovite-montmorillonite mixed-layer* from weathered ignimbrite, Kemmlitz, GDR

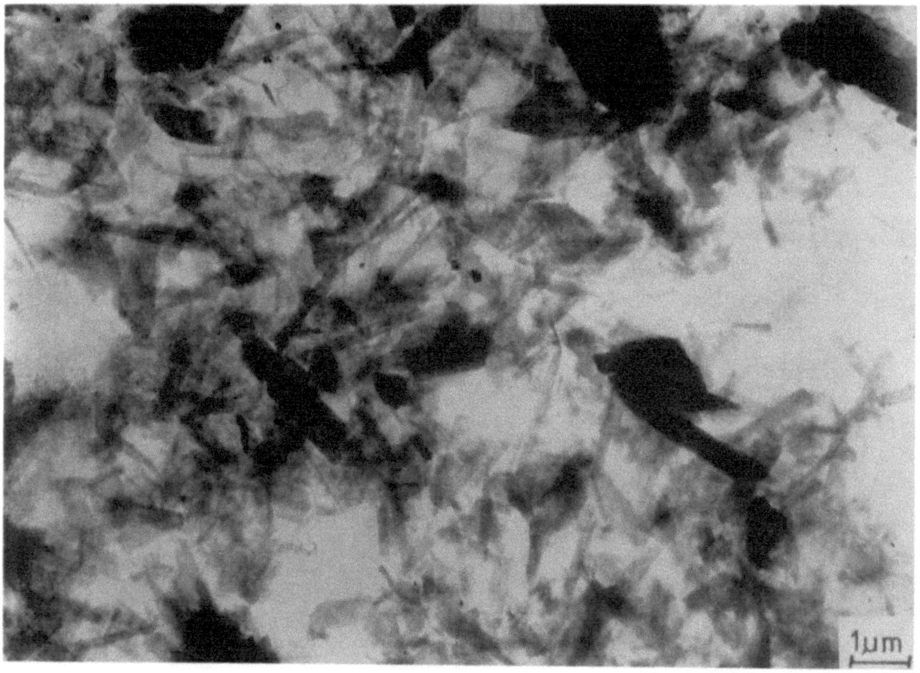

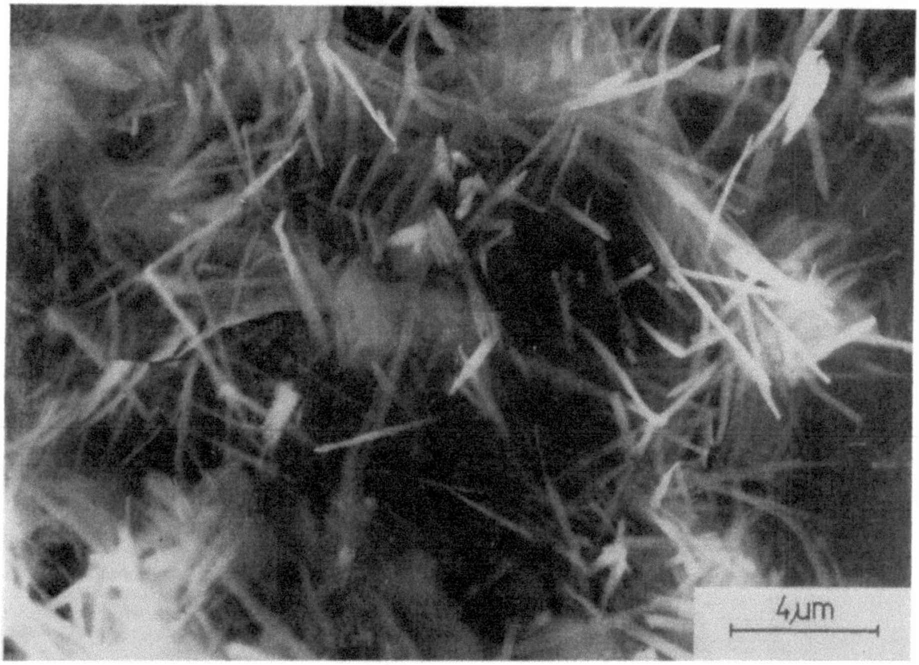

Fig. 317. TEM: *Muscovite-montmorillonite mixed-layer*, Kemmlitz, GDR

Fig. 318. SEM: *Muscovite-montmorillonite mixed-layer* from weathered ignimbrite, Kemmlitz, GDR

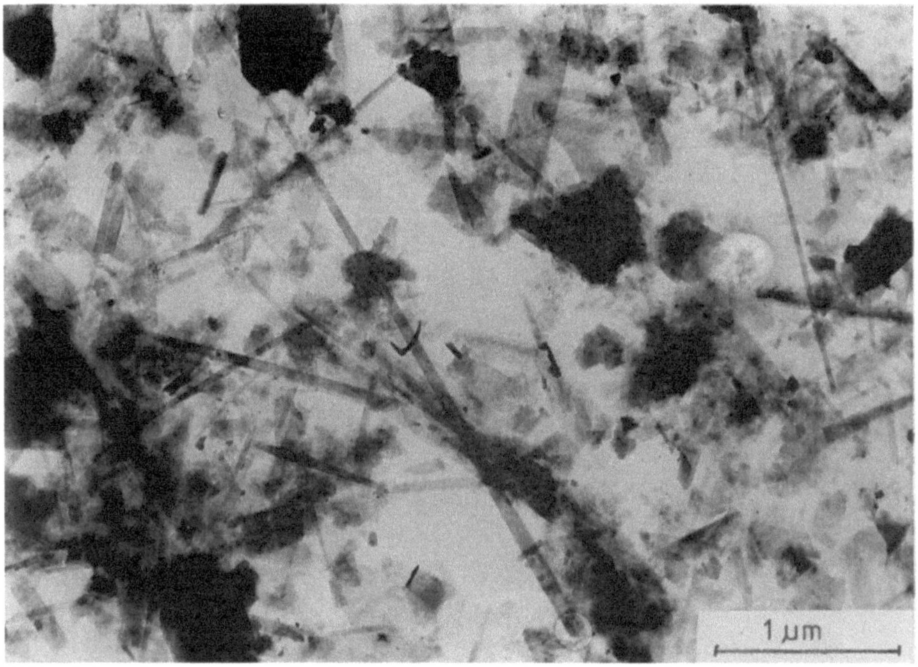

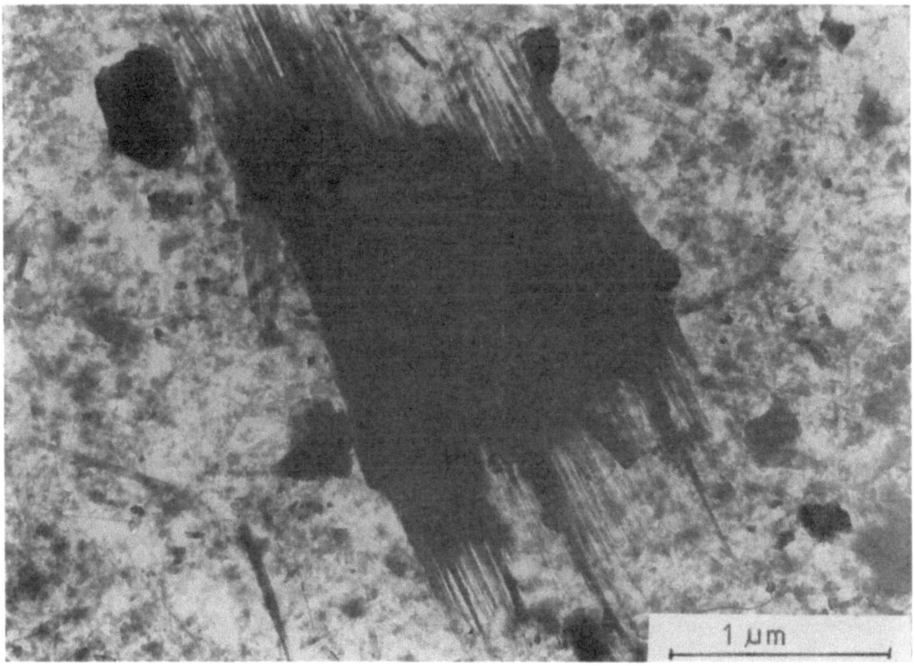

Fig. 319. TEM: *Muscovite-montmorillonite mixed-layer*, Neuhaus, GDR

Fig. 320. TEM: *Muscovite-montmorillonite mixed-layer* (Mu 80%, Mo 20%), from Triassic feldspathic sandstone, Kahla, GDR

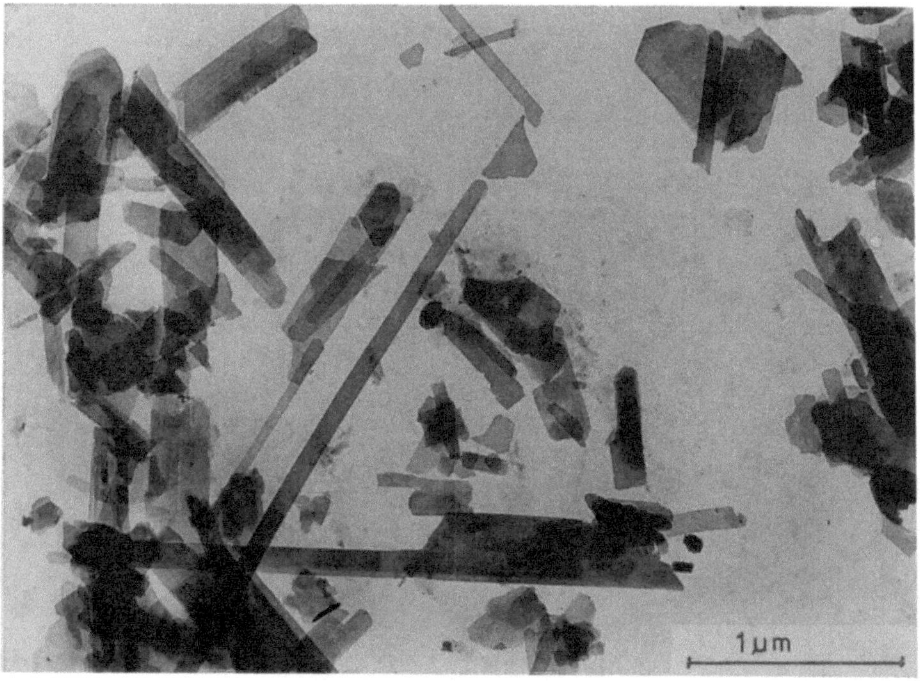

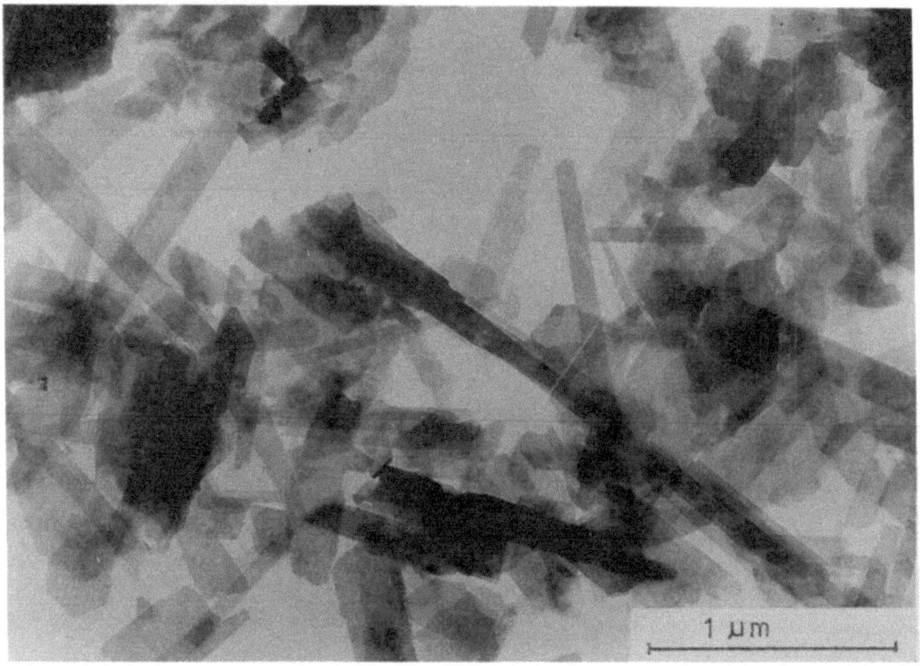

Figs. 321, 322. TEMs: Muscovite-montmorillonite mixed-layer (Mu 80%, Mo 20%) from altered tuff, Füzerradvany near Sarospatak, Hungary

Fig. 322.

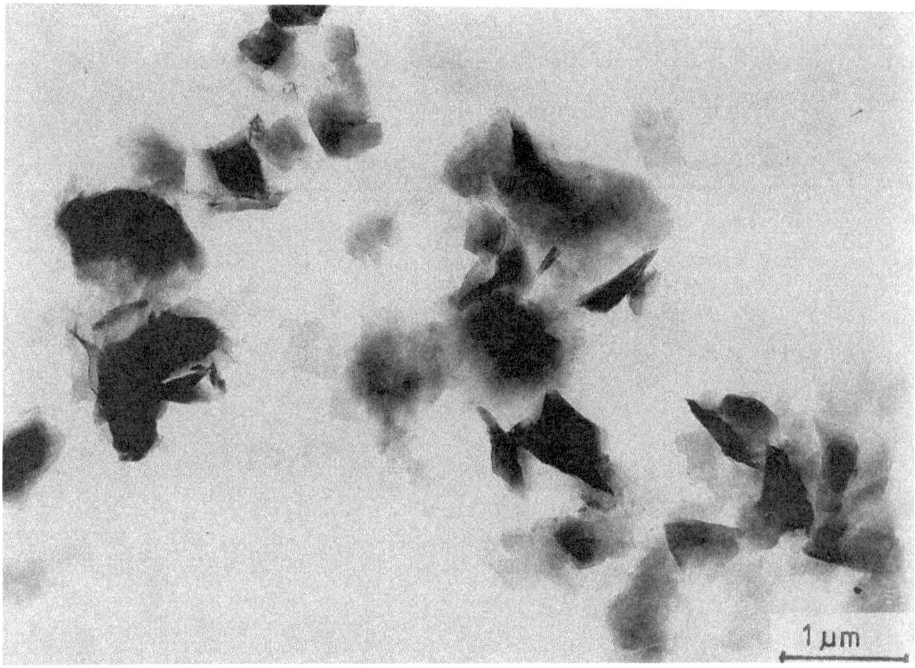

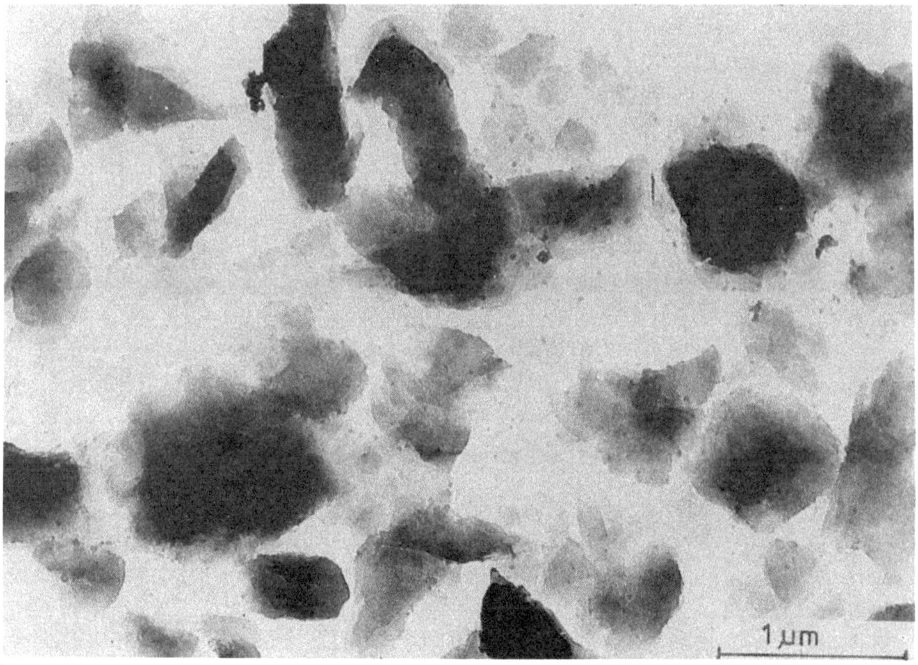

Figs. 323, 324. TEMs: *Muscovite-montmorillonite mixed-layer,* API No. 42, High Bridge, Kentucky, Arizona, USA

Fig. 324.

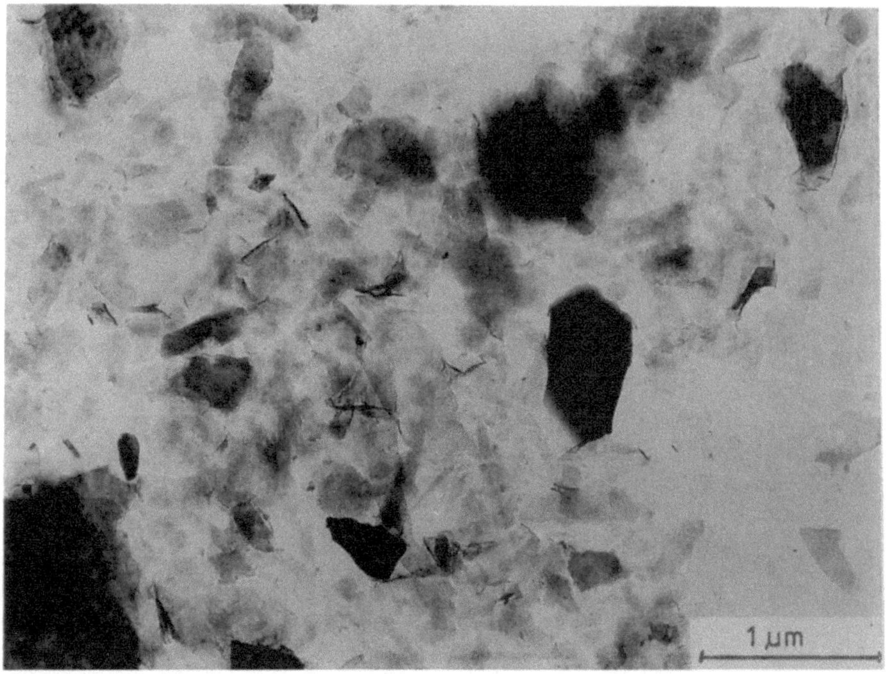

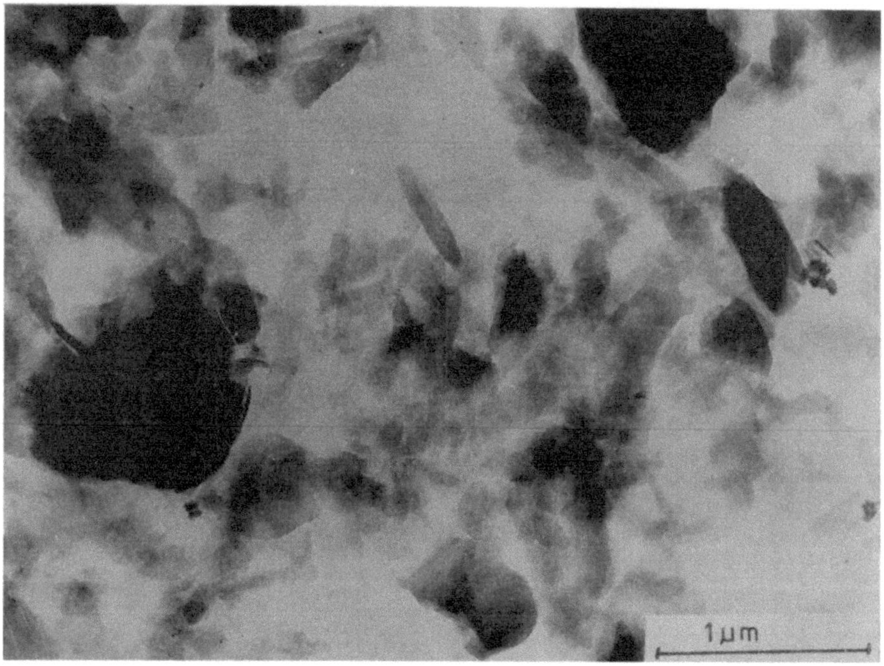

Figs. 325, 326. TEMs: *Muscovite-montmorillonite mixed-layer* from Ordovician bentonite, Kinnekulle, Sweden

Fig. 326.

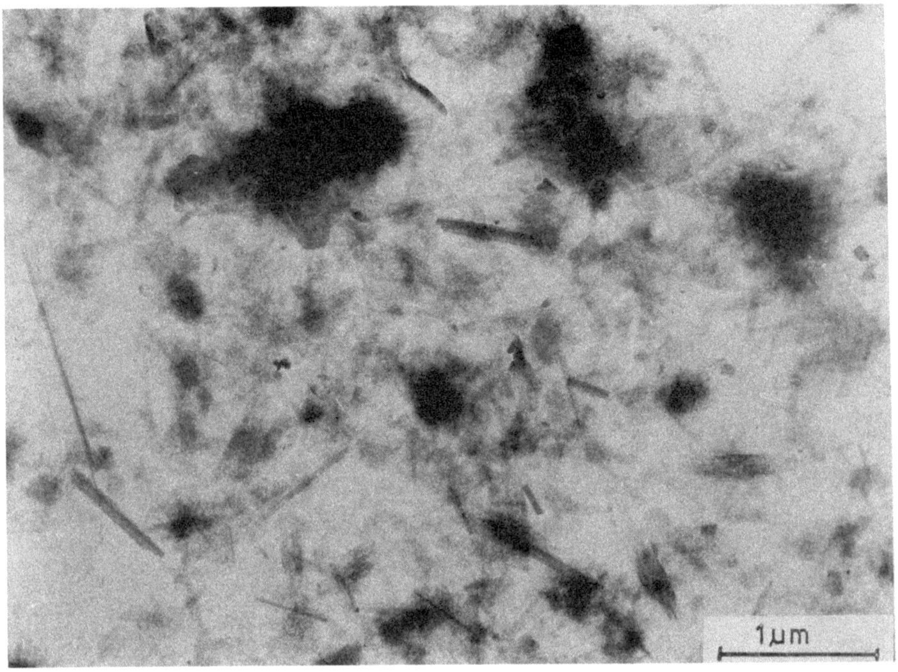

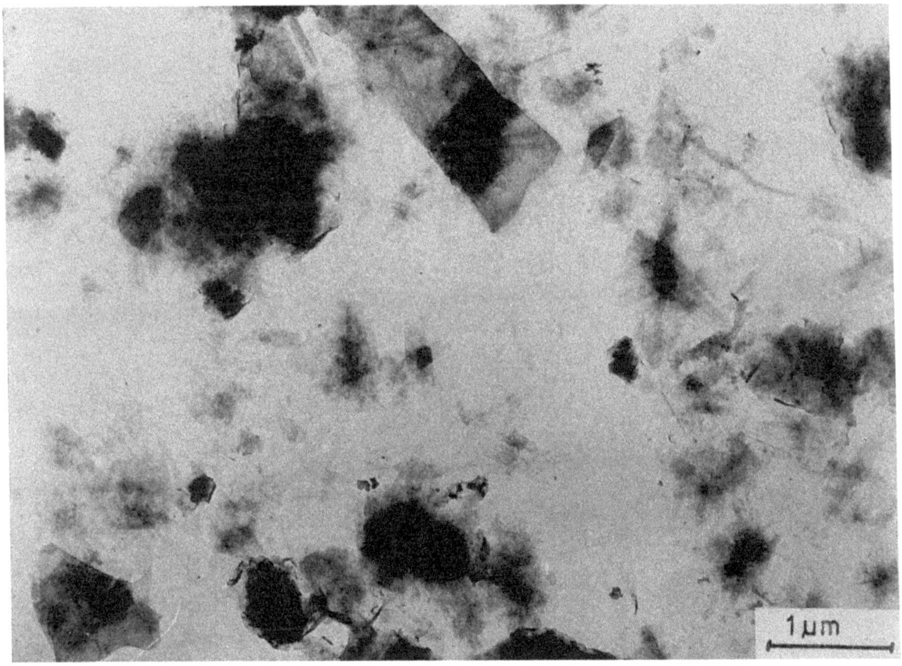

Figs. 327, 328. TEMs: *Muscovite-montmorillonite mixed-layer* (Mu 40%, Mo 60%) from Tertiary marine clay, near Friedland, GDR

Fig. 328.

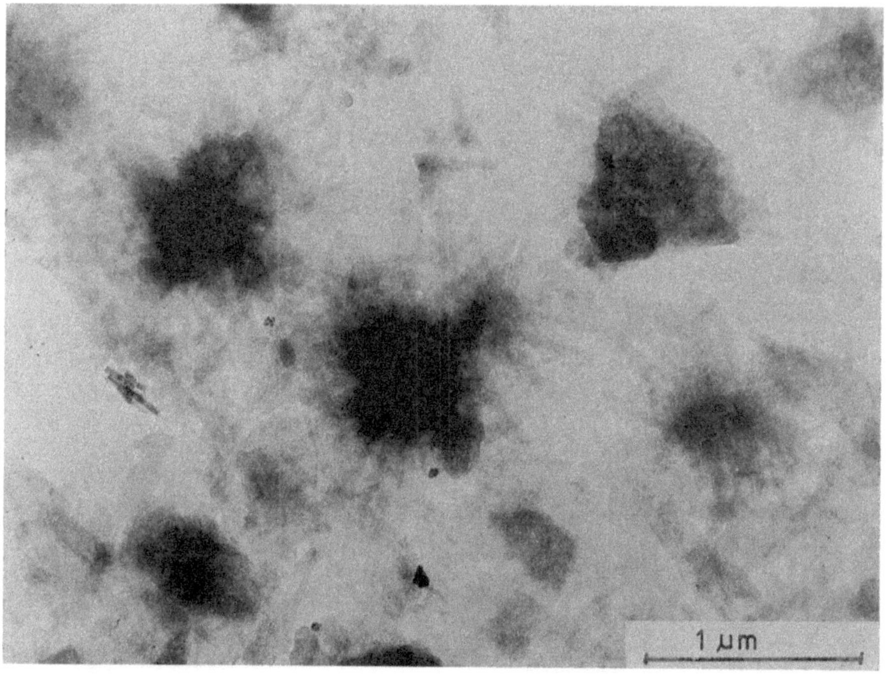

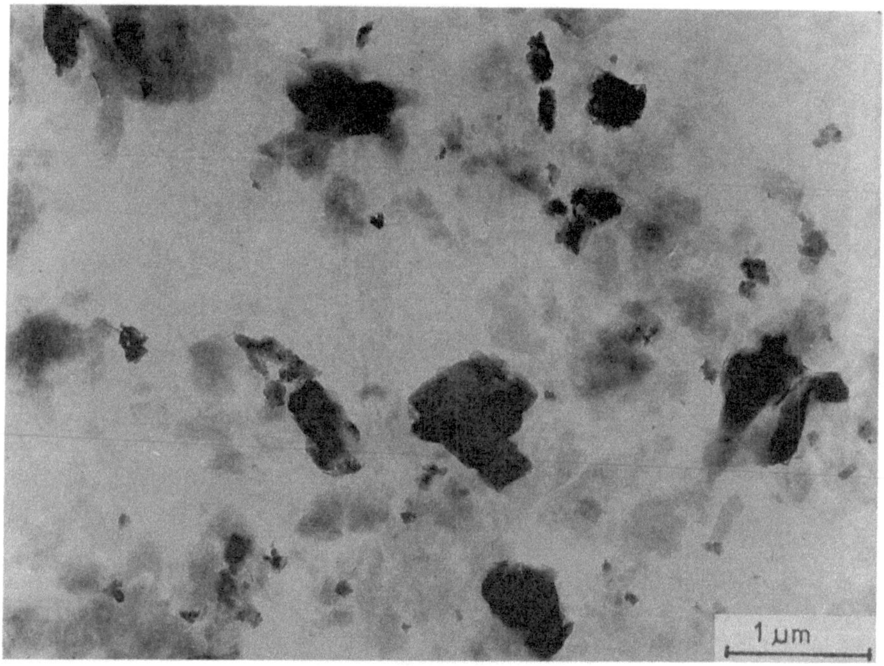

Fig. 329. TEM: *Muscovite-montmorillonite mixed-layer* (Mu 40%, Mo 60%) from drill core, Usedom, GDR

Fig. 330. TEM: *Muscovite-montmorillonite mixed-layer*, Nižne Hrabovec, Czechoslovakia

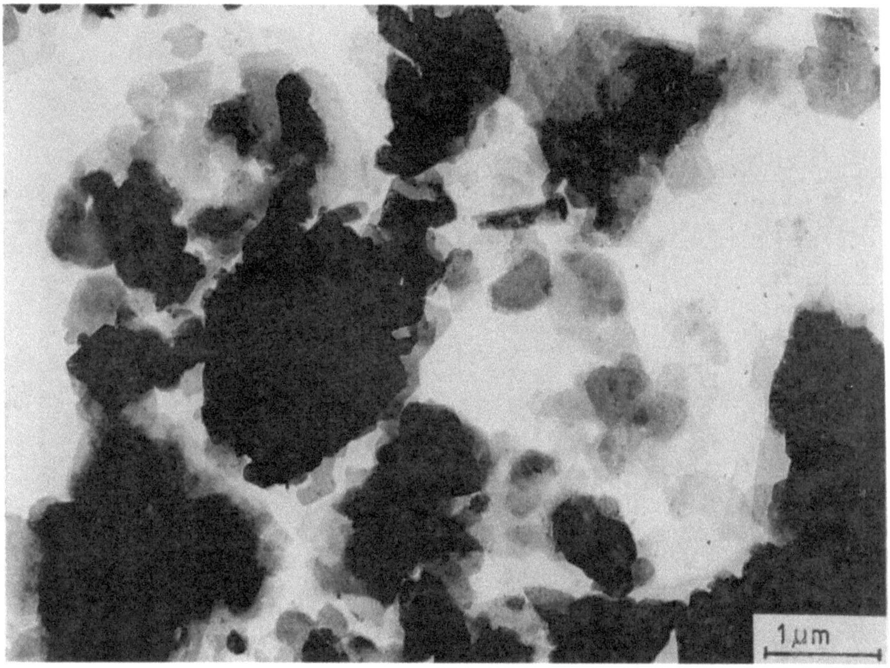

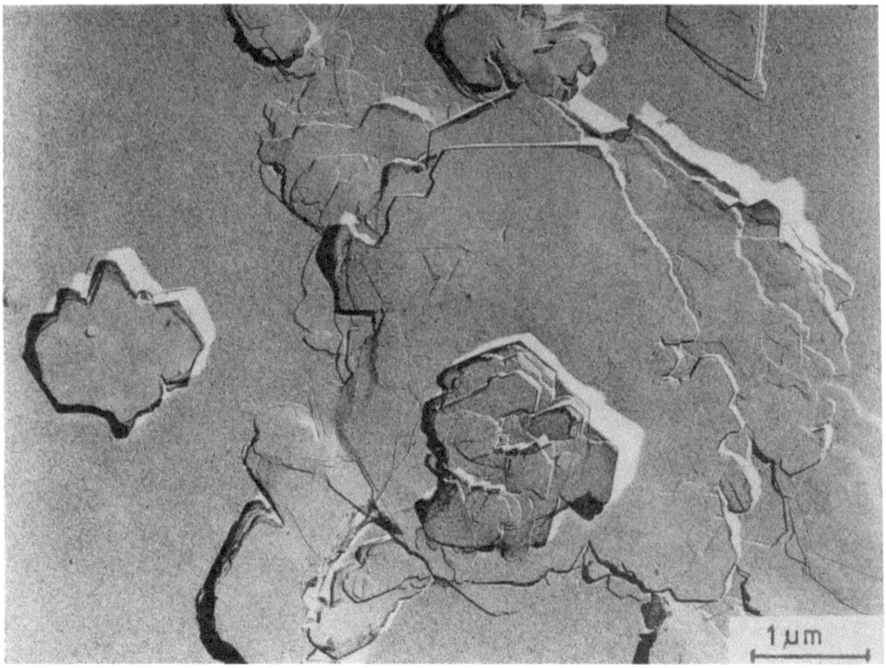

Fig. 331. TEM: *Kaolinite-montmorillonite mixed-layer*, coarse fraction, Jegłowa, Poland

Fig. 332. Replica: *Kaolinite-montmorillonite mixed-layer*, coarse fraction, Jegłowa, Poland

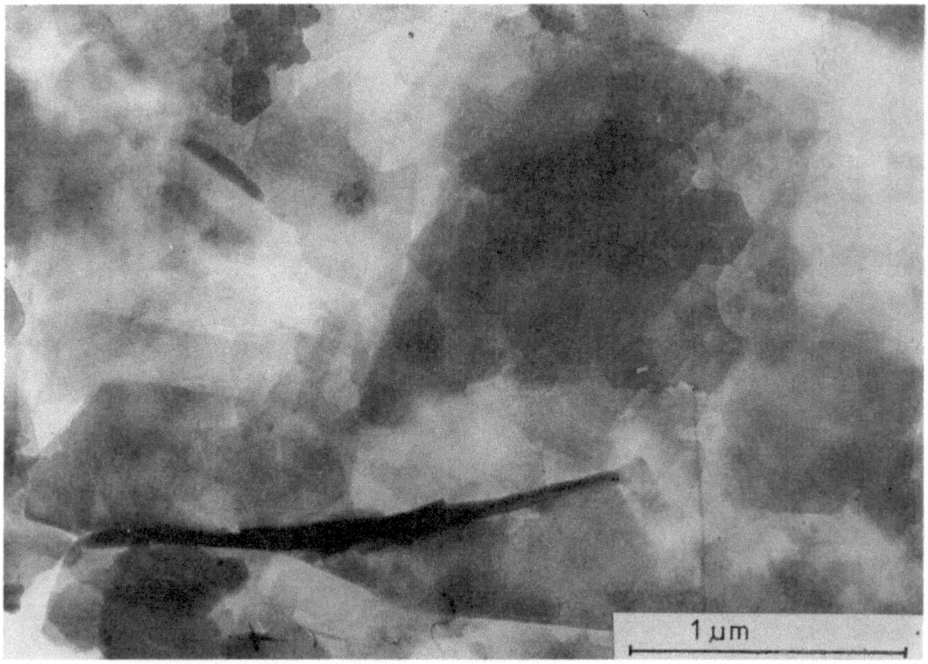

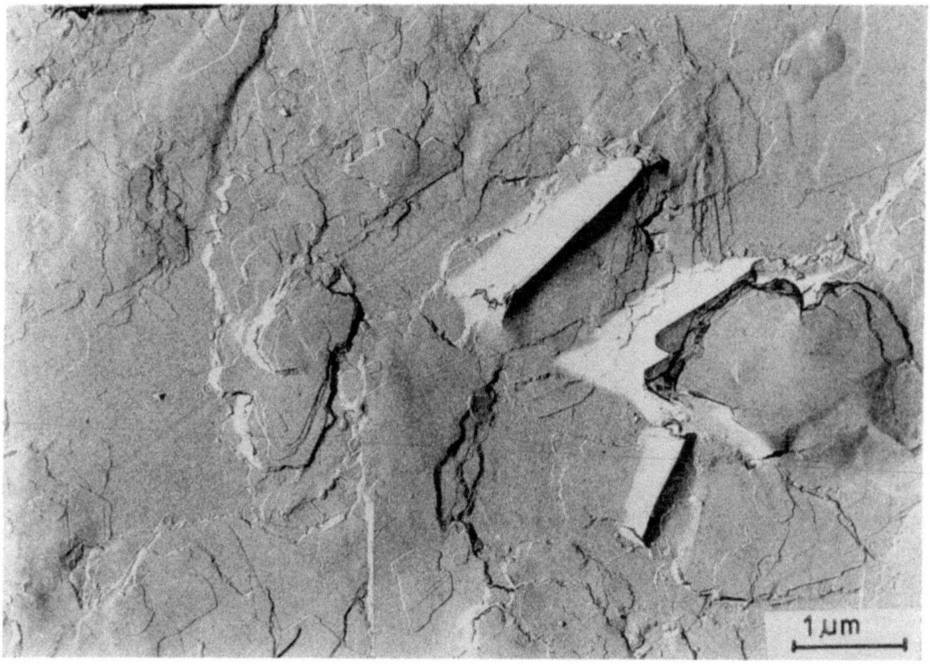

Fig. 333. TEM: *Kaolinite-montmorillonite mixed-layer*, fraction < 1 µm, Jegłowa, Poland

Fig. 334. Replica: *Kaolinite-montmorillonite mixed-layer*, fraction < 1 µm, Jegłowa, Poland

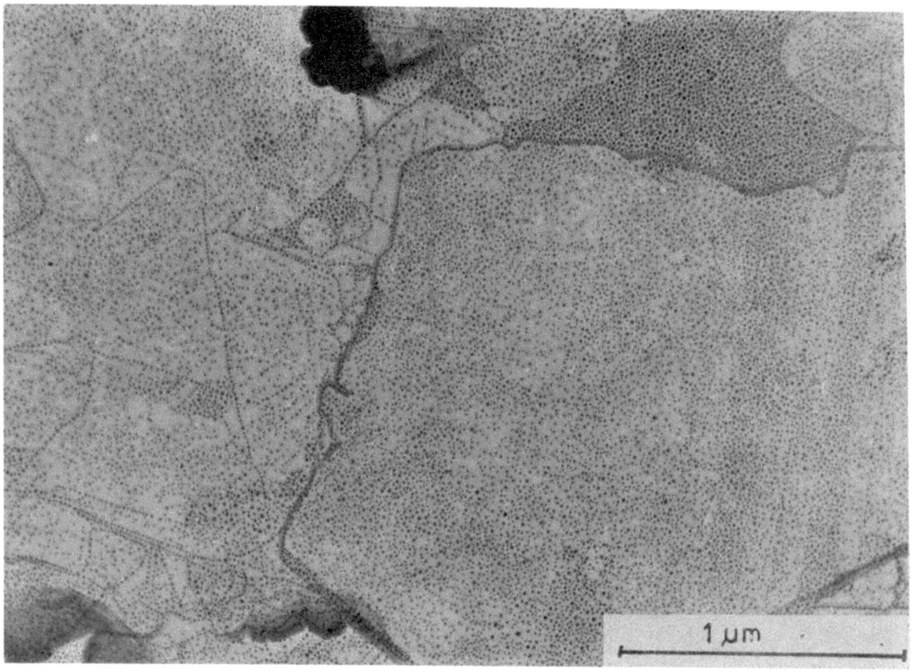

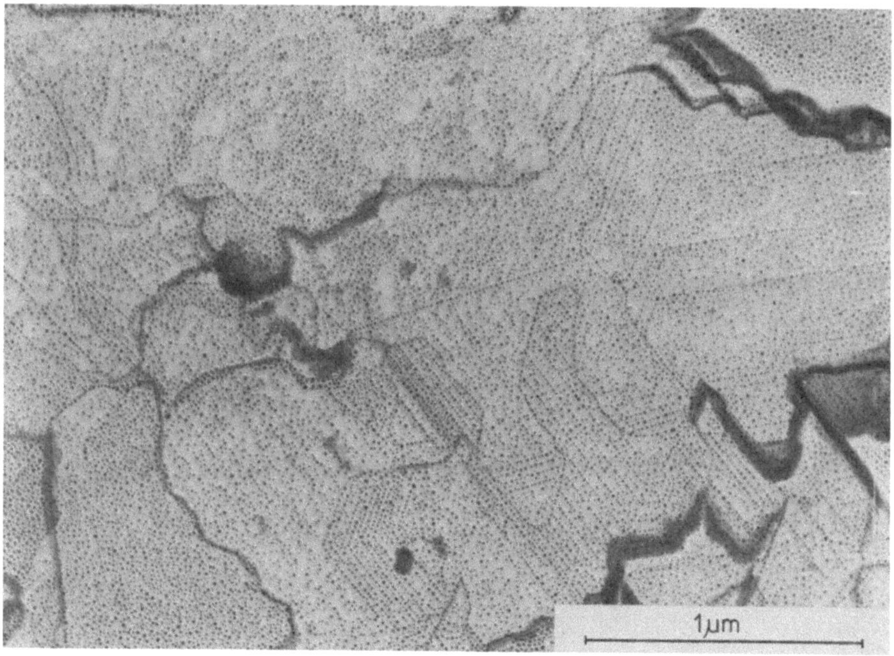

Figs. 335, 336. Replicas: *Kaolinite-montmorillonite mixed-layer*, gold decorated, Jegłowa, Poland

Fig. 336.

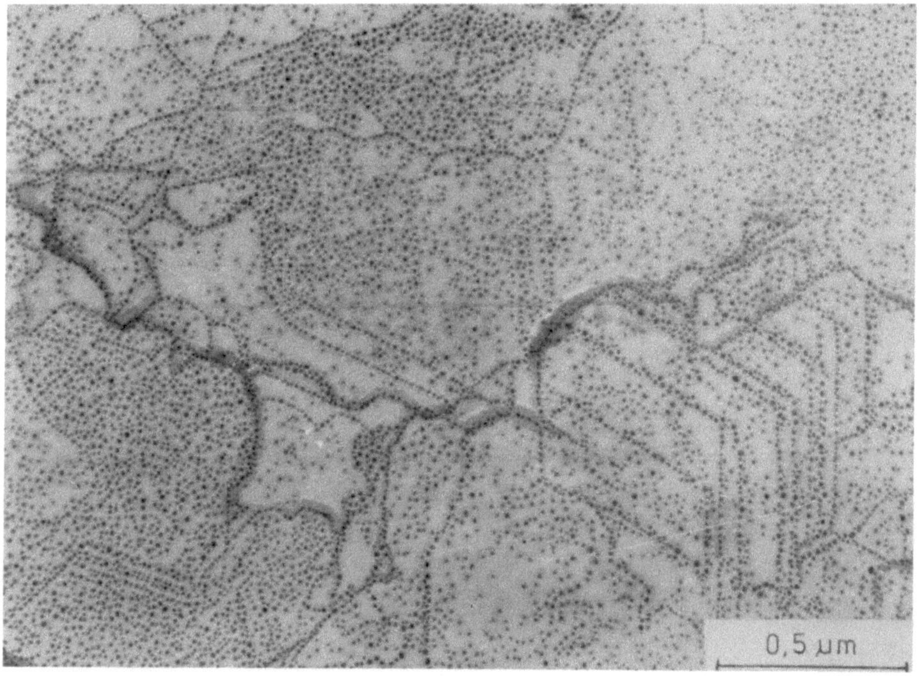

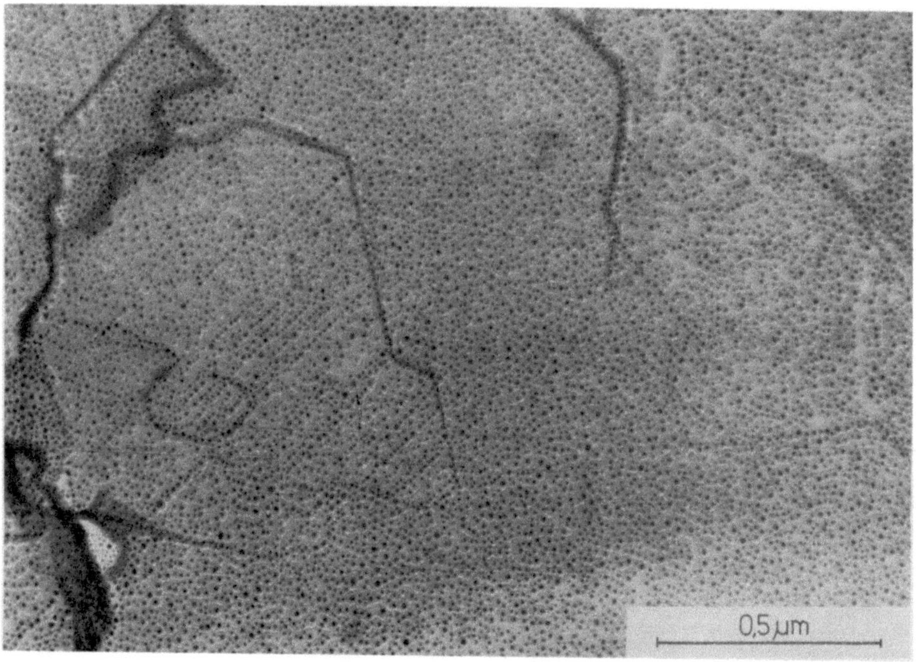

Figs. 337, 338. Replicas: *Kaolinite-montmorillonite mixed-layer*, gold decorated, Jegłowa, Poland

Fig. 338.

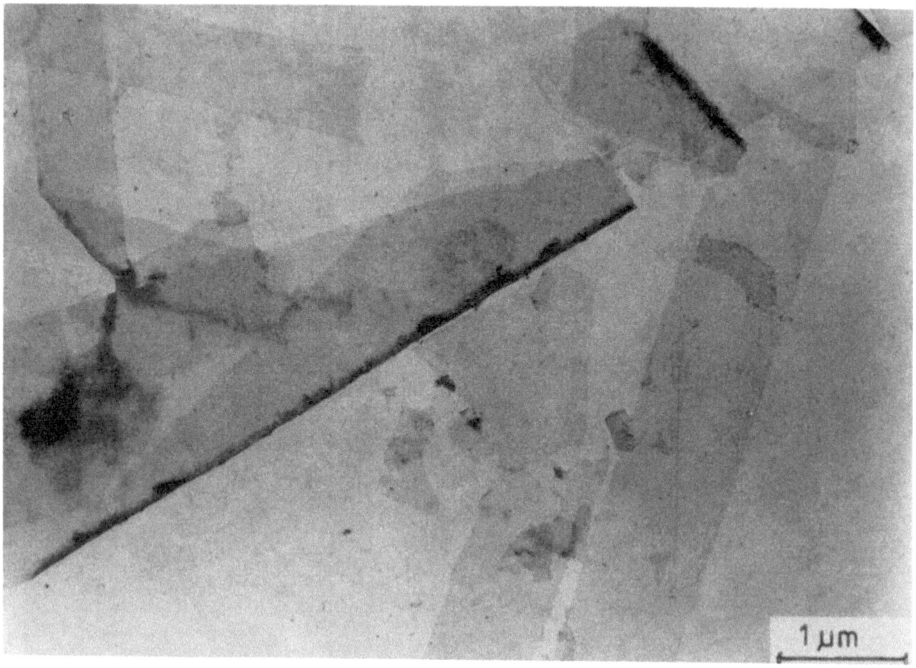

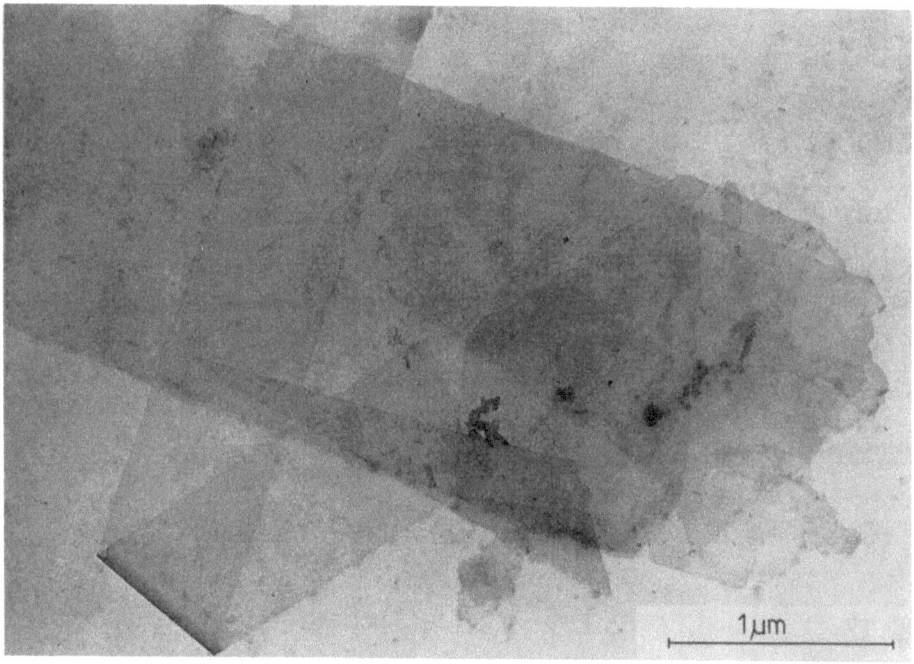

Figs. 339, 340. TEMs: *Rectorite*, New Little Rock, Arkansas, USA

Fig. 340.

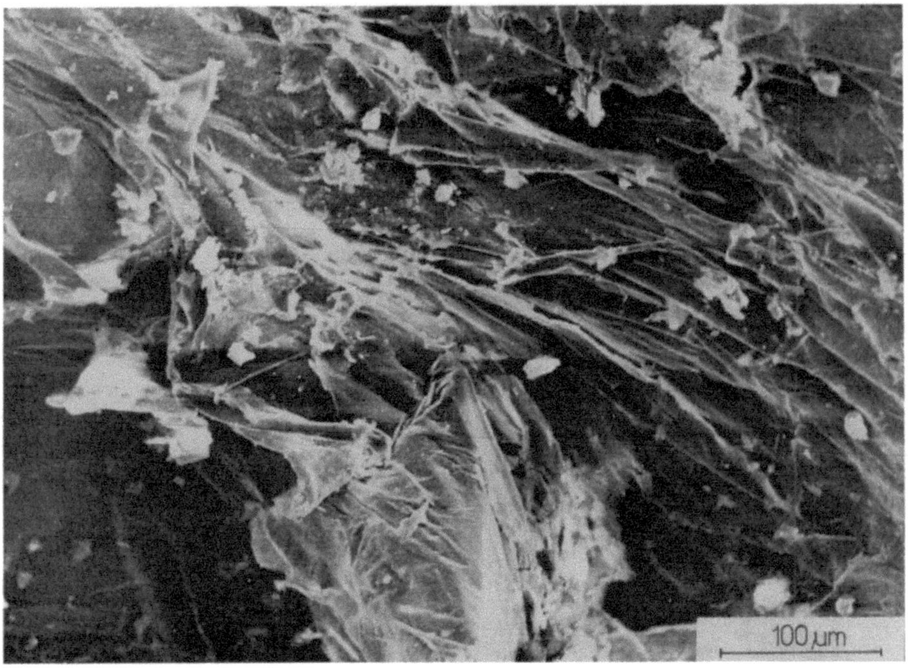

Fig. 341. Replica: *Rectorite*, New Little Rock, Arkansas, USA

Figs. 342, 343, 344. SEMs: *Rectorite*, New Little Rock, Arkansas, USA

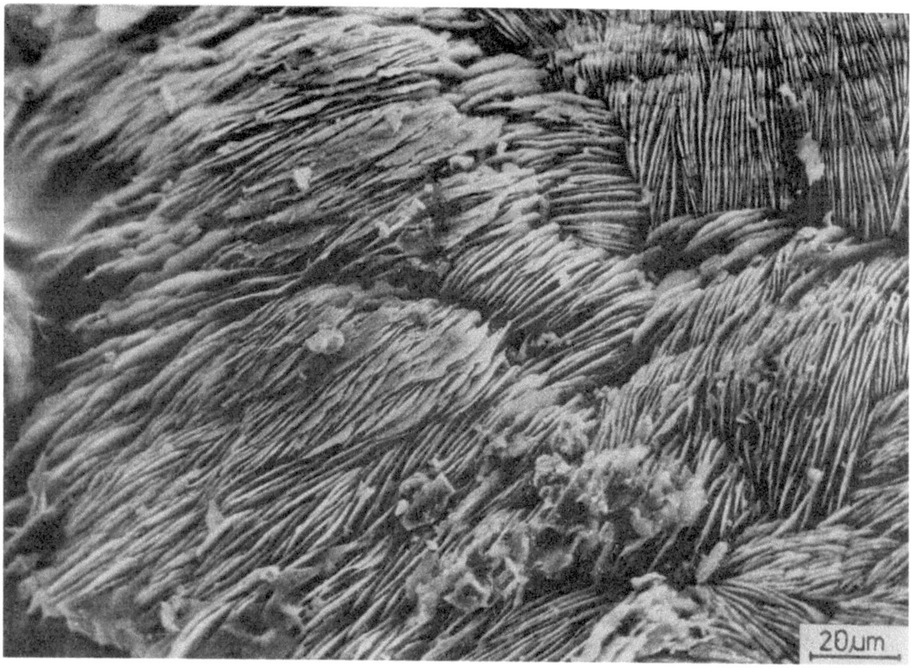

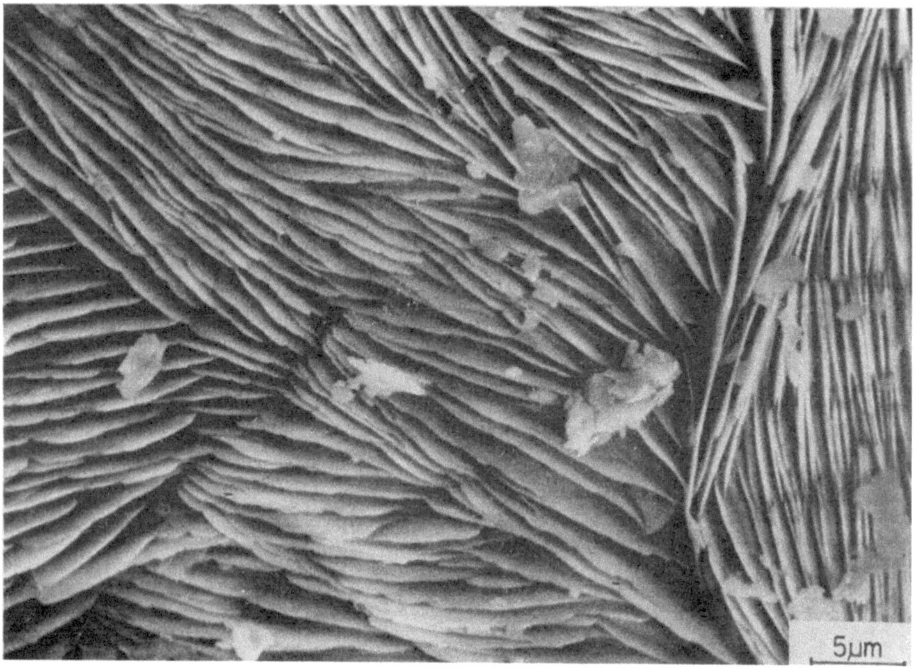

Fig. 343.

Fig. 344.

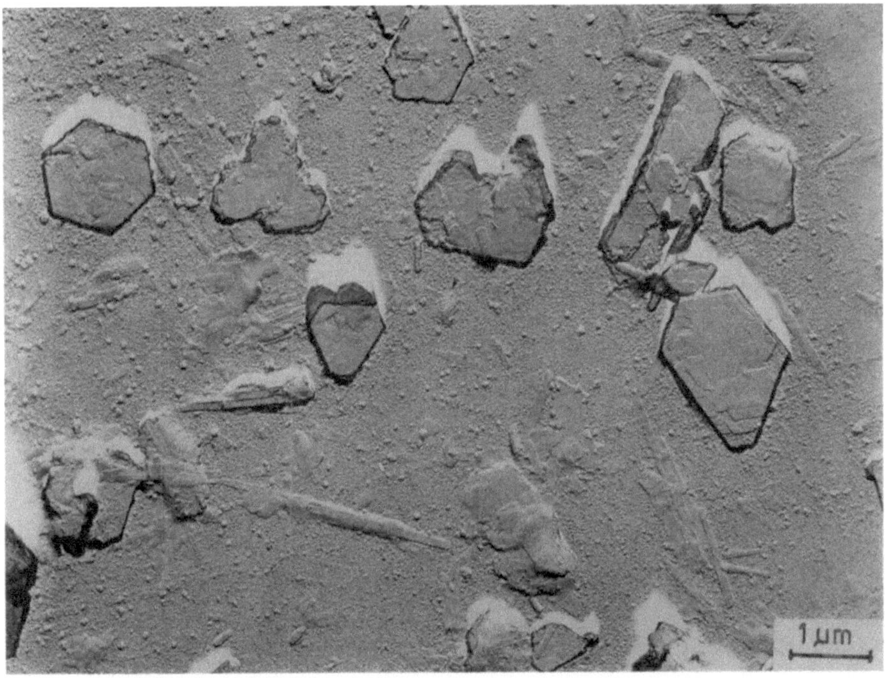

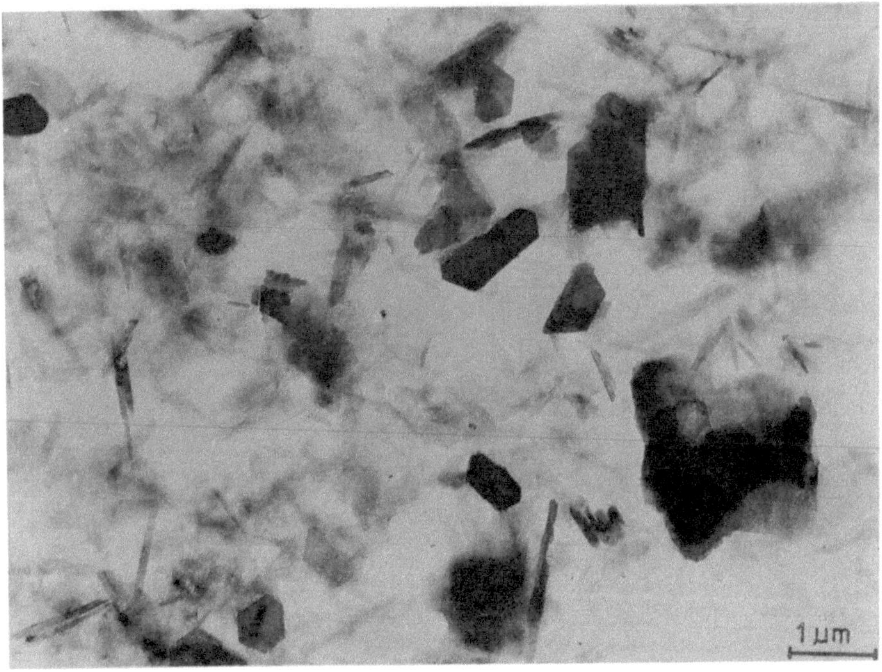

Fig. 345. TEM: *Allevardite* (laths) from altered tuff, Kiraly Hegy near Mad, Hungary

Fig. 346. Replica: *Allevardite* (laths) and kaolinite from altered tuff, Kiraly Hegy near Mad, Hungary

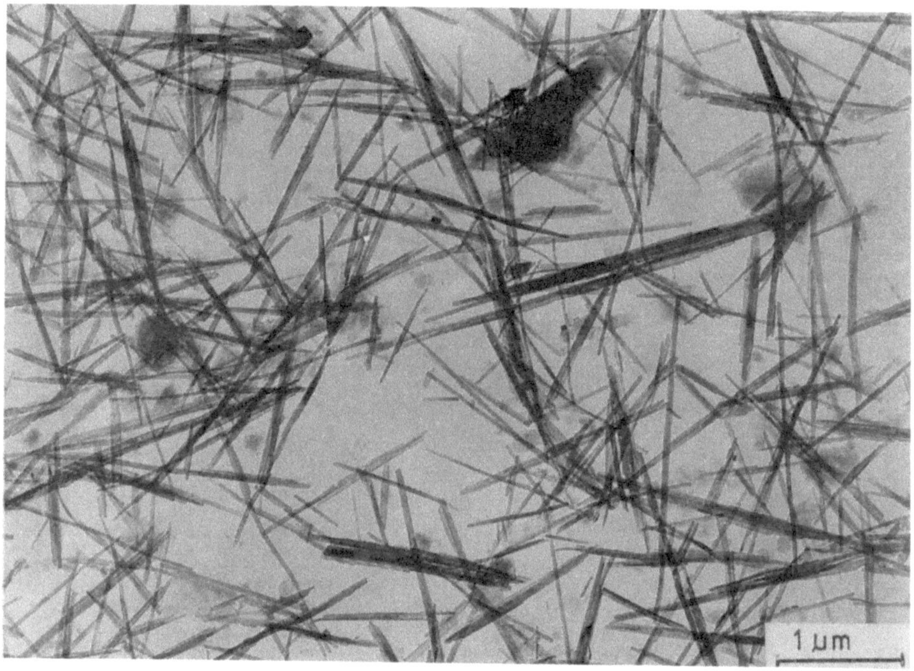

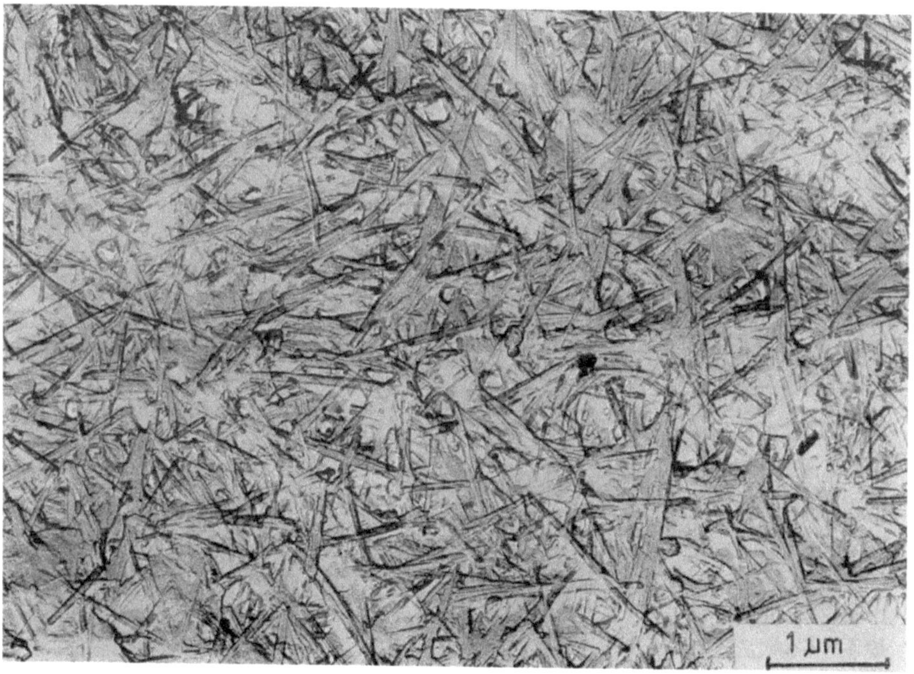

Fig. 347. TEM: *Palygorskite*, Attapulgus, Georgia, USA

Fig. 348. Replica: *Palygorskite*, Attapulgus, Georgia, USA

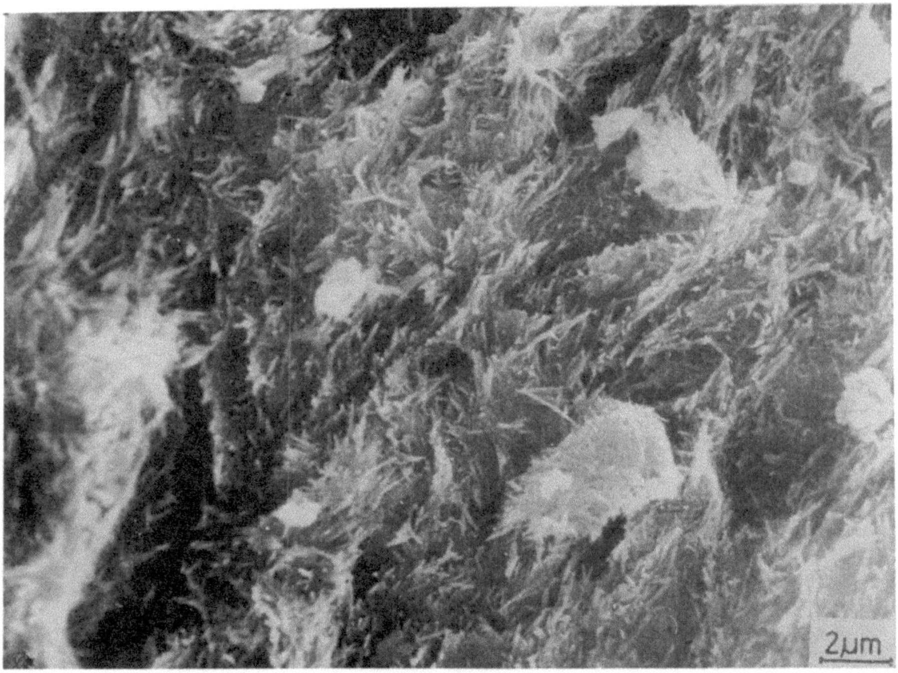

Figs. 349, 350. SEMs: *Palygorskite*, Attapulgus, Georgia, USA

Fig. 350.

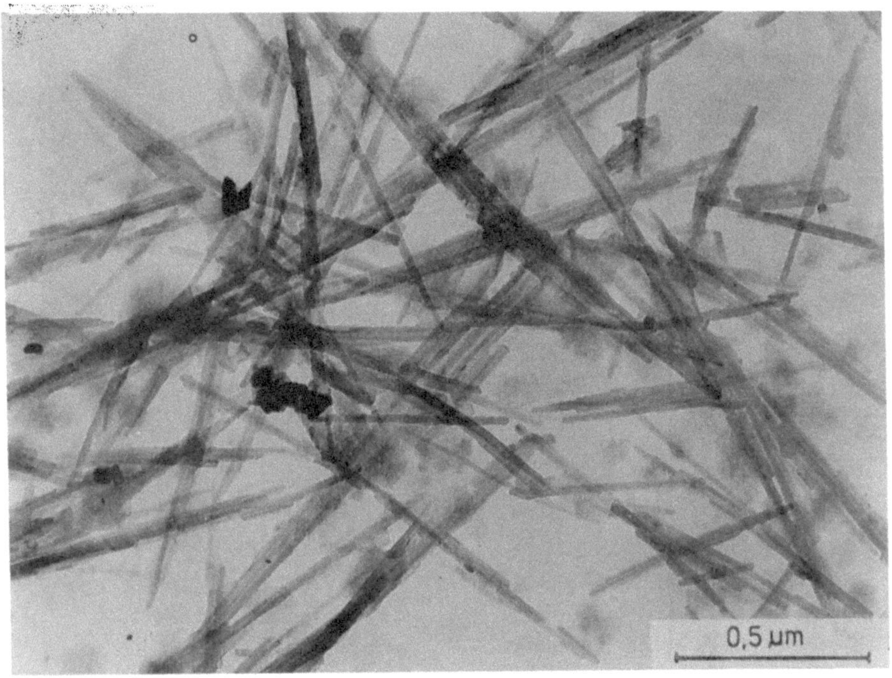

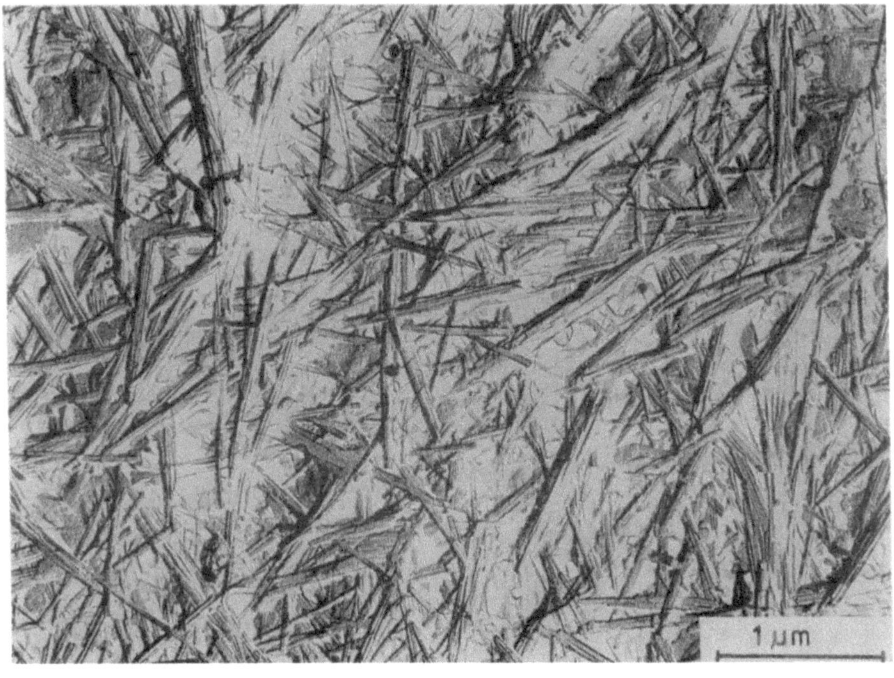

Fig. 351. TEM: *Palygorskite,* Quincy, Florida, USA

Fig. 352. Replica: *Palygorskite,* Quincy, Florida, USA

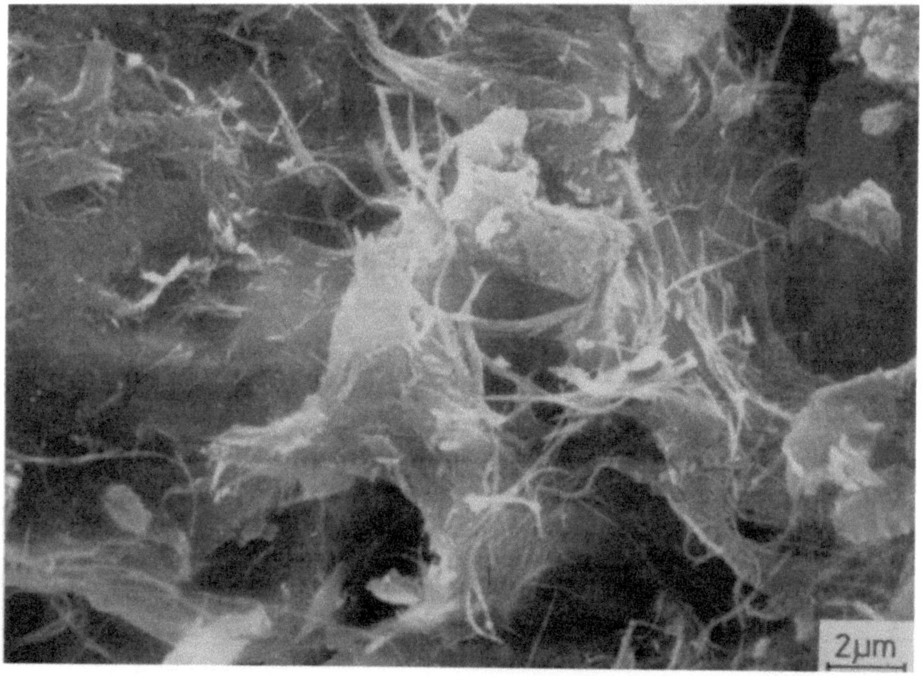

Fig. 353. SEM: *Palygorskite*, Quincy, Florida, USA

Fig. 354. TEM: *Palygorskite*, Ratkoszaha, Hungary

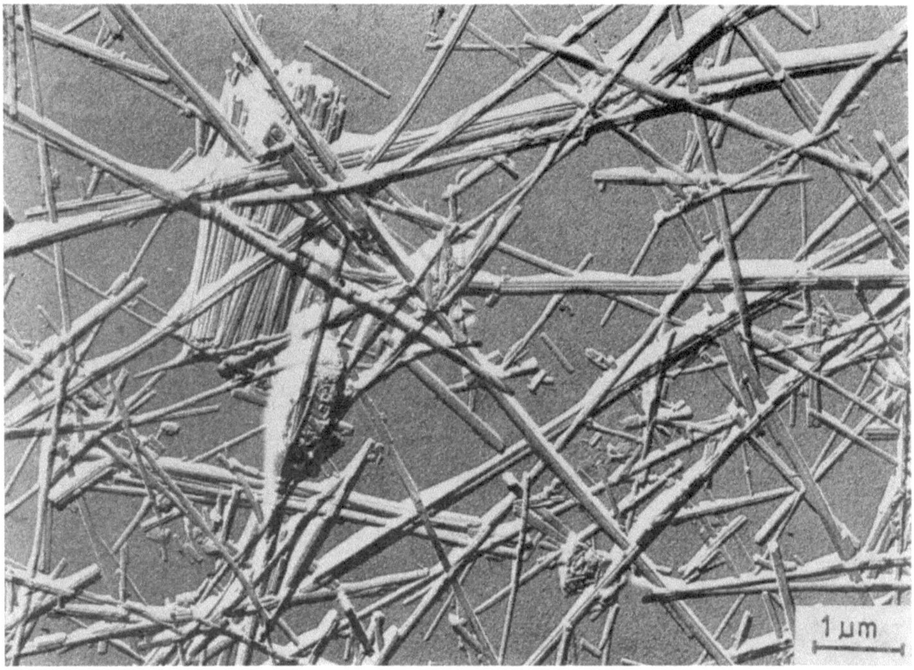

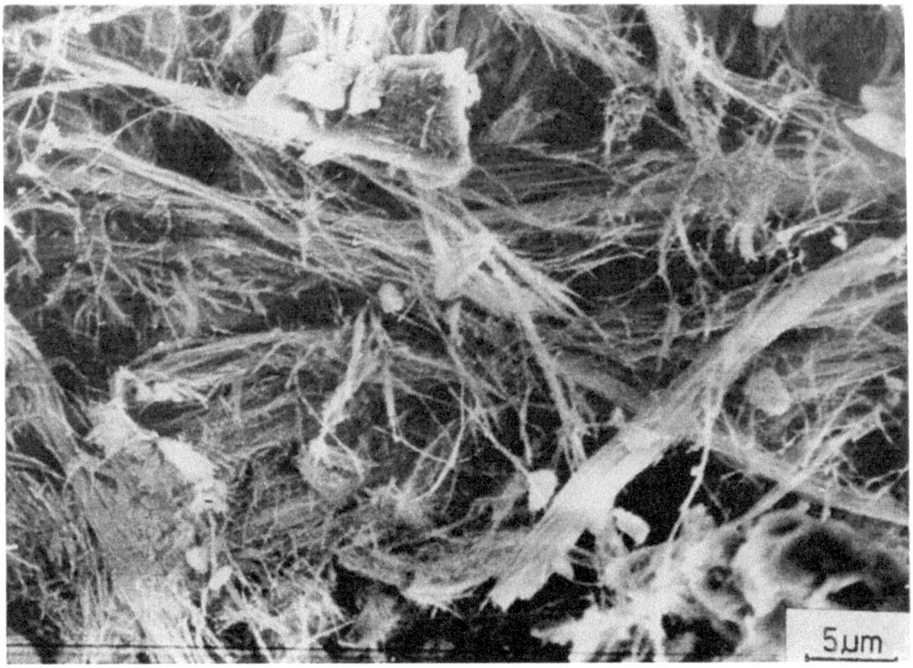

Fig. 355. Replica: *Palygorskite*, Ratkoszaha, Hungary

Fig. 356. SEM: *Palygorskite*, Ratkoszaha, Hungary

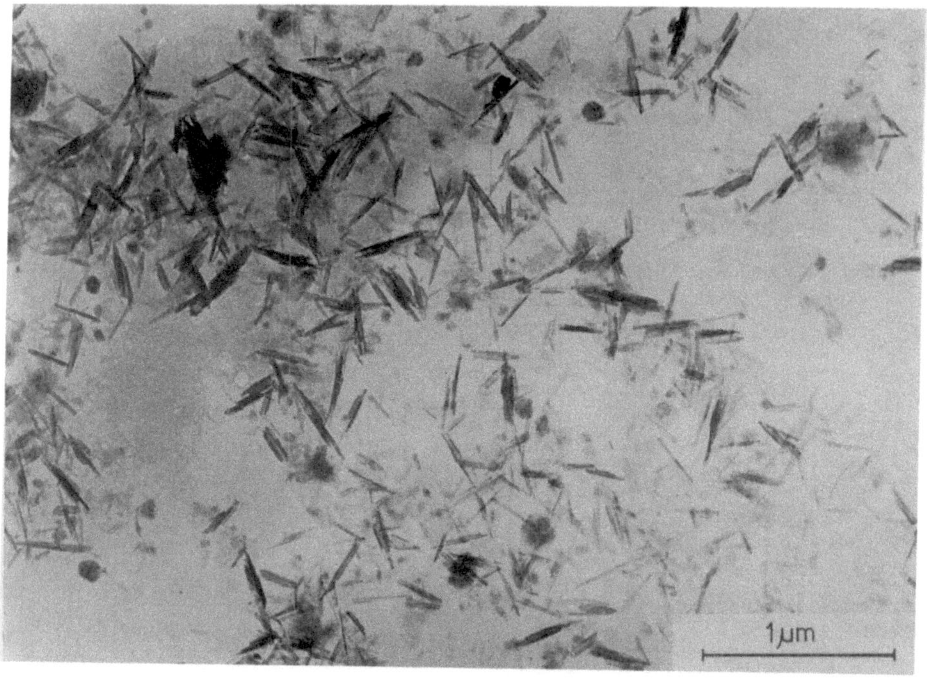

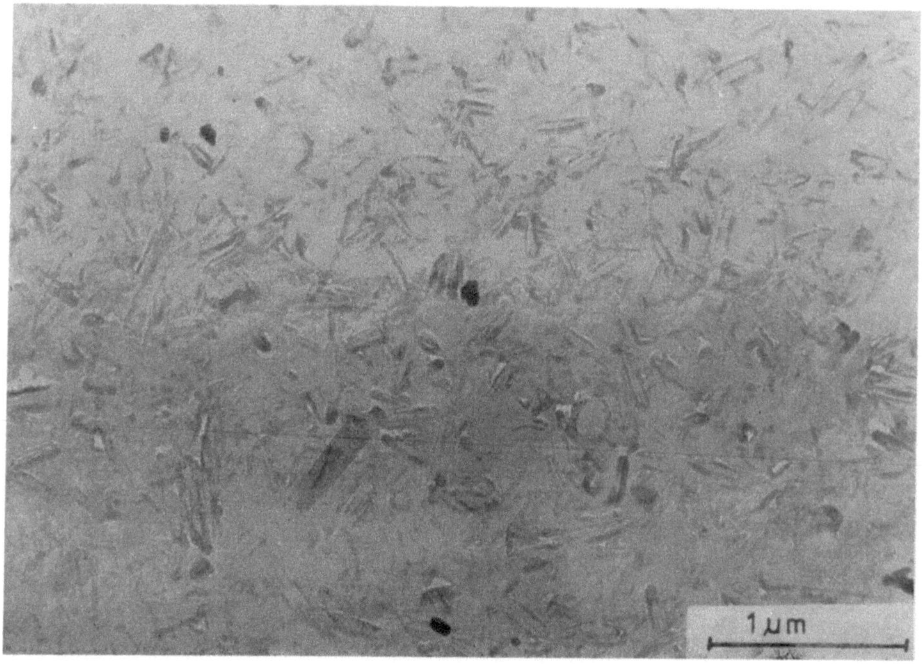

Fig. 357. TEM: *Palygorskite*, Palygorsk, Urals, USSR

Fig. 358. Replica: *Palygorskite*, Palygorsk, Urals, USSR

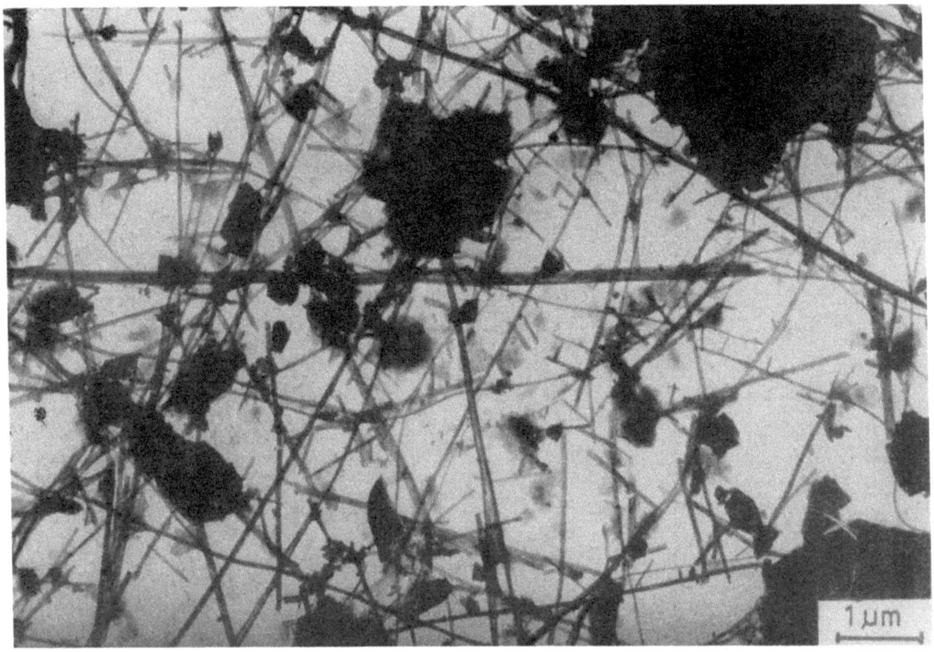

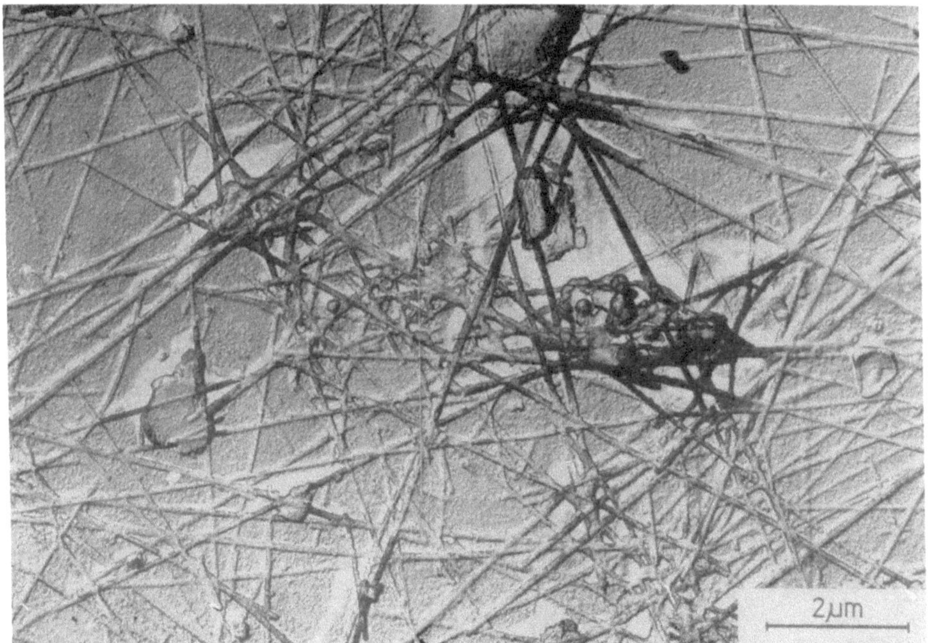

Fig. 359. TEM: *Palygorskite*, Beketavo, Hungary

Fig. 360. Replica: *Palygorskite*, Beketavo, Hungary

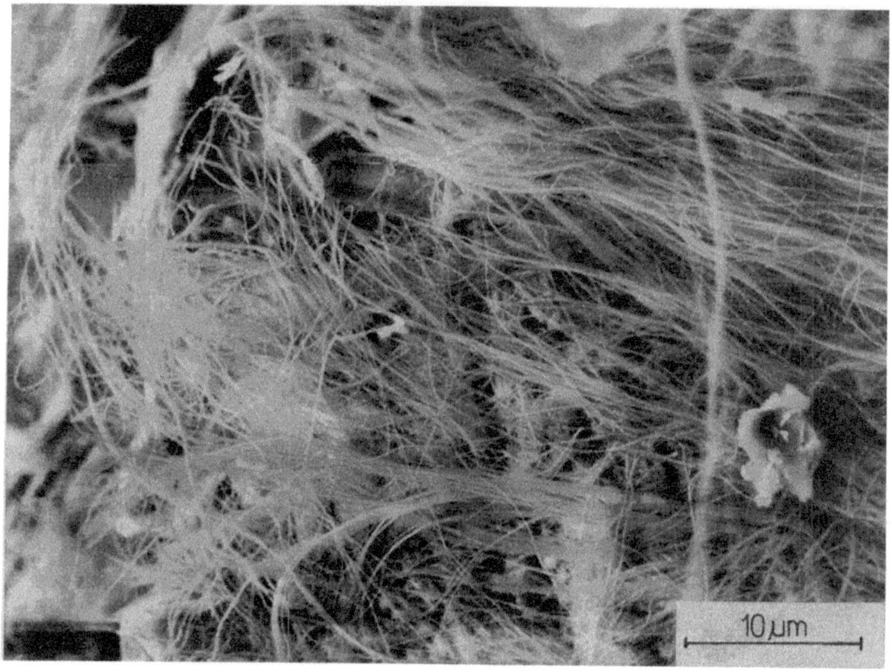

Figs. 361, 362. SEMs: *Palygorskite*, Beketavo, Hungary

Fig. 362.

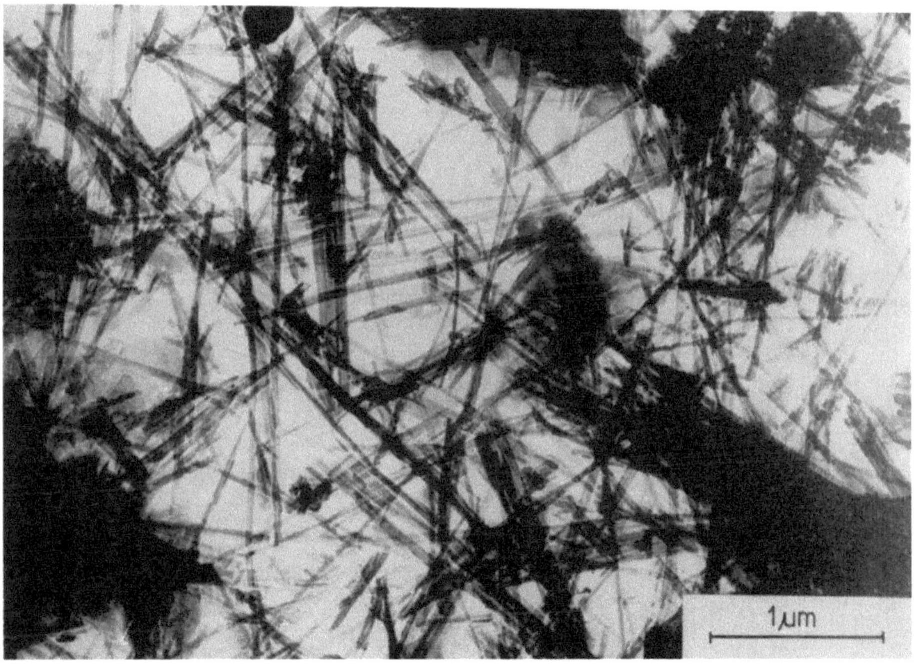

Figs. 363, 364. TEMs: *Palygorskite*, Tebel, Iraq

Fig. 364.

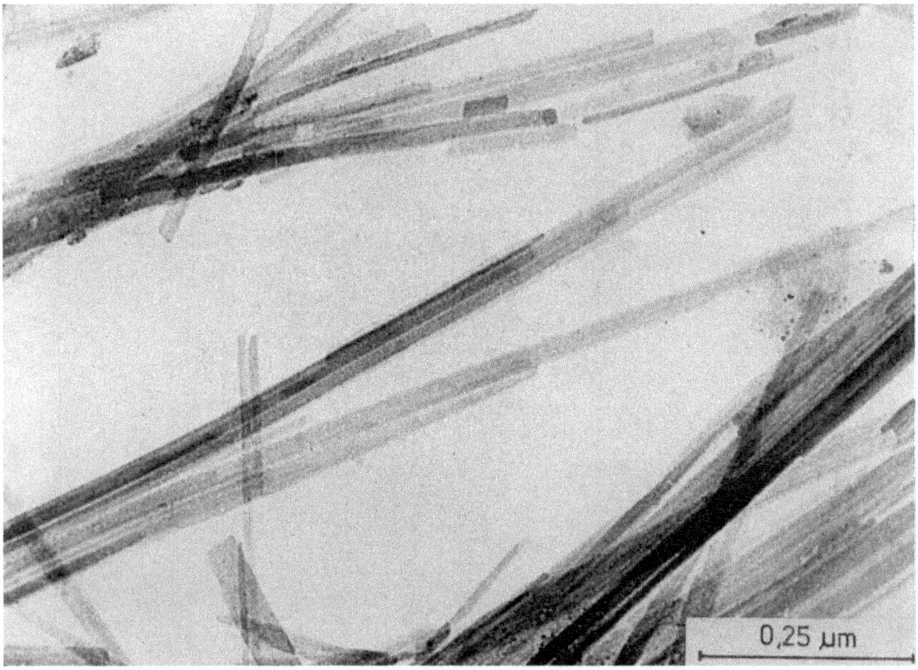

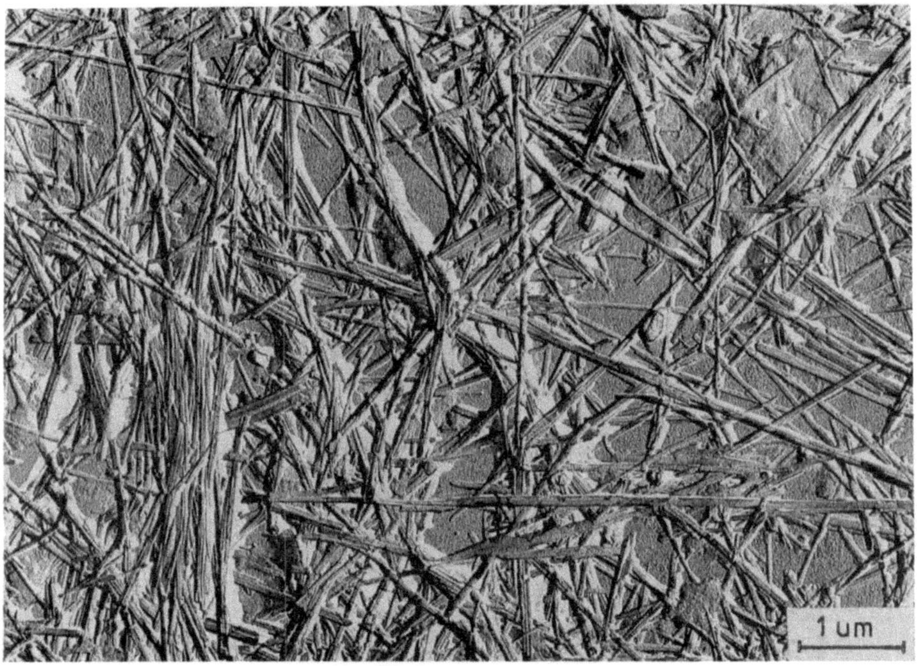

Fig. 365. TEM: *Palygorskite*, Tebel, Iraq

Fig. 366. Replica: *Palygorskite*, Tebel, Iraq

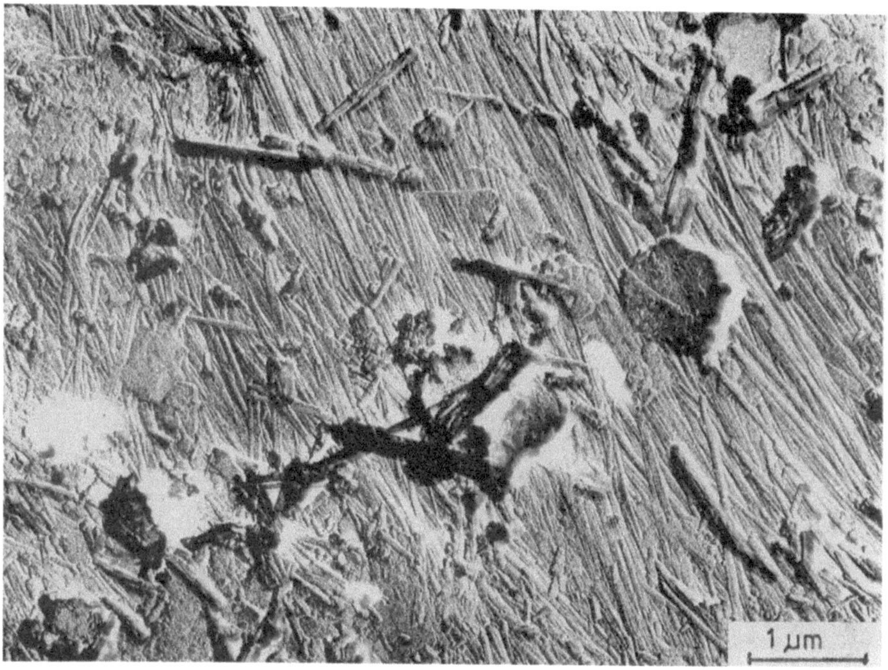

Fig. 367. Replica: *Palygorskite*, fracture parallel to fibres, Tebel, Iraq

Fig. 368. Replica: *Palygorskite*, fracture vertical to fibres, Tebel, Iraq

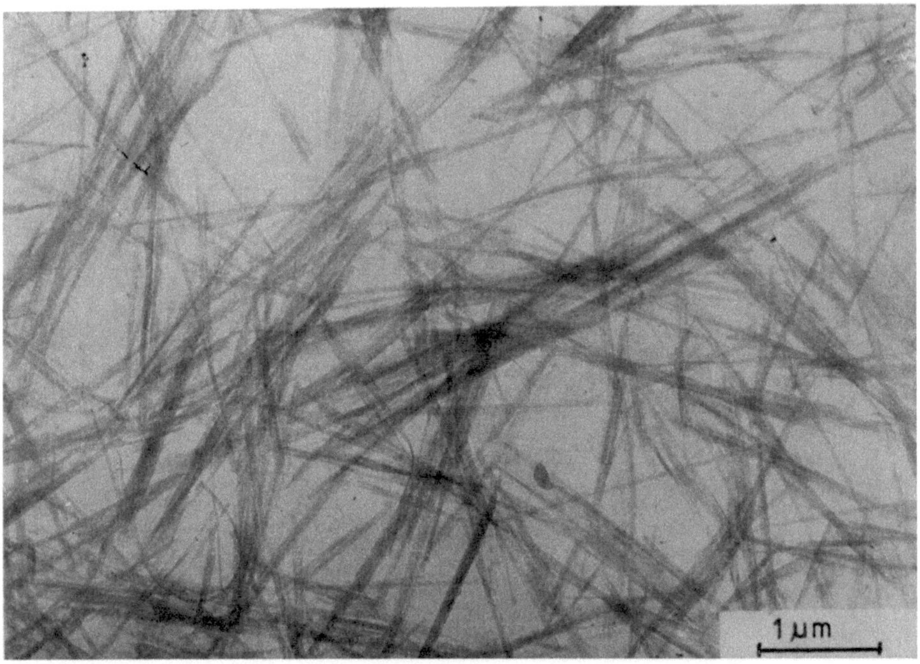

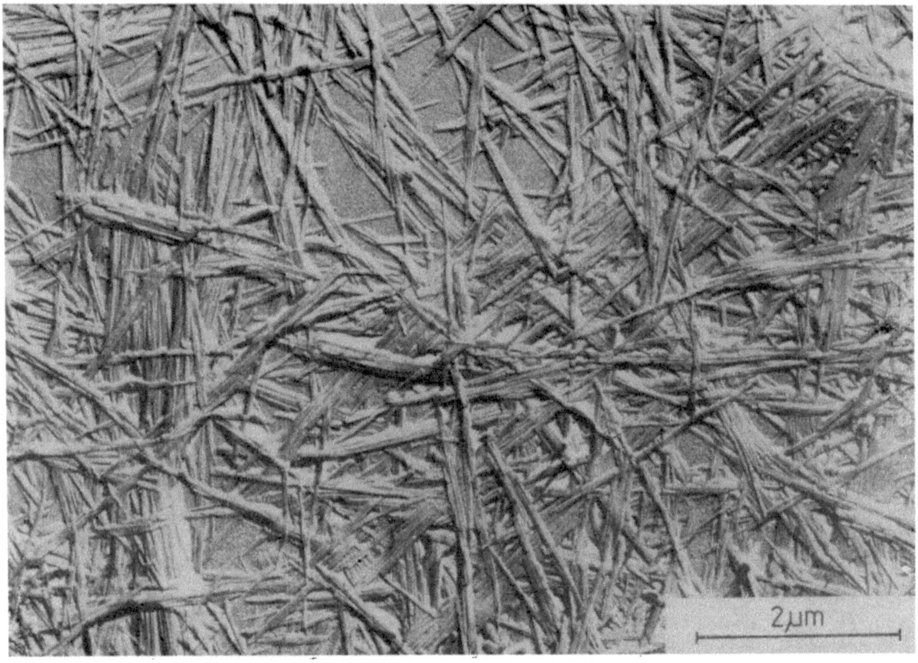

Fig. 369. TEM: *Sepiolite*, Vallecos, Spain

Fig. 370. Replica: *Sepiolite*, Vallecos, Spain

Figs. 371, 372. SEMs: *Sepiolite*, Vallecos, Spain

Fig. 372.

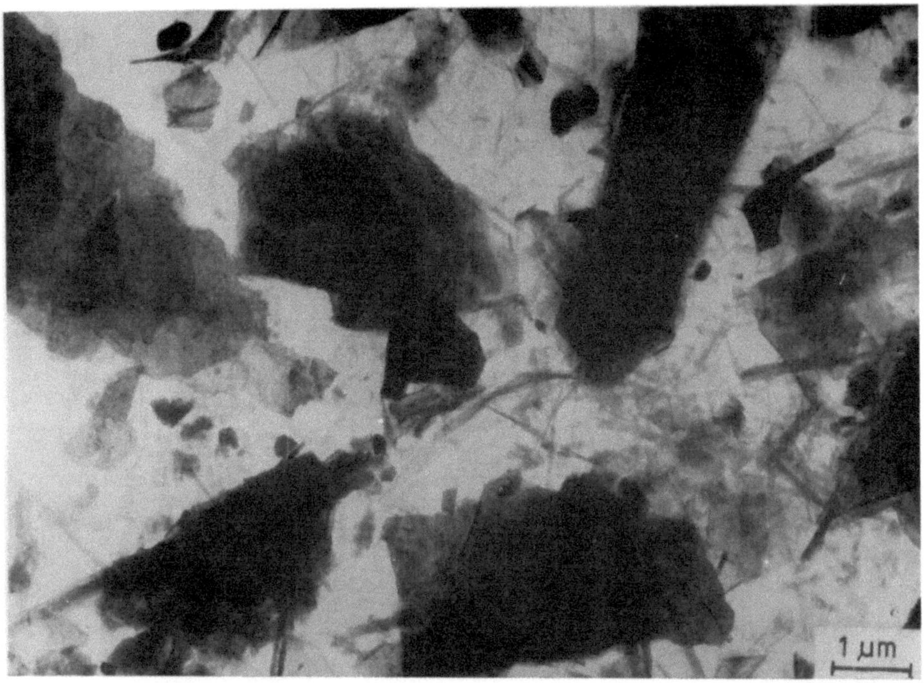

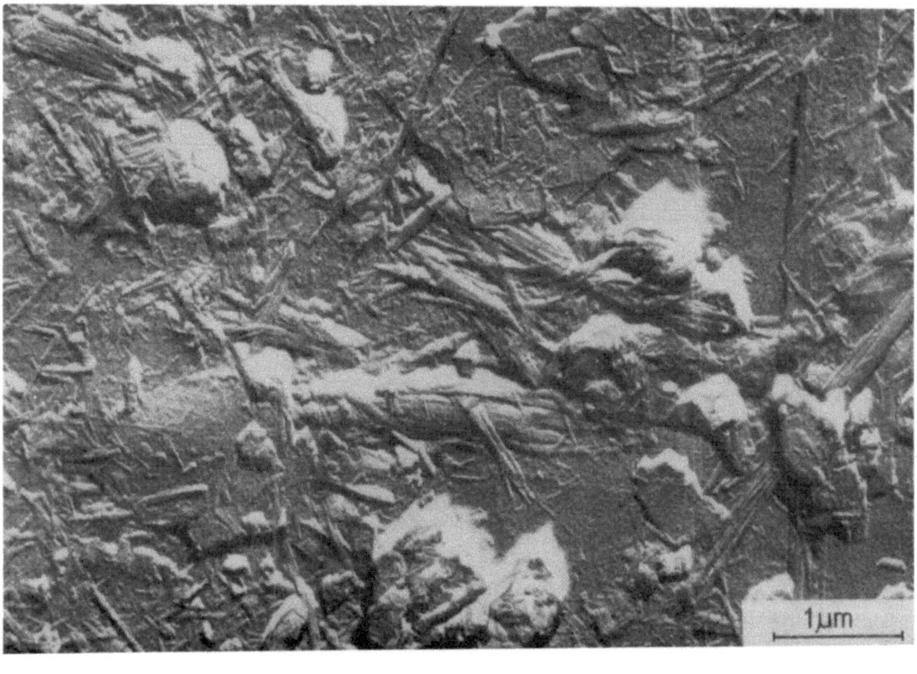

Fig. 373. TEM: *Sepiolite*, Jihočesky Kraj, Czechoslovakia

Fig. 374. Replica: *Sepiolite*, Jihočesky Kraj, Czechoslovakia

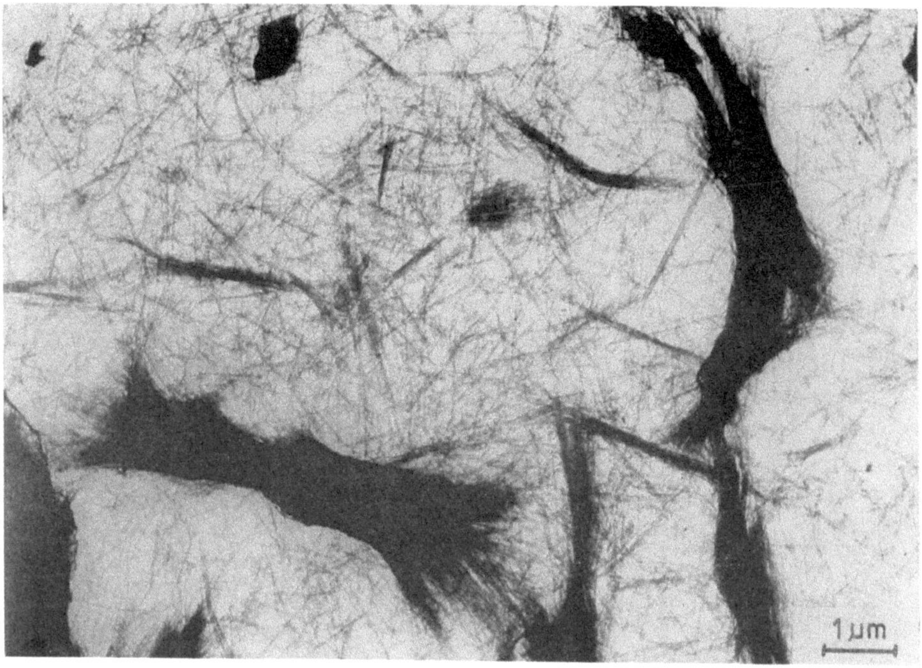

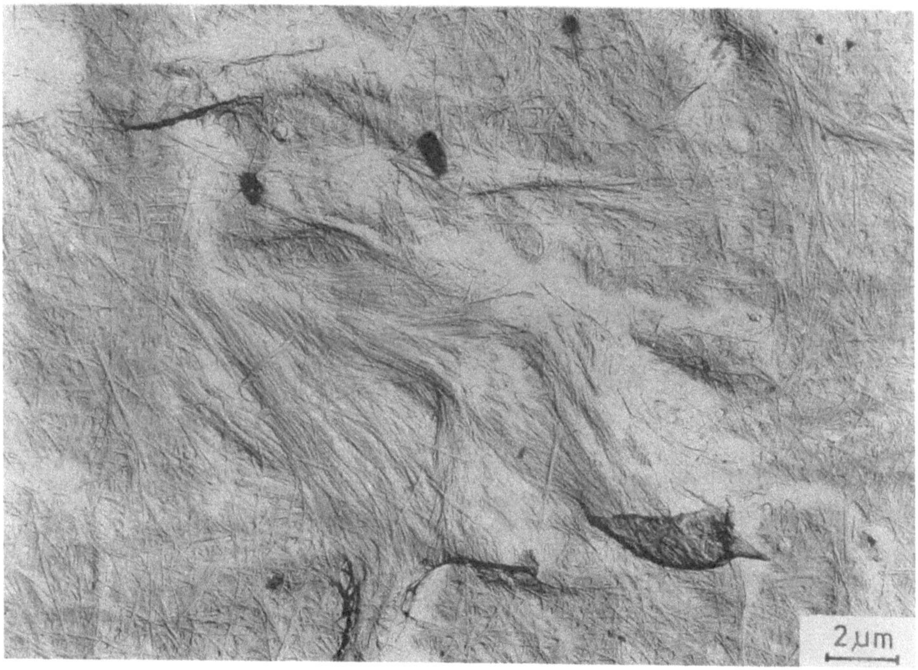

Fig. 375. TEM: *Sepiolite* from weathered serpentinite, Wiry (Szklary), Poland

Fig. 376. Replica: *Sepiolite* from weathered serpentinite, Wiry (Szklary), Poland

Fig. 377. TEM: *Goethite,* synthetic

Fig. 378. Replica: *Goethite,* synthetic

Fig. 379. TEM: *Goethite* from weathered serpentinite, Callenberg, GDR

Fig. 380. Replica: *Goethite* from weathered serpentinite, Callenberg, GDR

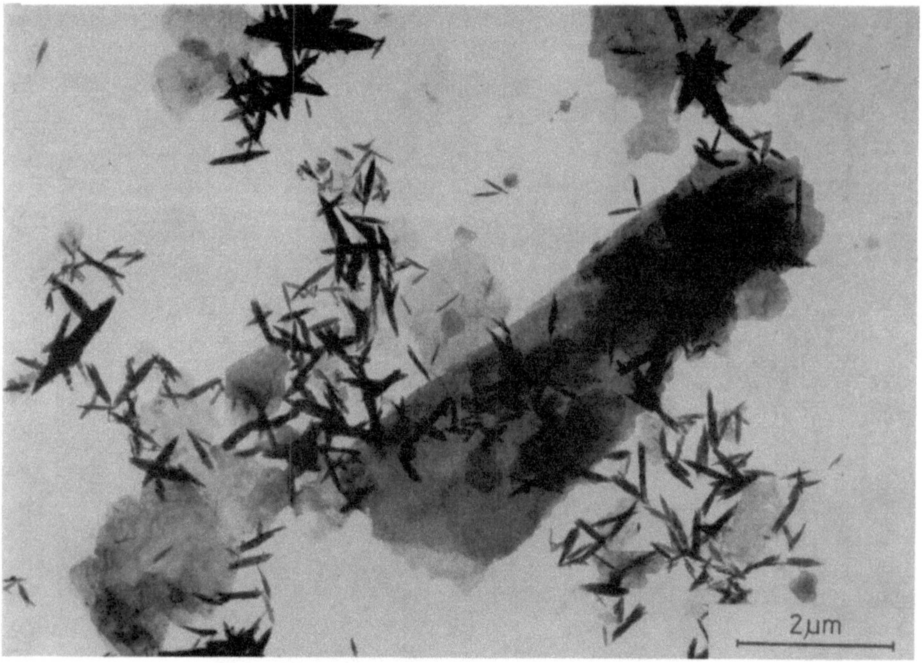

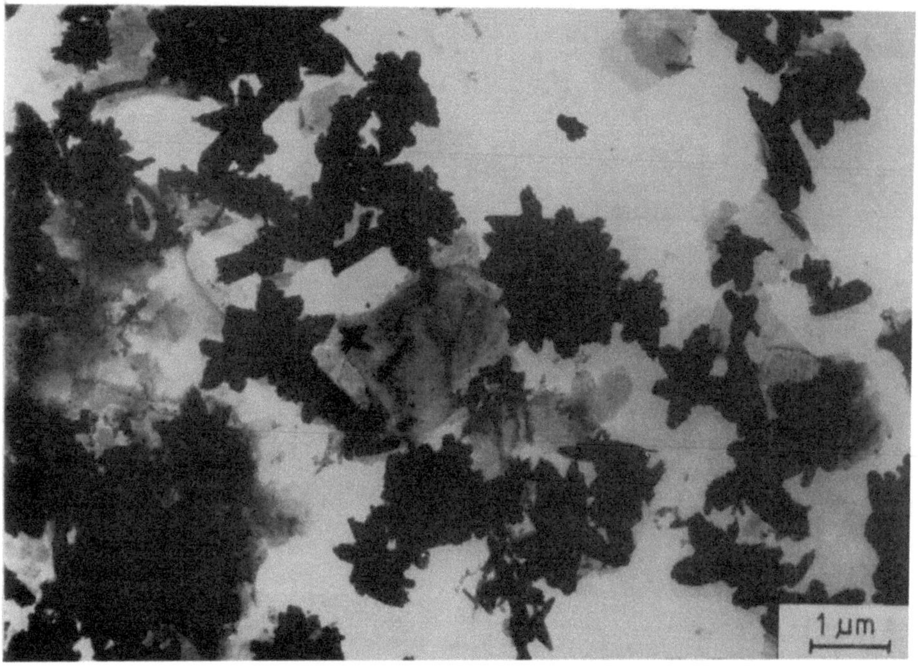

Fig. 381. TEM: *Goethite* from weathered serpentinite, Callenberg, GDR

Fig. 382. TEM: *Goethite* from weathered serpentinite, Callenberg, GDR

Fig. 383. TEM: *Goethite* from weathered serpentinite, Callenberg, GDR

Fig. 384. TEM: *Goethite* (star-shaped) from kaolin, Kemmlitz, GDR

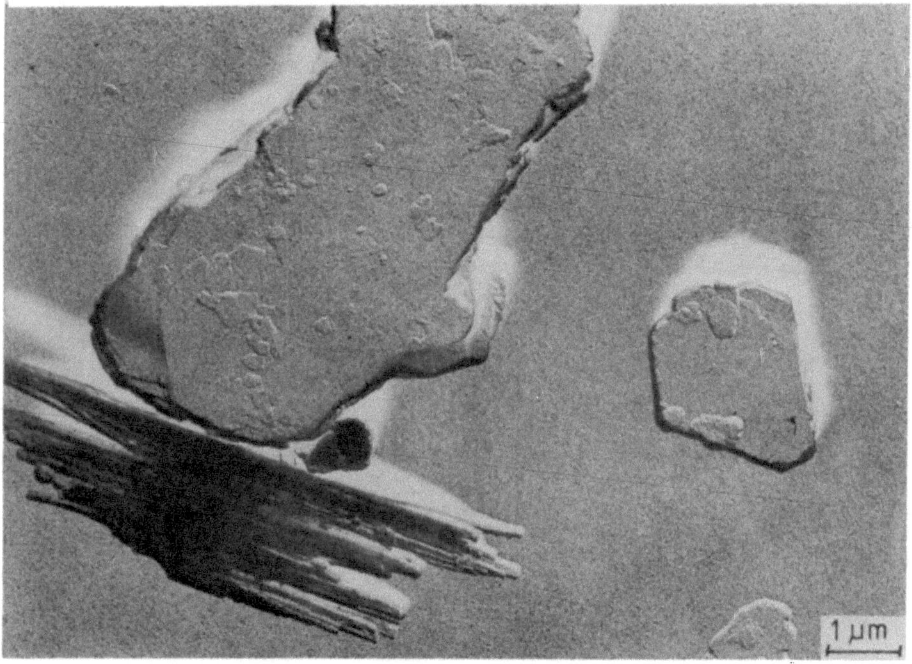

Fig. 385. Replica: *Goethite* from kaolin, Caminau, GDR

Fig. 386. Replica: *Goethite* from kaolin, Kemmlitz, GDR

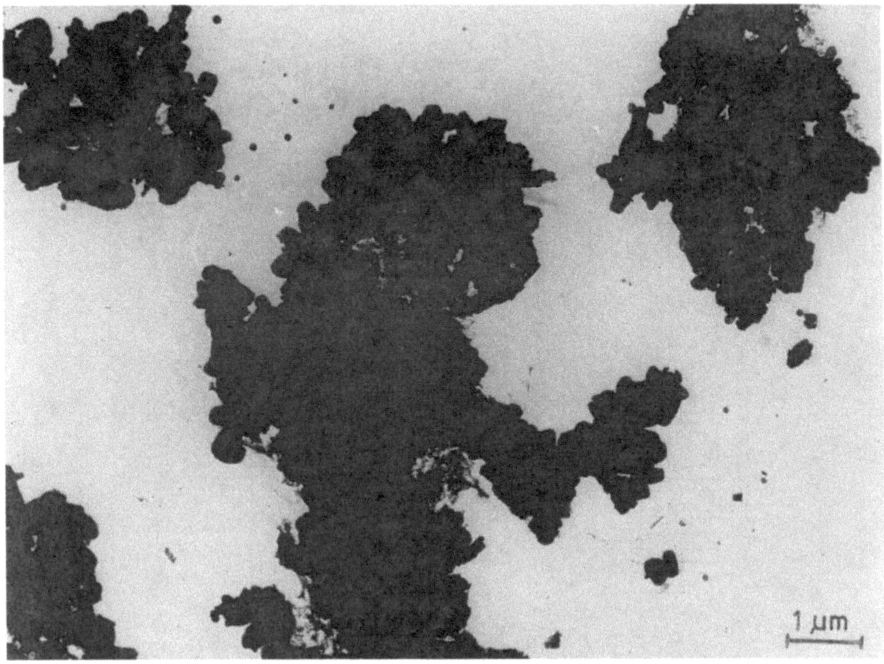

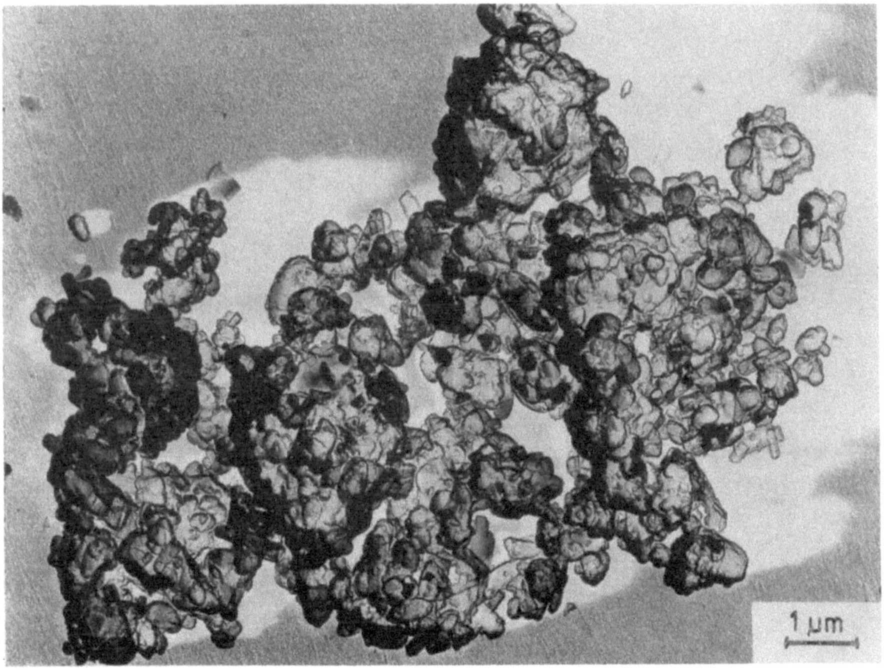

Fig. 387. TEM: *Hematite*, synthetic

Fig. 388. Replica: *Hematite*, synthetic

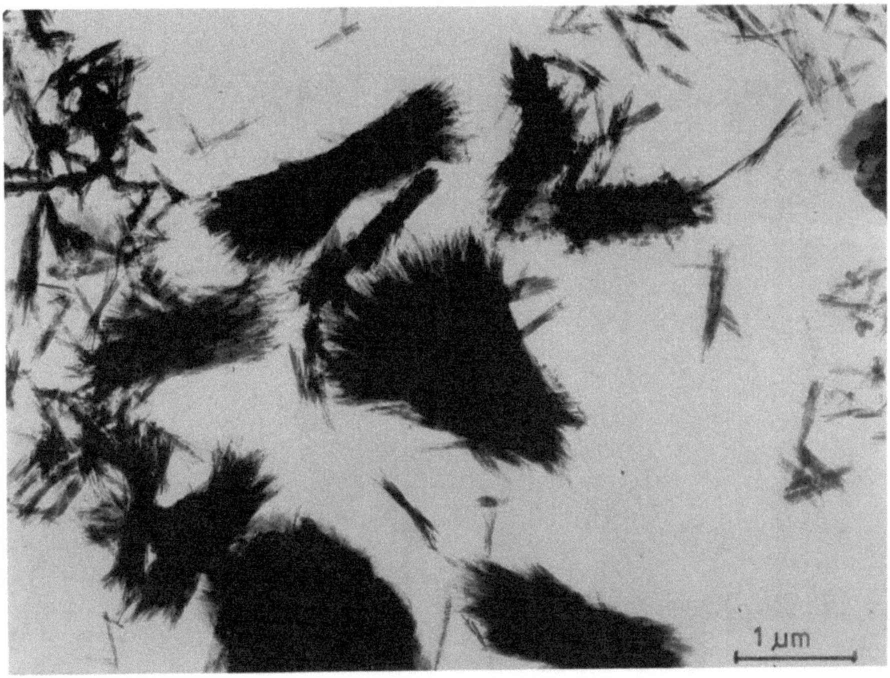

Figs. 389, 390, 391. TEMs: *Hematite* from weathered serpentinite, Callenberg, GDR

Fig. 390.

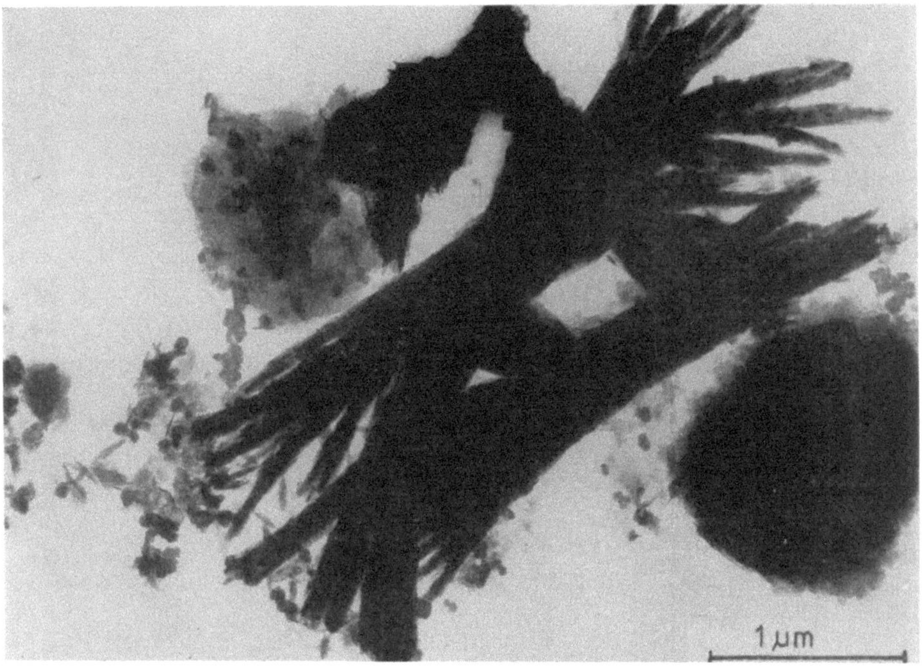

Fig. 391.

Fig. 392. TEM: *Hematite* (spheres and needles) from weathered serpentinite, Callenberg, GDR

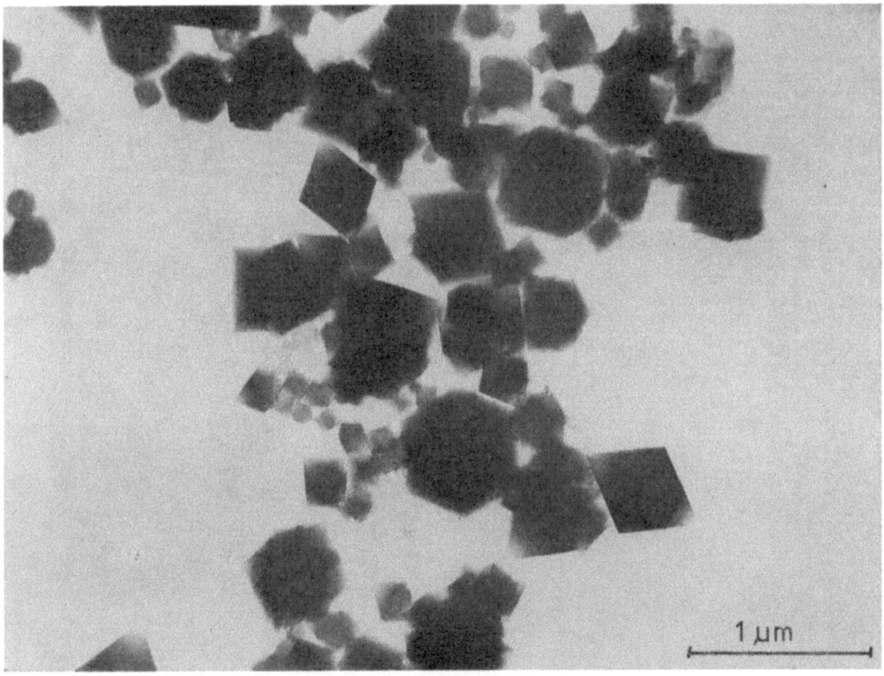

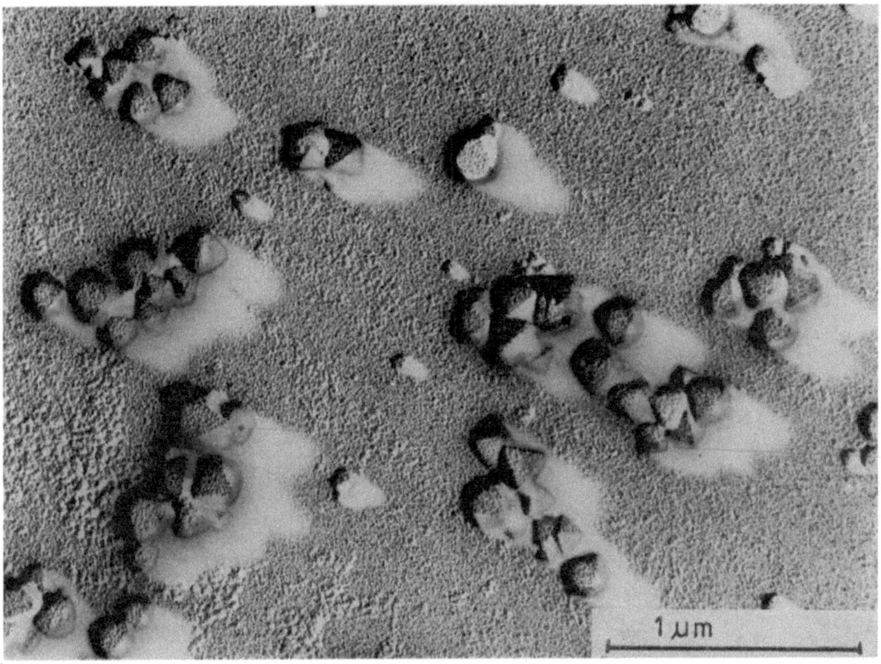

Fig. 393. TEM: *Anatase* (bipyramids) from granodiorite kaolin, Caminau, GDR

Fig. 394. Replica: *Anatase* (bipyramids) from granodiorite kaolin, Caminau, GDR

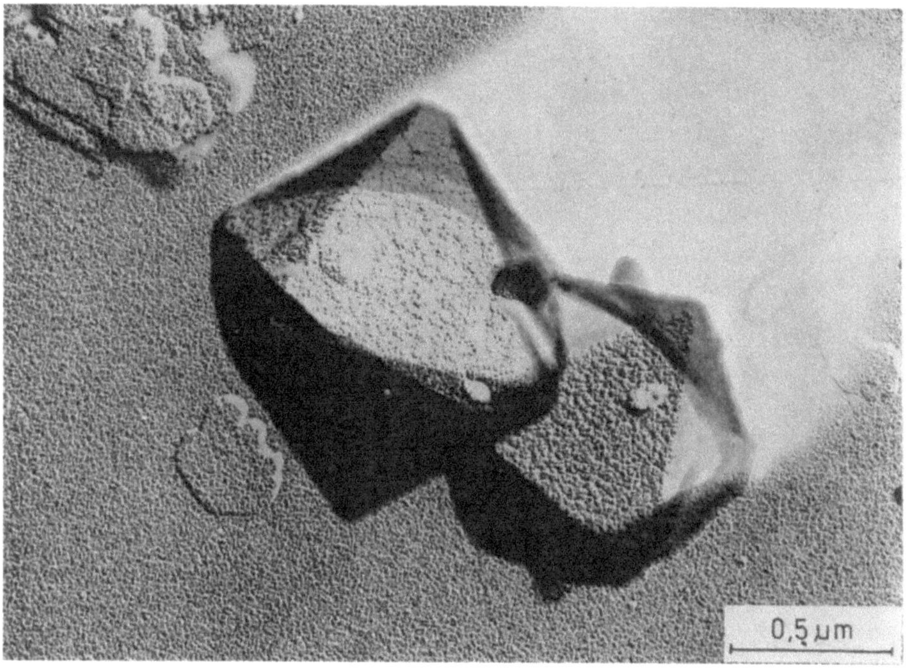

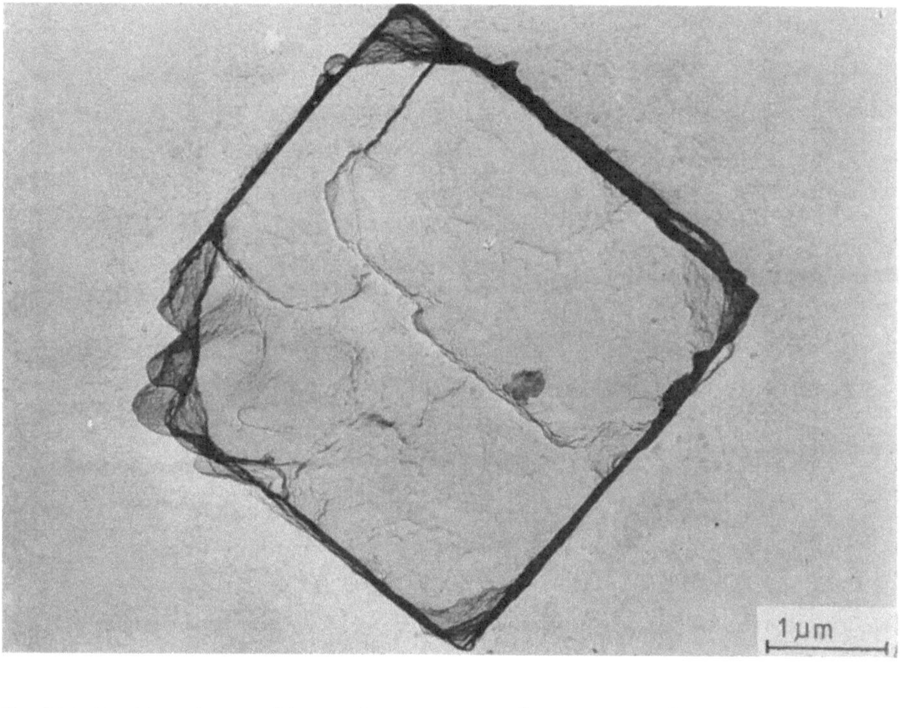

Fig. 395. Replica: *Anatase* (bipyramids) from granodiorite kaolin, Caminau, GDR

Fig. 396. Replica: *Crandallite* (which can be mixed up with Anttase) from sandstone kaolin, Spergau, GDR

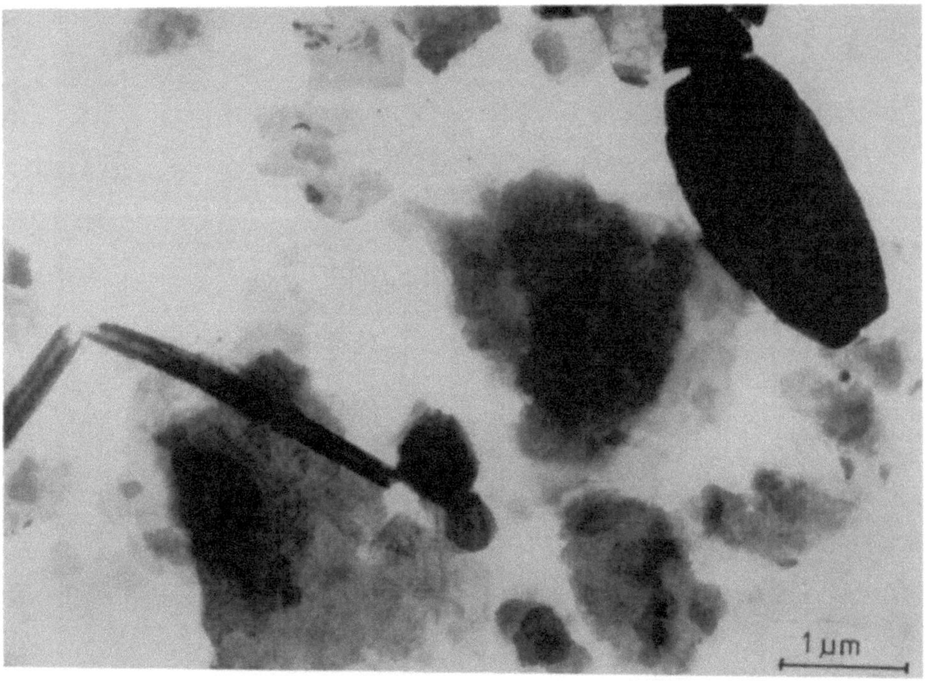

Fig. 397. TEM: *Anatase* (barrel-shaped) from granodiorite kaolin, Caminau, GDR

Fig. 398. Replica: *Anatase* (barrel-shaped) from granodiorite kaolin, Caminau, GDR

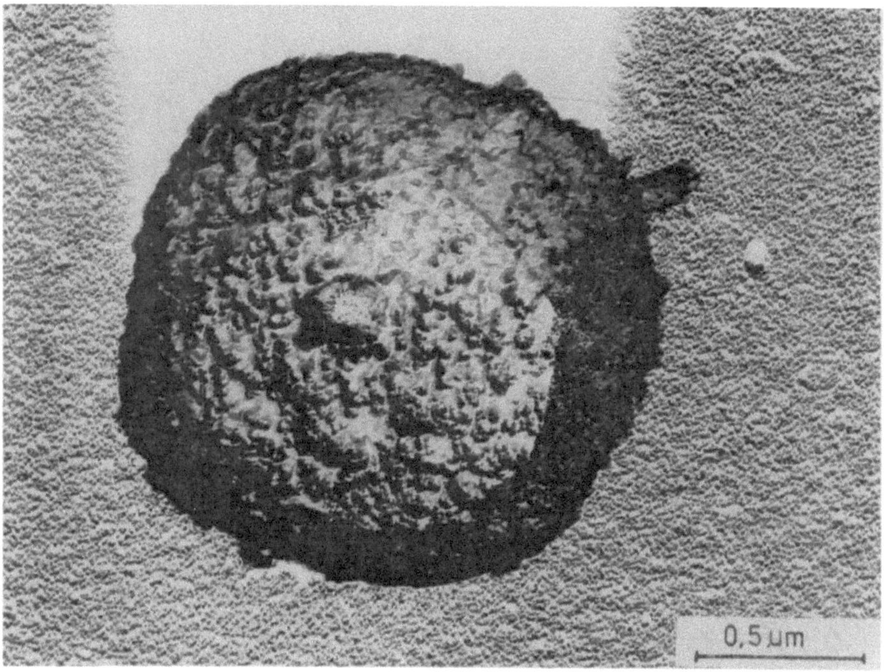

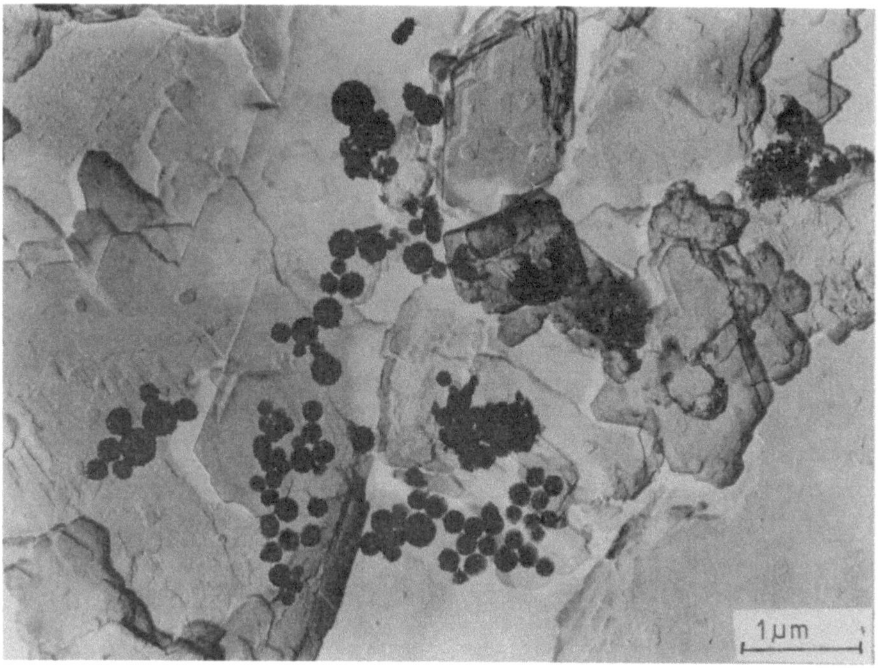

Fig. 399. Replica: *Anatase* (globular aggregate) from porphyry kaolin, Kemmlitz, GDR

Fig. 400. TEM: *Anatase* (spheres) from sandstone kaolin, Röblingen, GDR

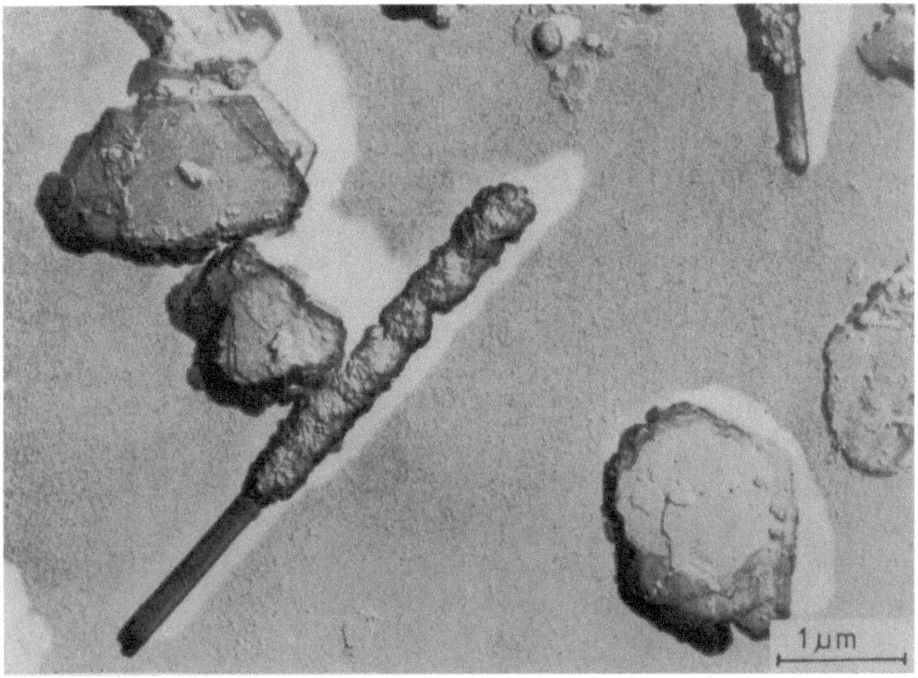

Fig. 401. Replica: *Rutile*, partly anatase-coated, from porphyry kaolin, Kemmlitz, GDR

Fig. 402. Replica: *Rutile* (needle) and *anatase* (aggregates) from granite kaolin, Otovice, Czechoslovakia

Fig. 403. Replica: *Rutile* from sandstone kaolin, Spergau, GDR

Fig. 404. Replica: *Rutile* (needle) and anatase (bipyramid and aggregates) from sandstone kaolin, Röblingen, GDR

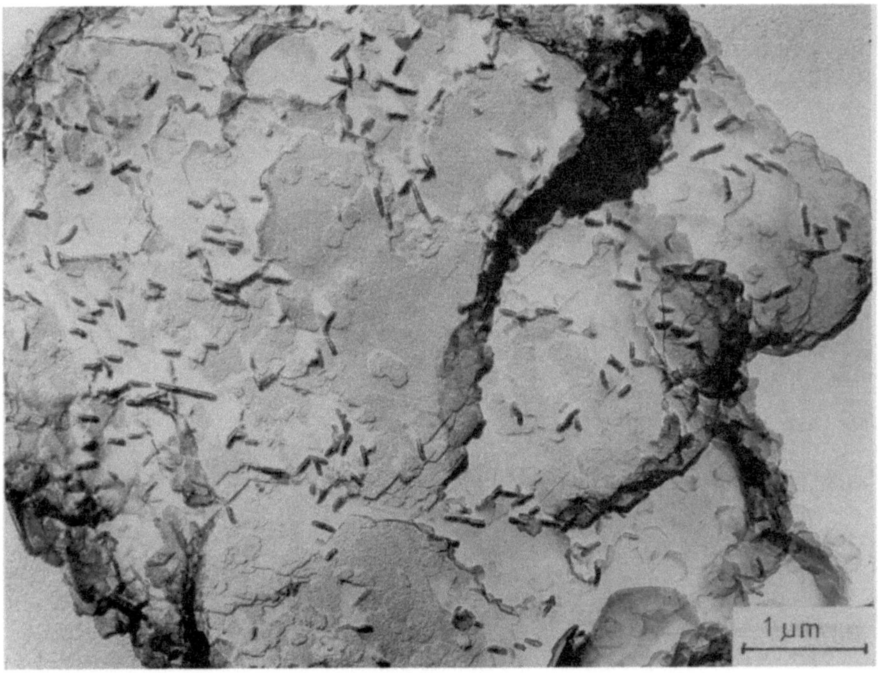

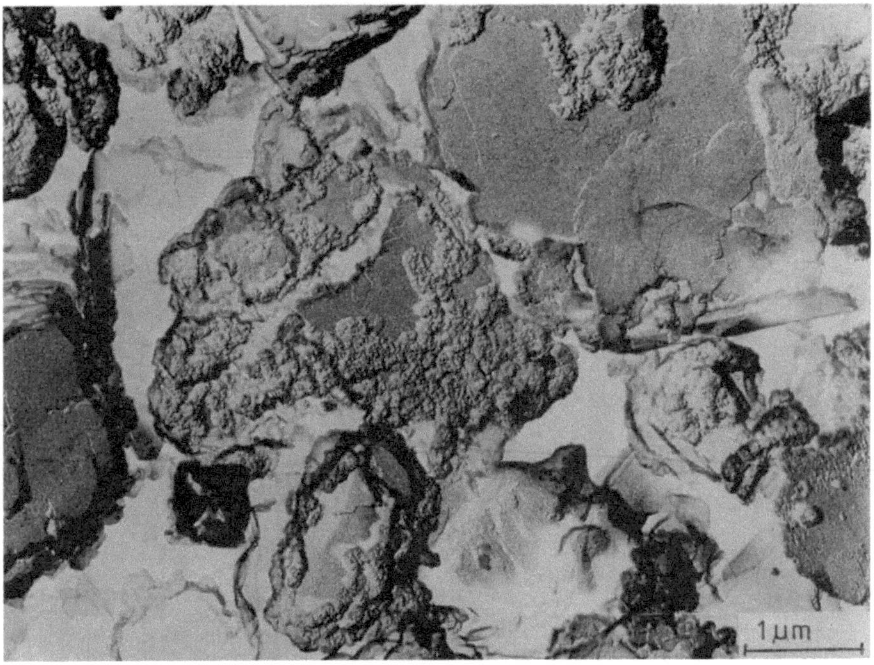

Fig. 405. Replica: *Iron minerals* on *kaolinite* from granodiorite kaolin, Caminau, GDR

Fig. 406. Replica: *Lepidocrocite* on *kaolinite* from porphyry kaolin, Kemmlitz, GDR

Fig. 407. Replica: *Iron minerals* on *kaolinite* from granodiorite kaolin, Caminau, GDR

Fig. 408. Replica: *Iron minerals* on *kaolinite* from sandstone kaolin, Röblingen, GDR

Figs. 409, 410. SEMs: *Pyrite* from porphyry kaolin, Fuchsberg near Halle, GDR

Fig. 410.

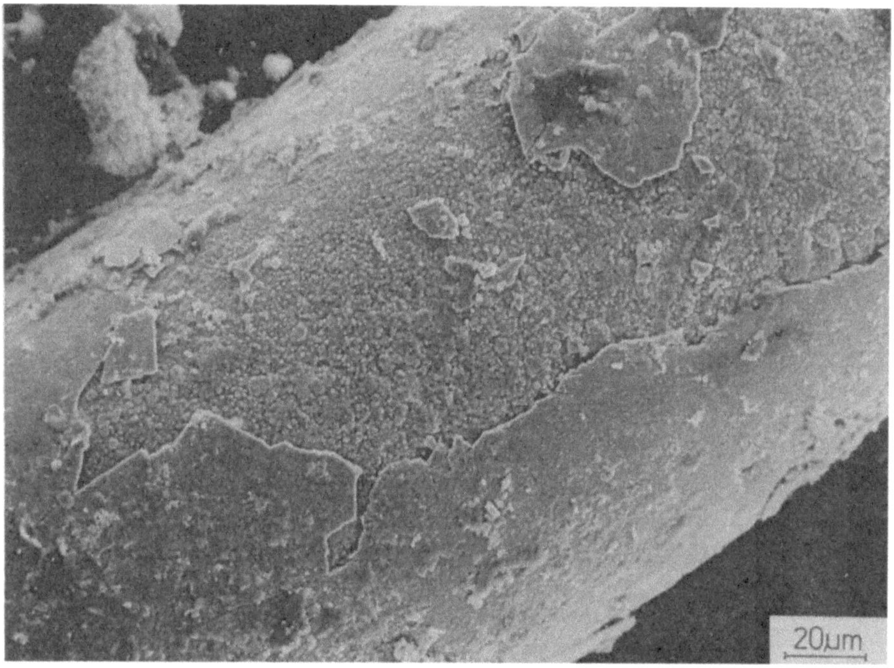

Fig. 411. SEM: *Pyrite* from marine Eocene clay, Friedland, GDR

Fig. 412. SEM: *Pyrite*, detail of Fig. 411

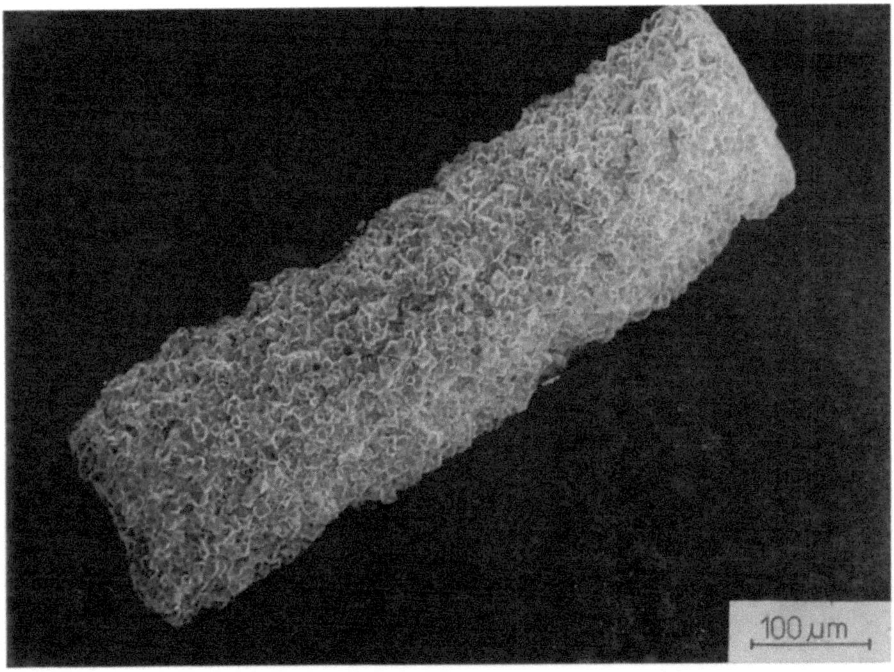

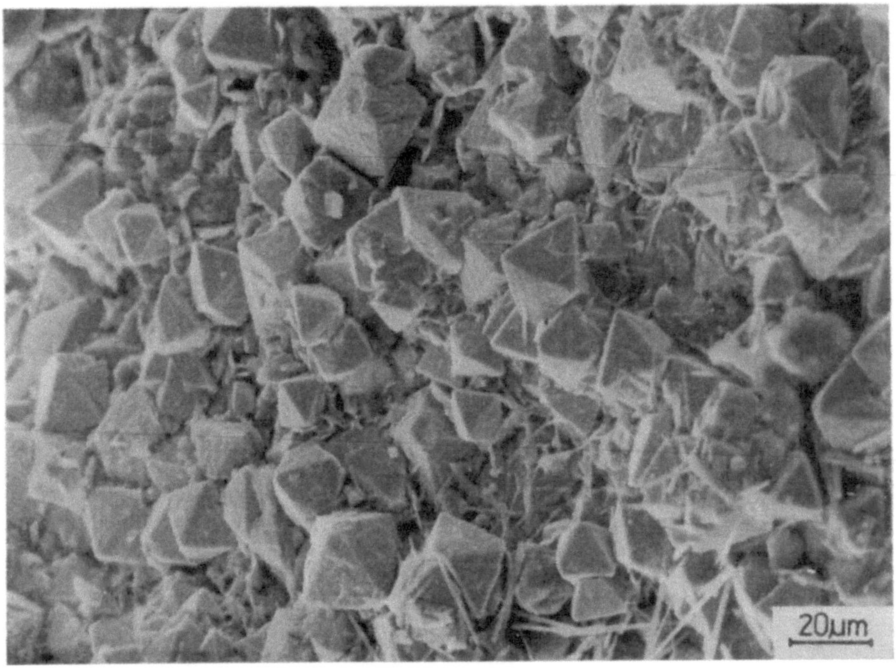

Fig. 413. SEM: *Pyrite* from marine Eocene clay, Friedland, GDR

Fig. 414. SEM: *Pyrite*, detail of Fig. 413

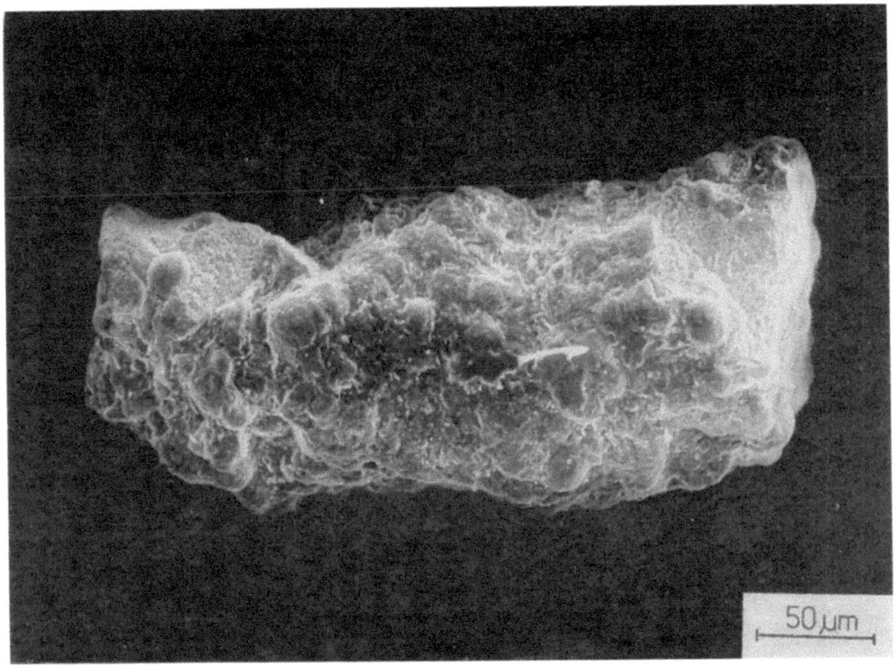

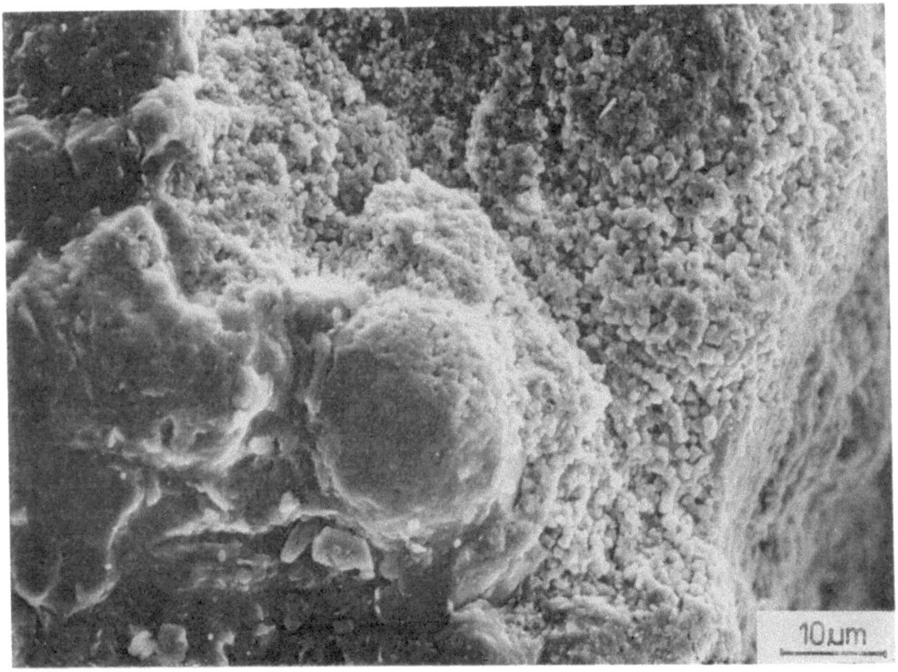

Fig. 415. SEM: Framboidal *pyrite* from marine Eocene clay, Friedland, GDR

Fig. 416. SEM: *Pyrite*, detail of Fig. 415

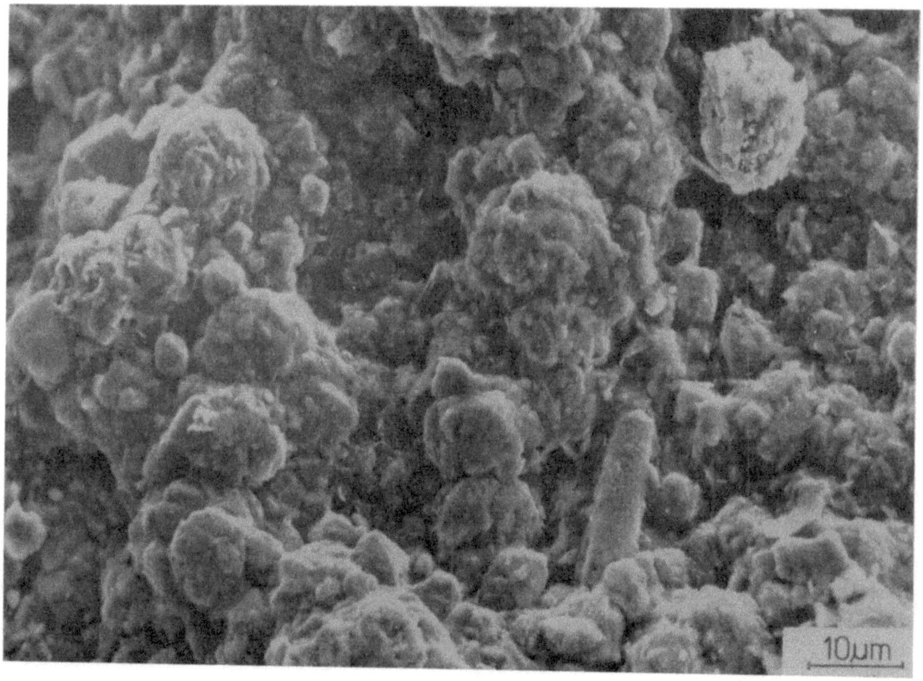

Fig. 417. SEM: Framboidal *pyrite* from marine Eocene clay, Friedland, GDR

Fig. 418. SEM: *Pyrite*, detail of Fig. 417

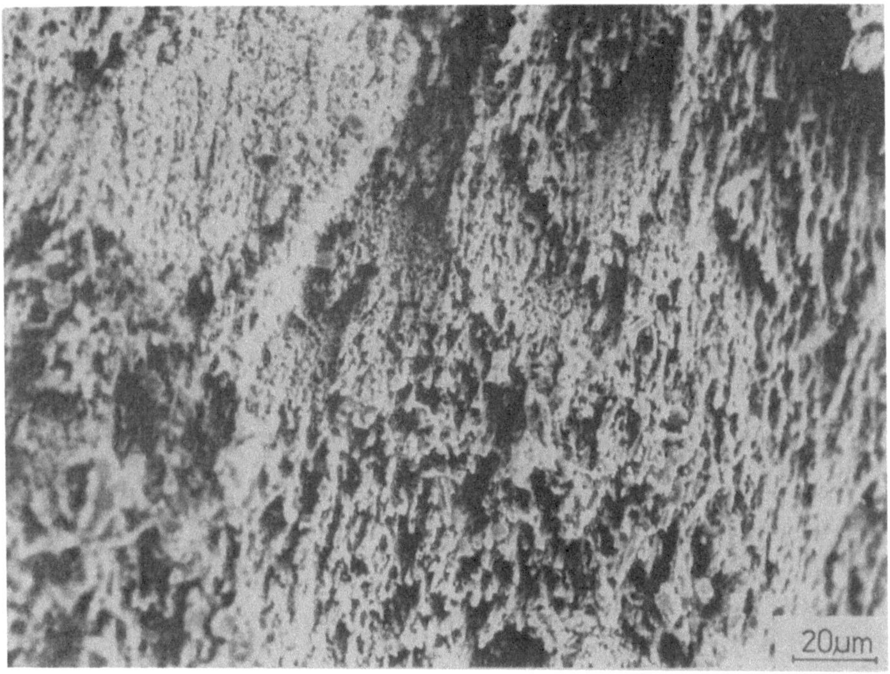

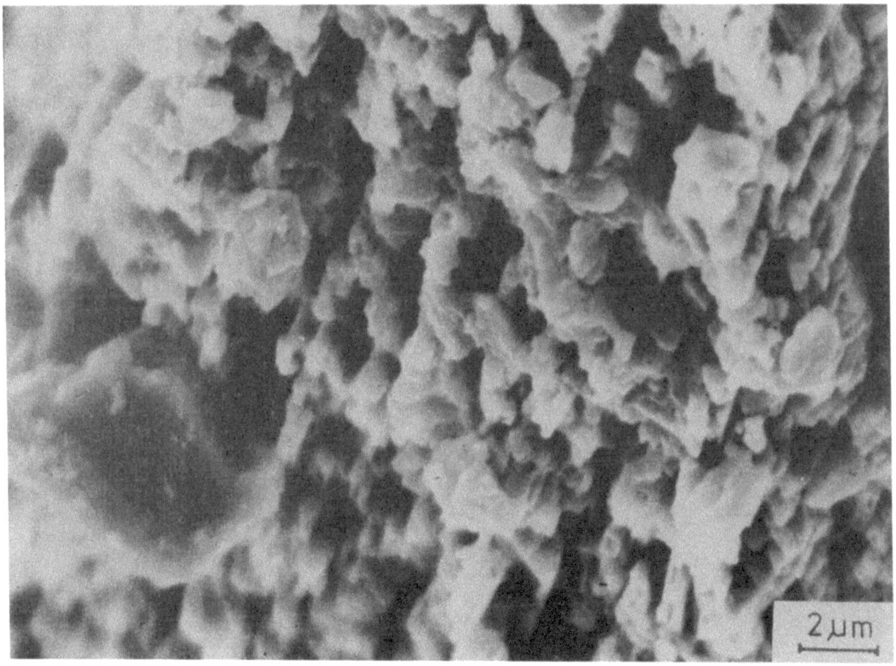

Fig. 419. SEM: *Pyrite* on a crack in a siderite concretion from marine Eocene clay, Friedland, GDR

Fig. 420. SEM: *Pyrite*, detail of Fig. 419

Fig. 421. SEM: *Siderite* and *pyrite* from marine Eocene clay, Friedland, GDR

Fig. 422. SEM: *Siderite* and *pyrite*, detail of Fig. 421

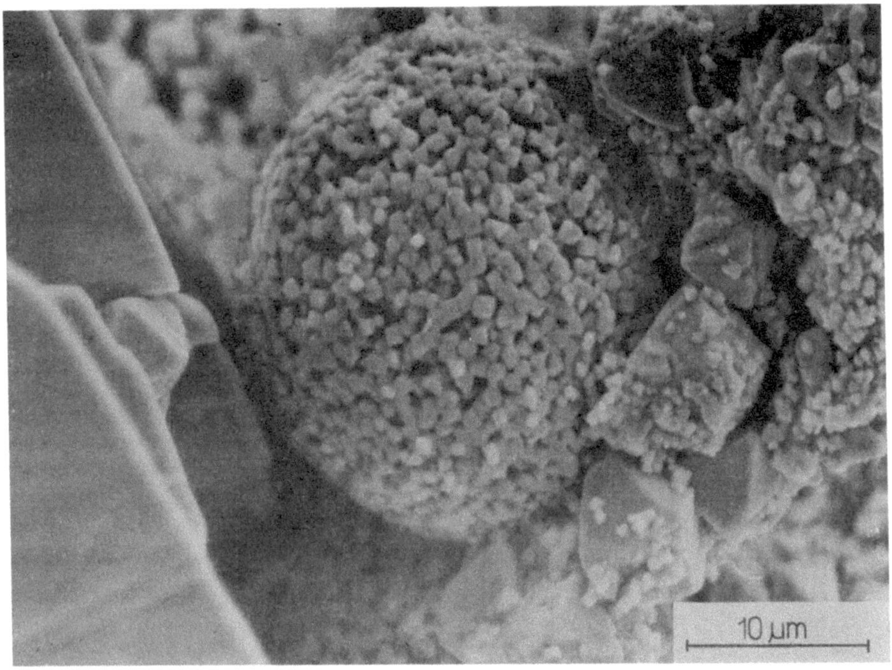

Fig. 423. SEM: *Pyrite,* detail of Fig. 422

Fig. 424. SEM: *Pyrite,* detail of Fig. 423

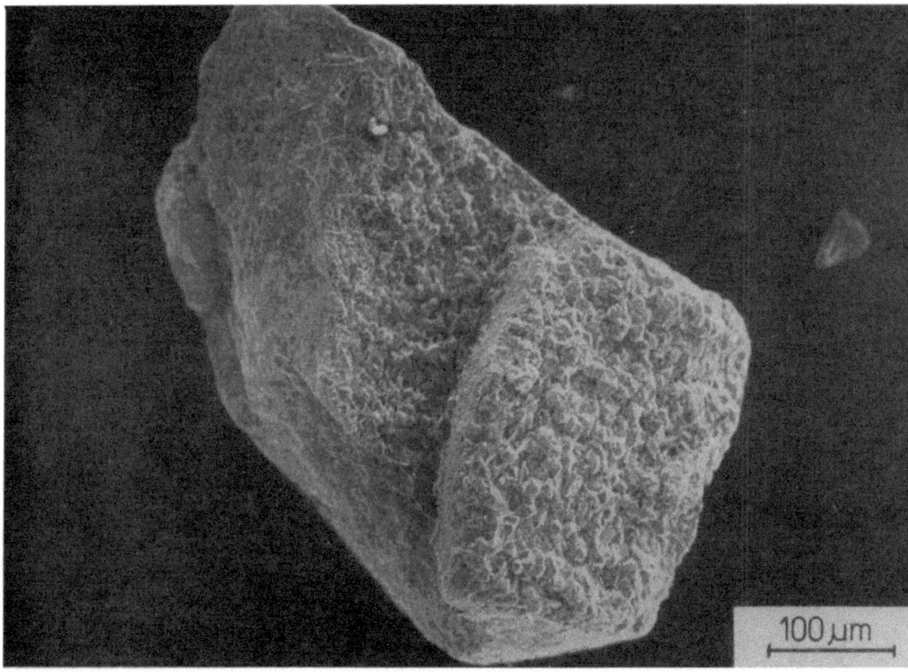

Fig. 425. SEM: Weathered *pyrite* with *iron sulfate* on its surface, Friedland, GDR

Fig. 426. SEM: Detail of Fig. 425

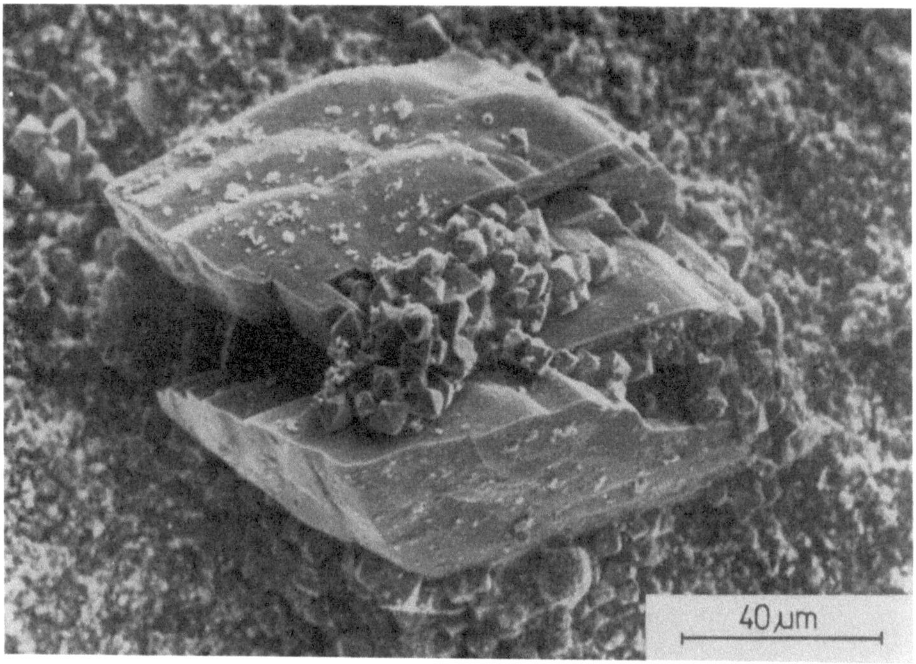

Fig. 427. SEM: *Siderite* and *pyrite* from marine Eocene clay, Friedland, GDR

Fig. 428. SEM: *Pyrite*, detail of Fig. 427

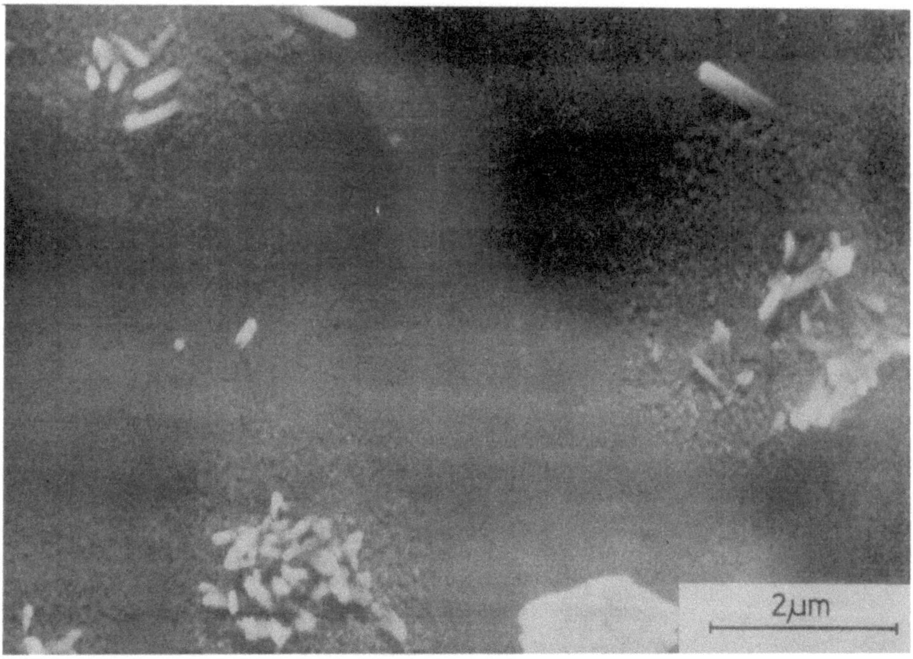

Fig. 429. SEM: *Sulfate minerals* (decomposed pyrite) on *siderite*, Friedland ,GDR

Fig. 430. SEM: Detail of Fig. 429

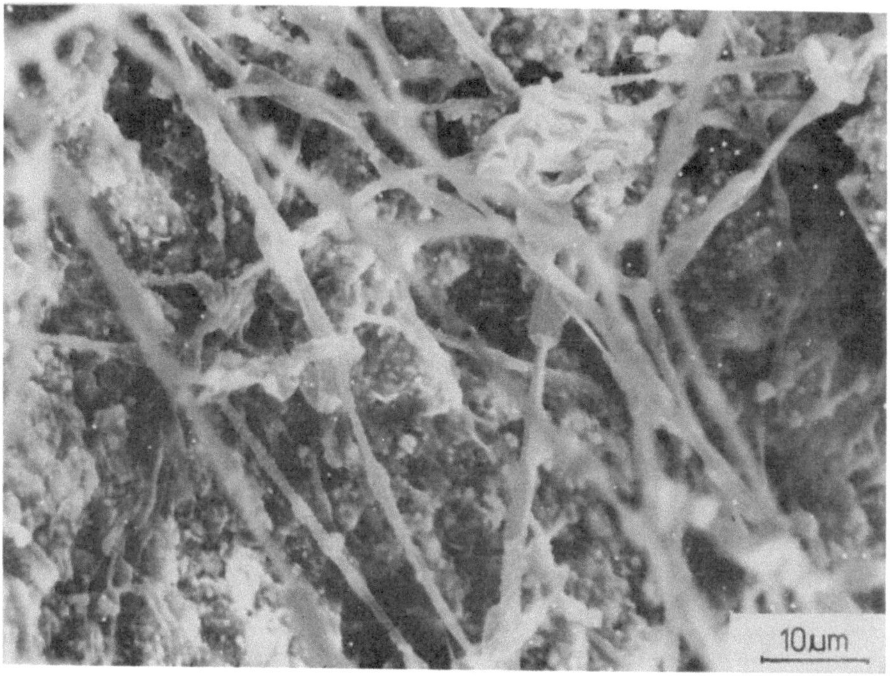

Fig. 431. SEM: *Thenardite* from Newcastle Sandstone, Crook County, Wyoming, USA

Fig. 432. SEM: *Thenardite* from porphyry kaolin, Fuchsberg near Halle, GDR

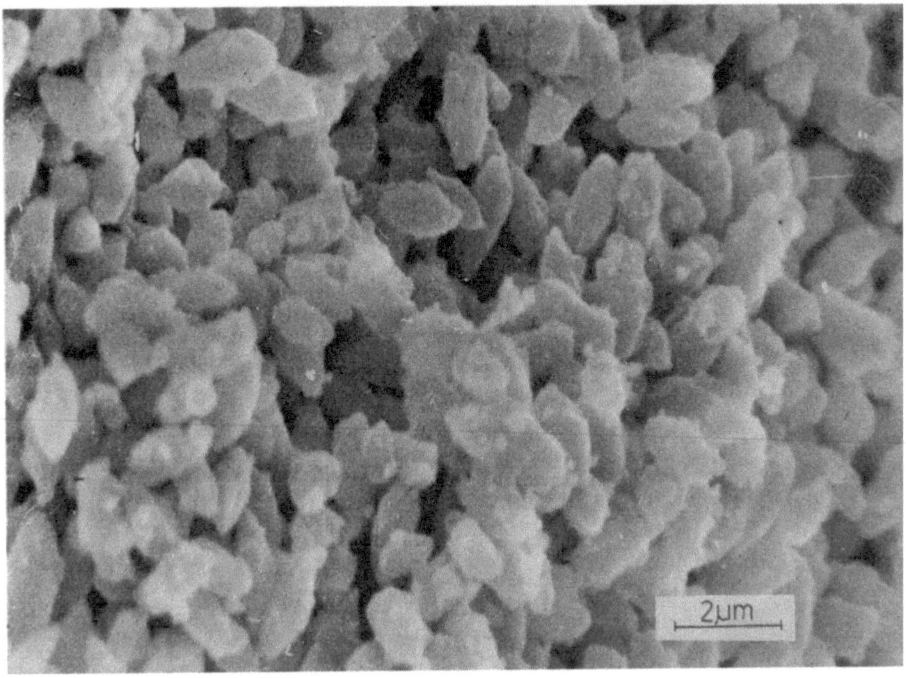

Fig. 433. Replica: *Alunite* from porphyry kaolin, Fuchsberg near Halle, GDR

Fig. 434. SEM: *Alunite* from a nodule of porphyry kaolin, Altenbach near Wurzen, GDR

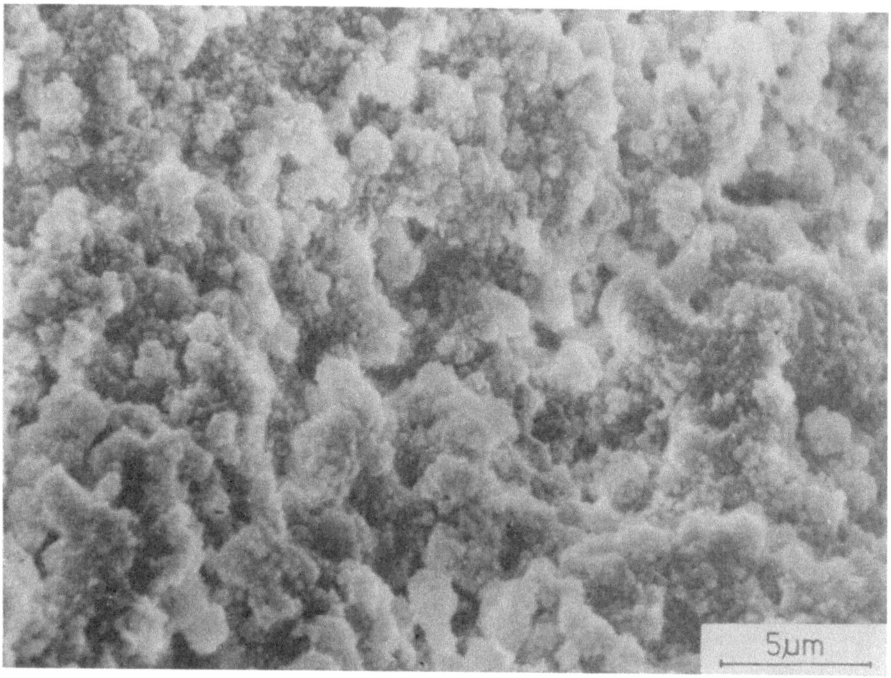

Fig. 435. SEM: *Jarosite*, yellow crust from the clay surface, Friedland, GDR

Fig. 436. SEM: *Jarosite*, yellow powder in joints of the Mowry Shale, Crook County, Wyoming, USA

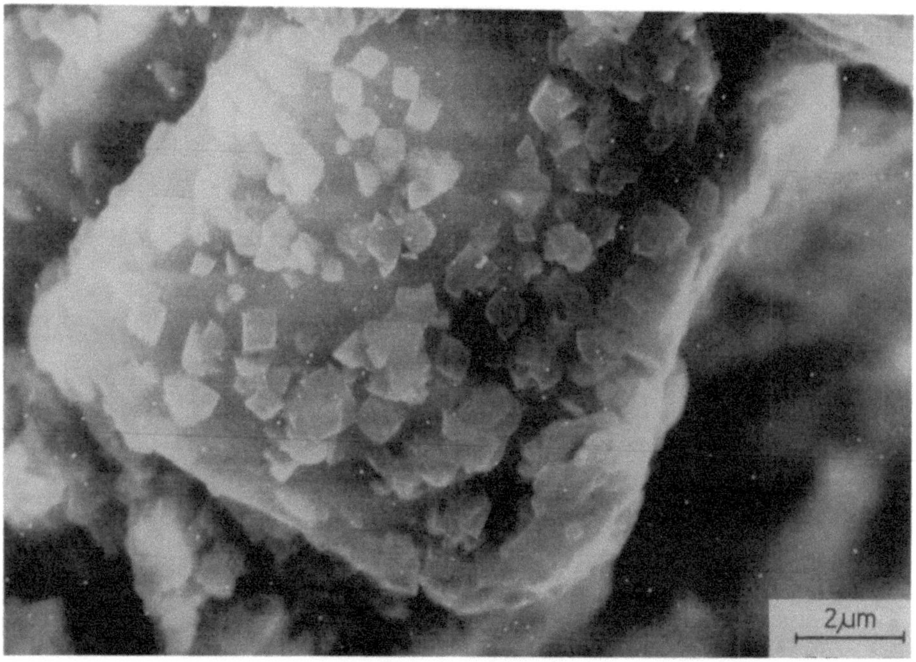

Fig. 437. SEM: *Jarosite* in a pore of Mowry Shale, Crook County, Wyoming USA

Fig. 438. SEM: *Jarosite* on kaolinite from porphyry kaolin, Fuchsberg near Halle, GDR

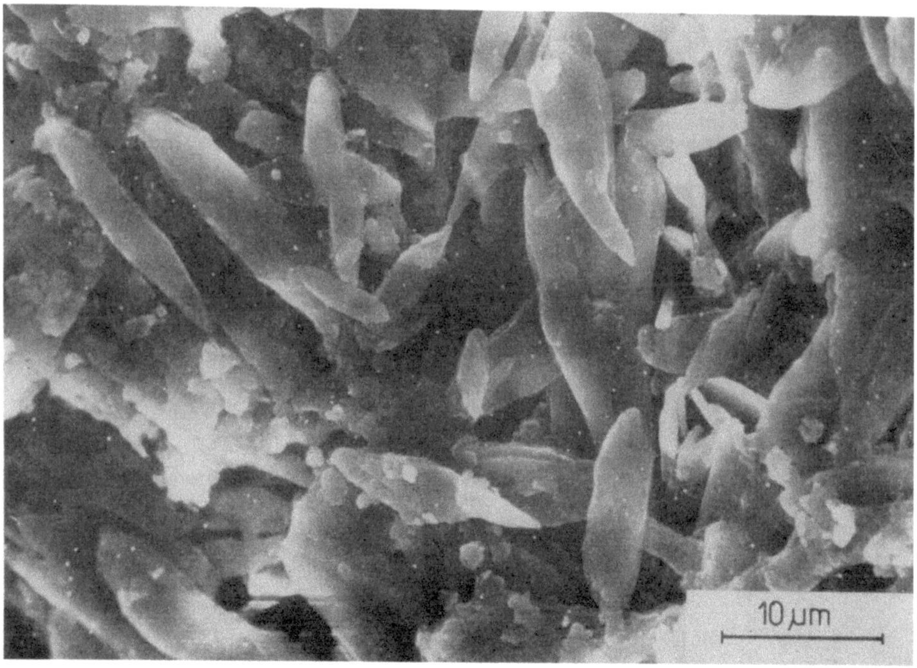

Fig. 439. SEM: *Jarosite,* joint filling in Eocene clay, interbedded with tuff layers, Fur, Denmark

Fig. 440. SEM: *Jarosite* from a brown crust on porphyry kaolin, Fuchsberg near Halle, GDR

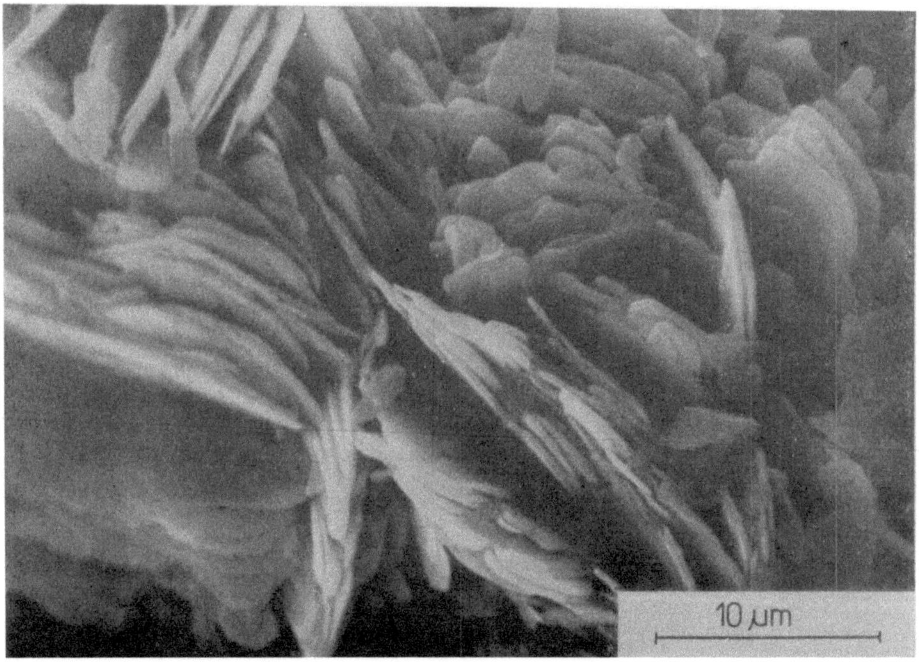

Fig. 441. SEM: *Jarosite* from yellow crusts on the clay surface, Friedland, GDR

Fig. 442. SEM: *Jarosite* from a joint filling in clay, Friedland, GDR

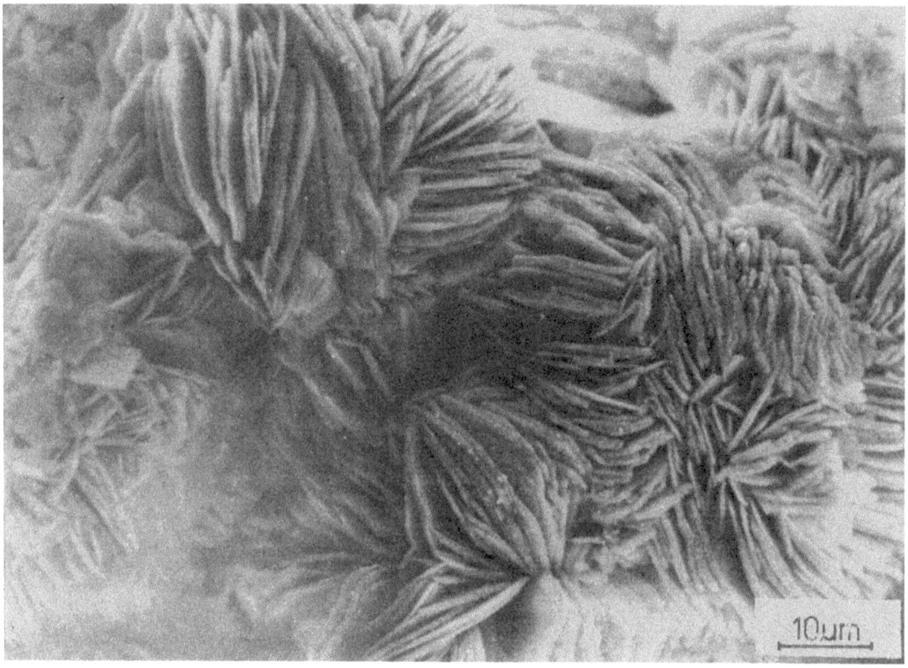

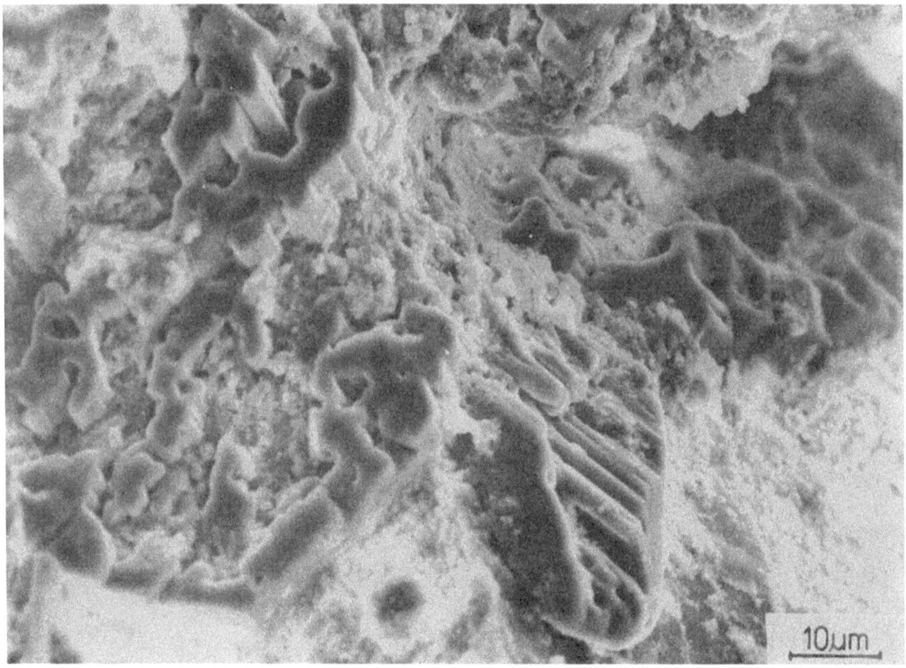

Fig. 443. SEM: *Jarosite*, yellow dots on the clay surface, Friedland, GDR

Fig. 444. SEM: *Jarosite* (grains) and *gypsum* from Silurian pyritic slate, Gräfenthal, GDR

Figs. 445, 446. SEMs: *Szomolnokite* on fissures in clay, Friedland, GDR

Fig. 446.

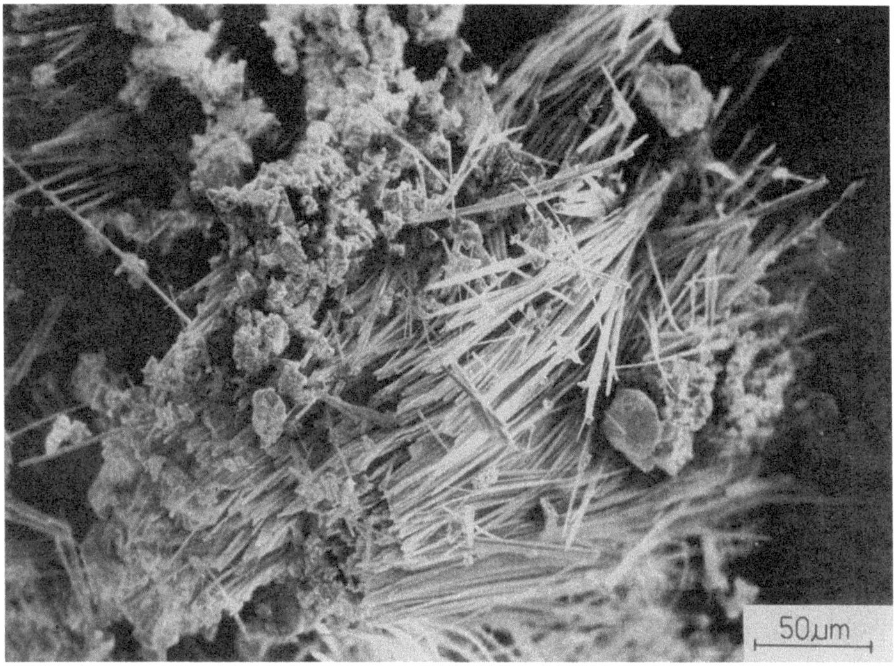

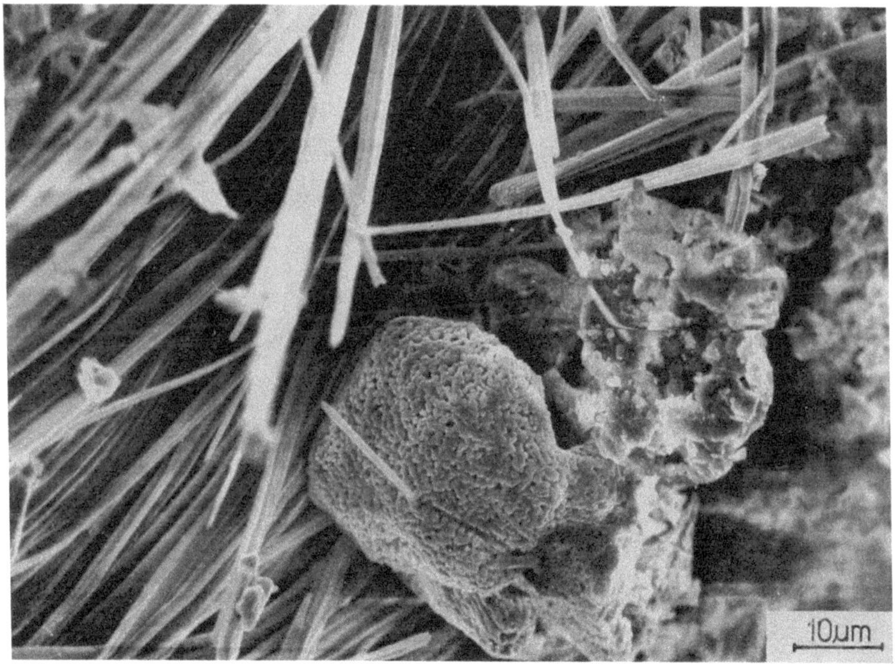

Figs. 447, 448, 449, 450. SEMs: *Szomolnokite* (needles and grains) from the surface of weathered pyrite in clay, Friedland, GDR

Fig. 448.

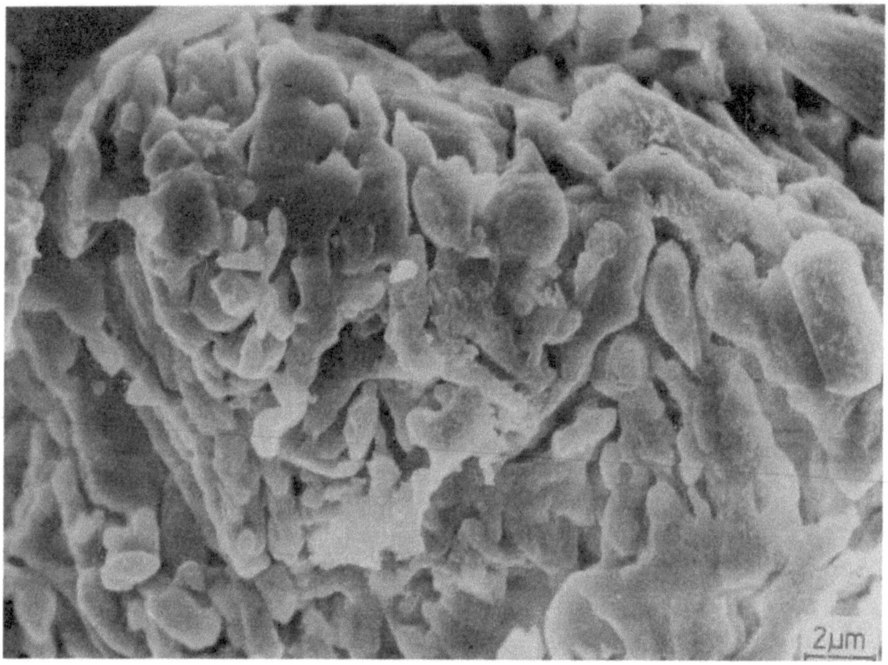

Fig. 449.

(explained at p. 265)

Fig. 450.

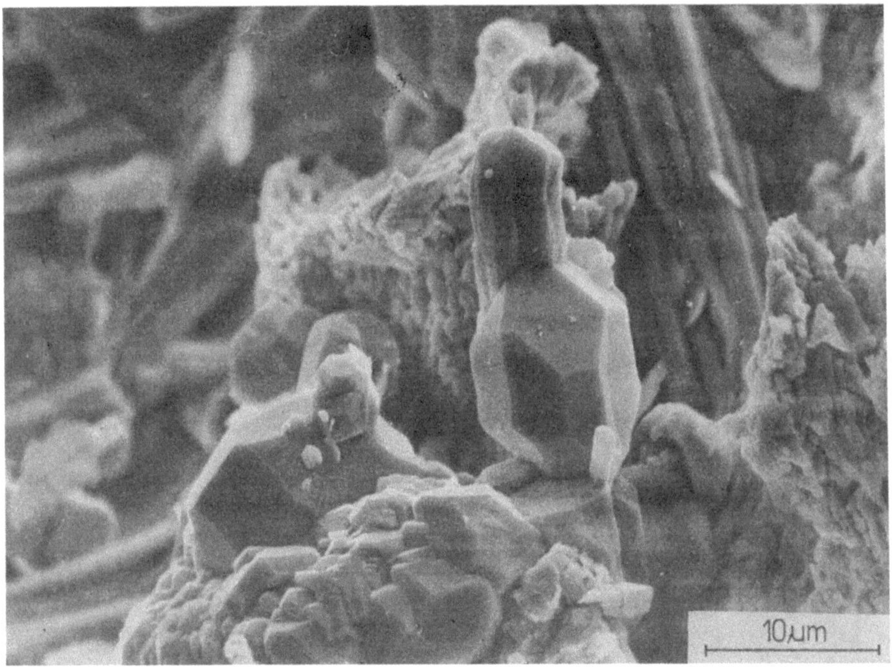

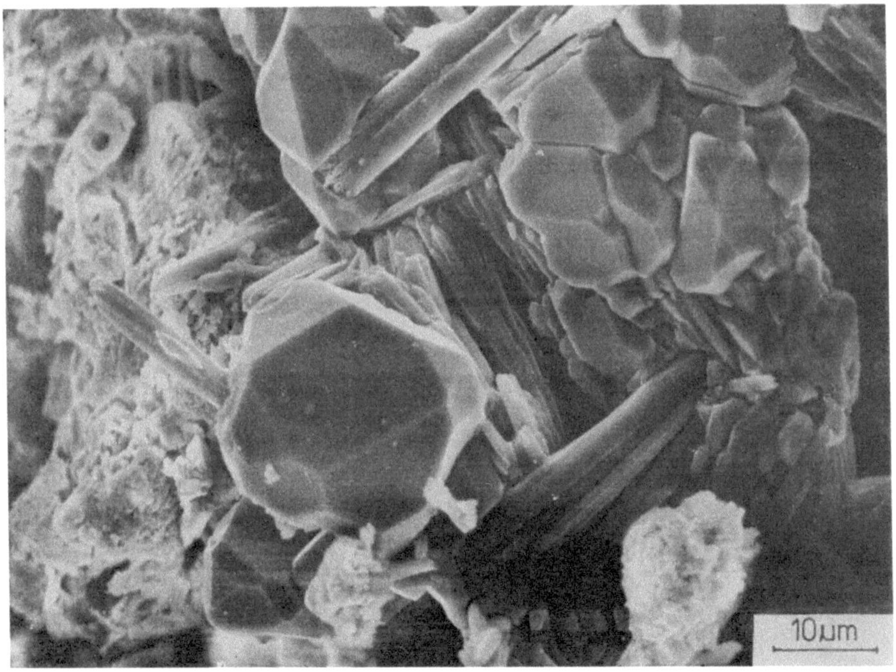

Figs. 451, 452. SEMs: *Szomolnokite* from the surface of weathered pyrite in clay, Friedland, GDR

Fig. 452.

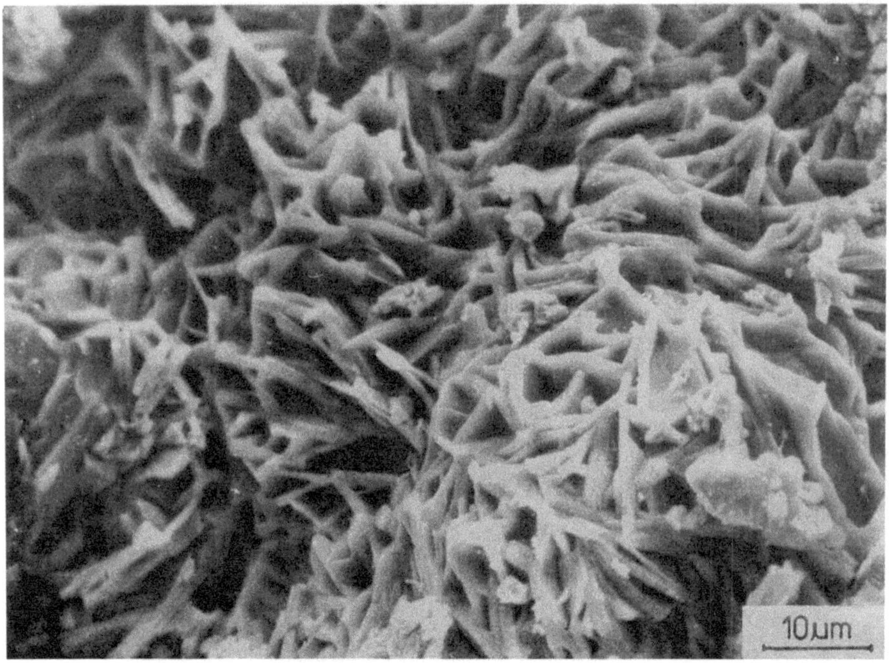

Fig. 453. SEM: *Rozenite* and *szomolnokite* (needles) on weathered pyrite from clay, Friedland, GDR

Fig. 454. SEM: Detail of Fig. 453

Figs. 455, 456. SEMs: *Coquimbite* from kaolin, Fuchsberg near Halle, GDR

Fig. 456.

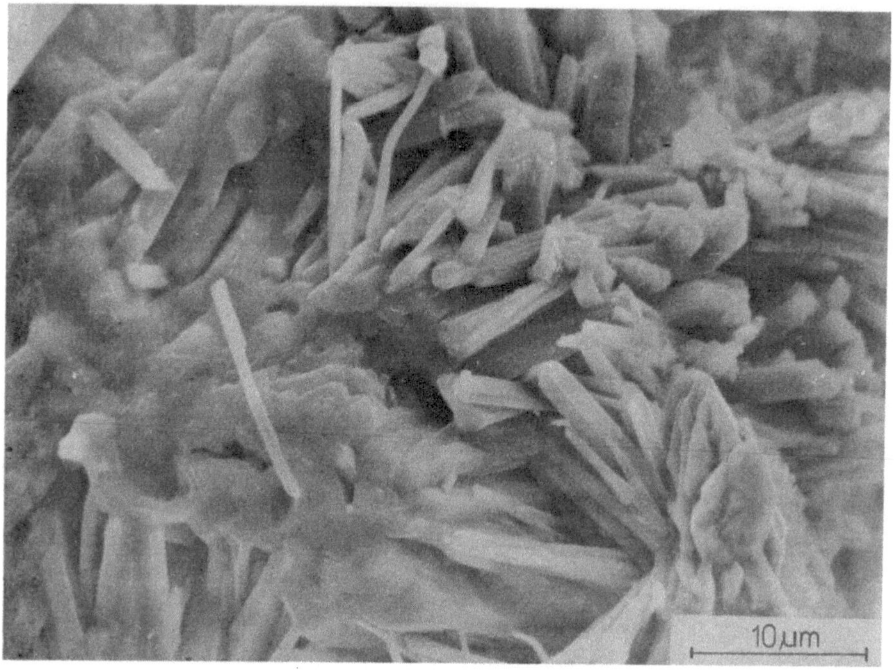

Fig. 457. SEM: *Coquimbite* (needles) and *copiapite* (grains) from kaolin, Fuchsberg near Halle,. GDR

Fig. 458. SEM: *Copiapite,* detail of Fig. 457

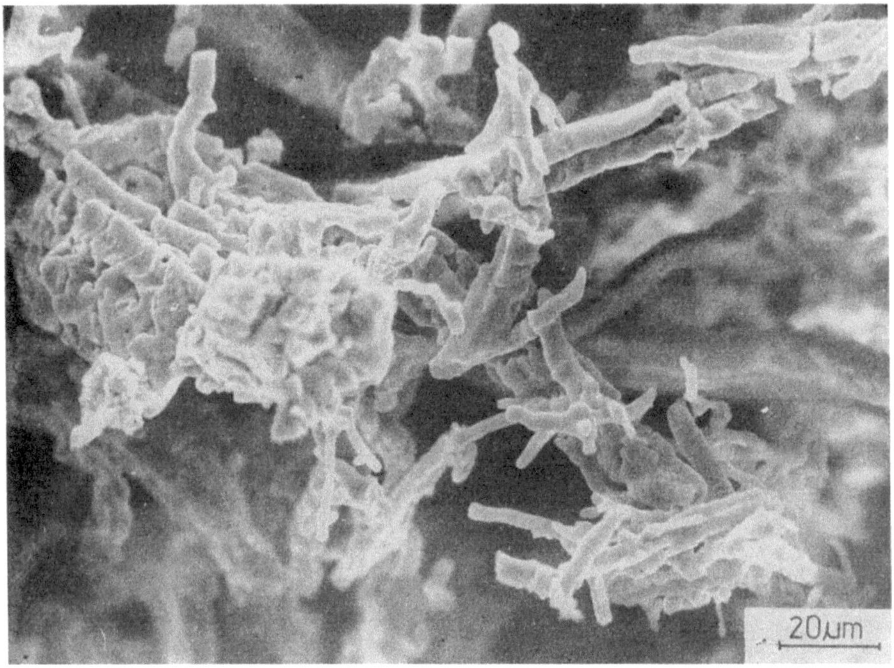

Figs. 459, 460, 461, 462. SEMs: *Rozenite* from fissures in clay, Friedland, GDR

Fig. 460.

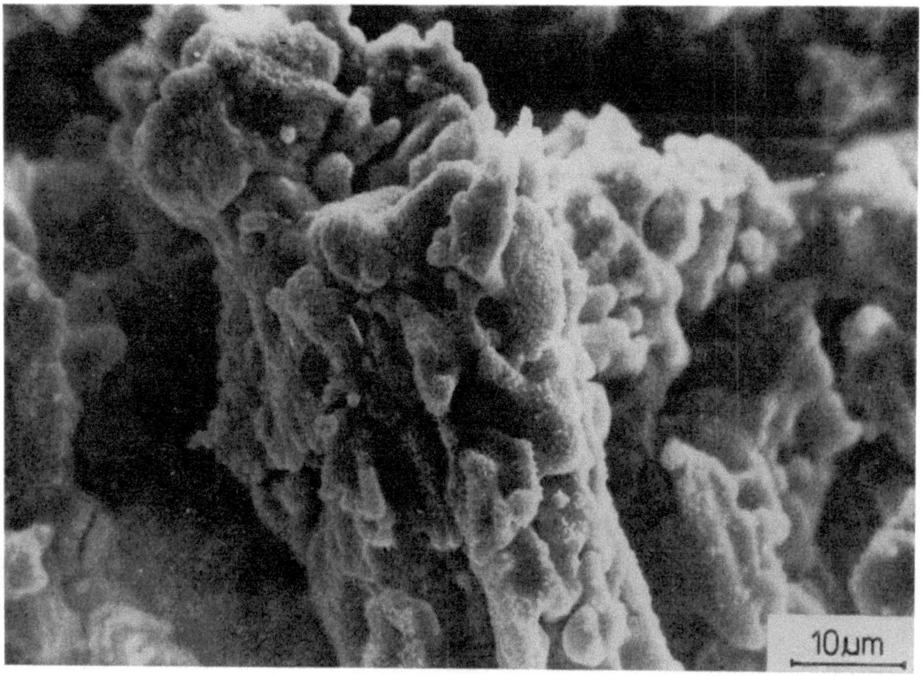

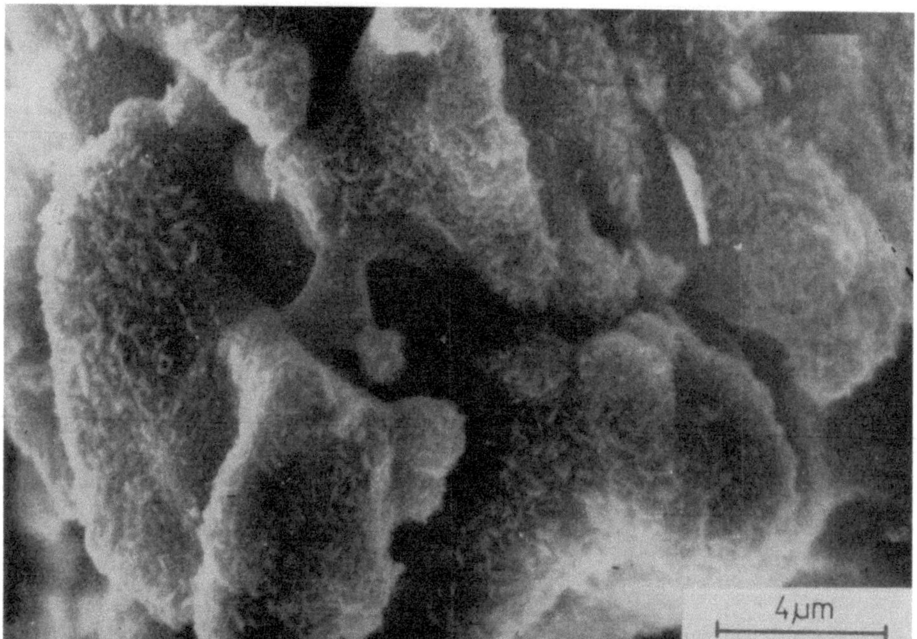

Fig. 461.

(explained at p. 271)

Fig. 462.

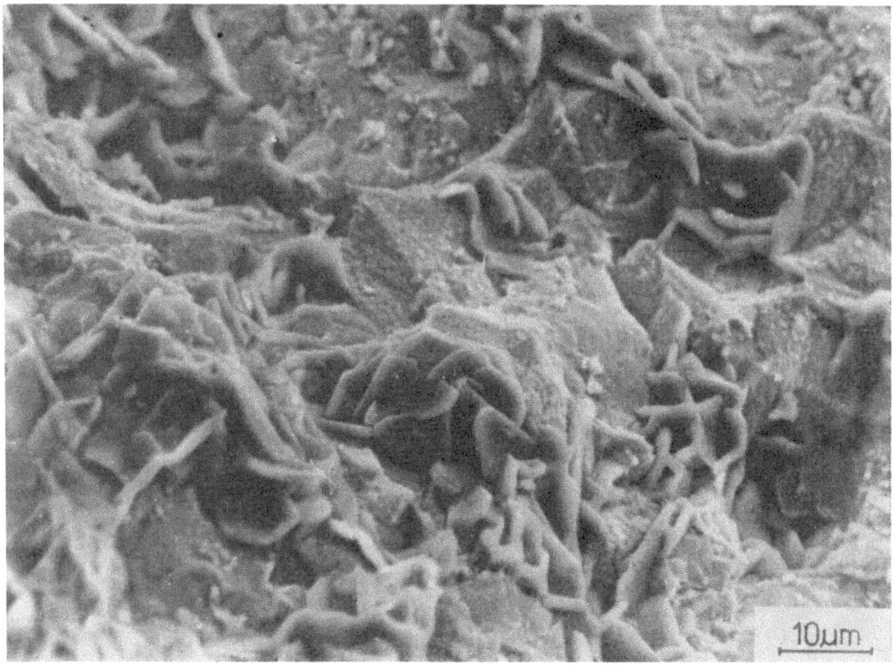

Figs. 463, 464. SEMs: *Rozenite* on the surface of decomposed pyrite in clay, Friedland, GDR

Fig. 464.

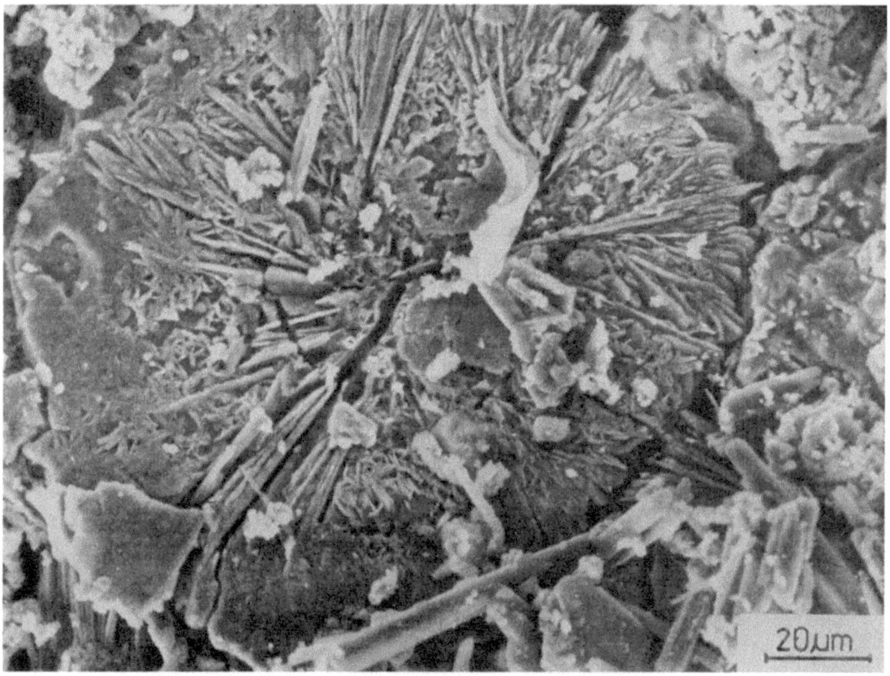

Fig. 465. SEM: *Melanterite* ("blue stone") from kaolin, Podlesi, Czechoslovakia

Fig. 466. SEM: Detail of Fig. 465

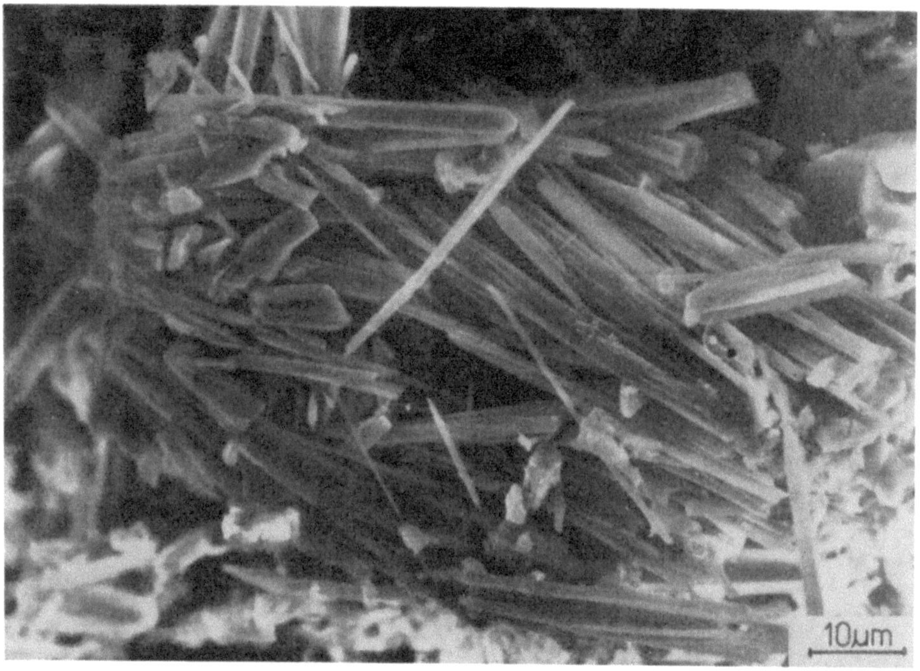

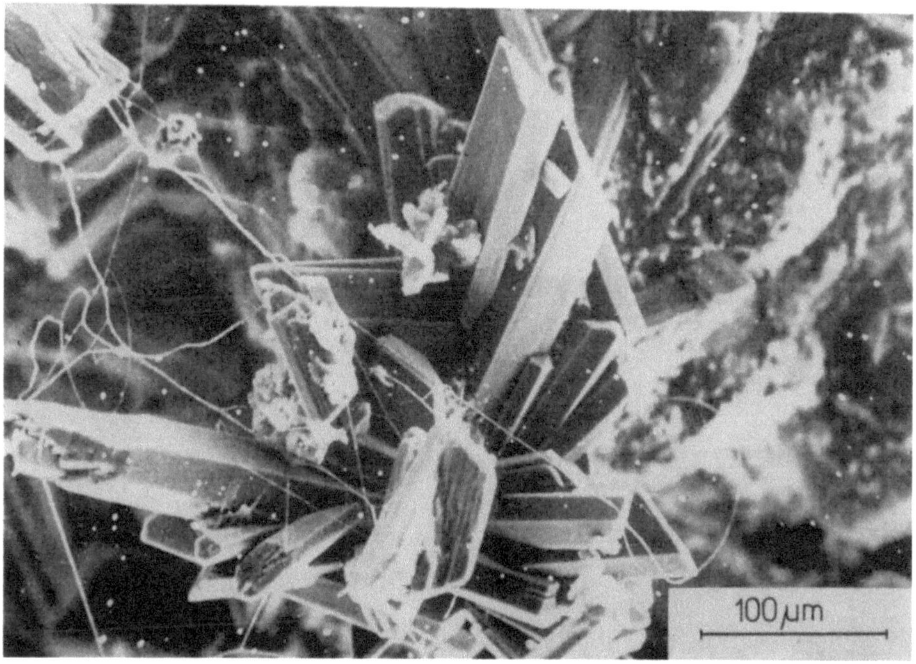

Fig. 467. SEM: *Roemerite*, white crust on decomposed pyrite from clay, Friedland, GDR

Fig. 468. SEM: *Gypsum* with fibres of *thenardite* from the surface of clay, Friedland, GDR

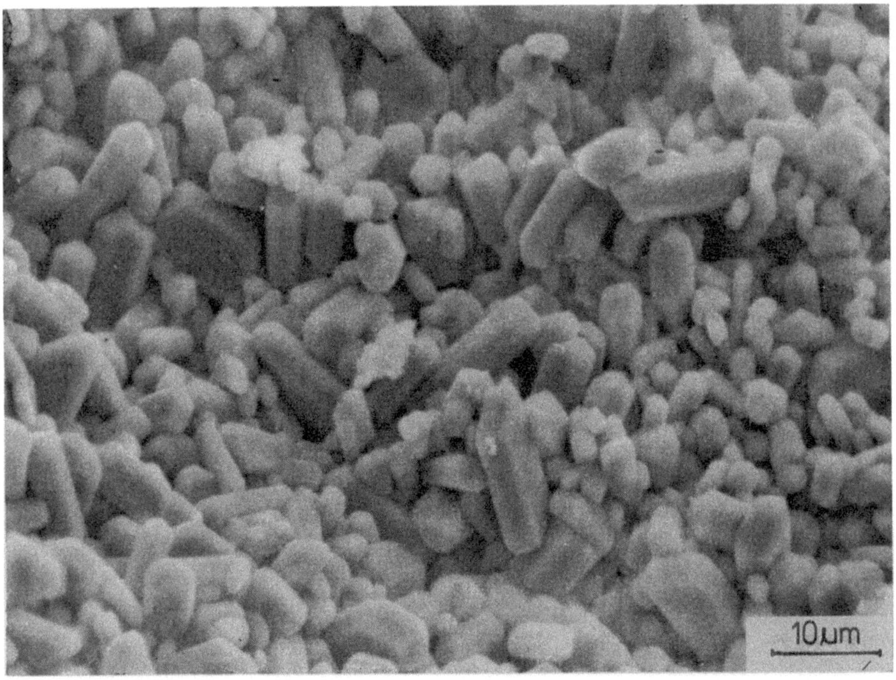

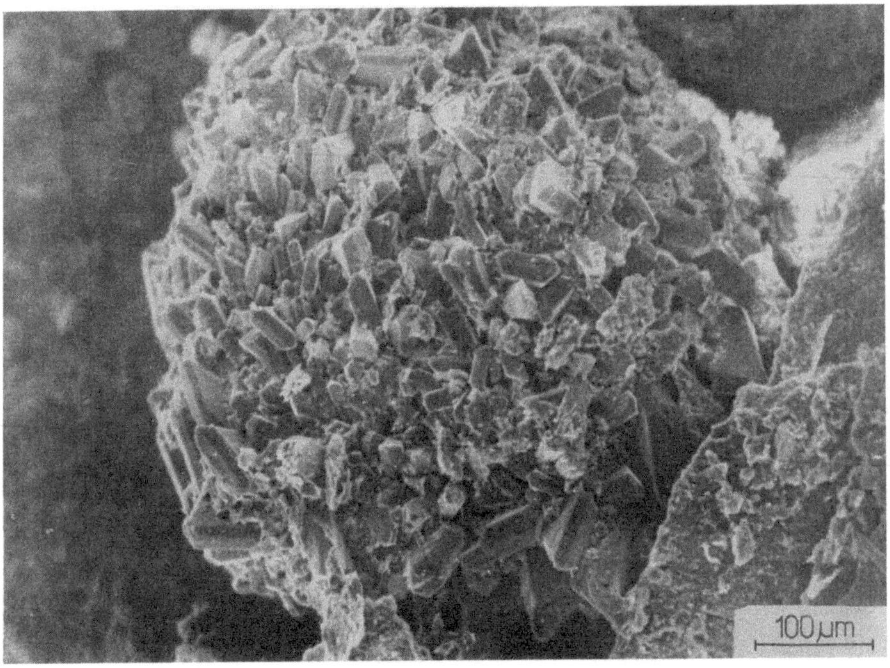

Figs. 469, 470, 471. SEMs: *Gypsum* from the surface of clay, Friedland, GDR

Fig. 470.

Fig. 471.

Fig. 472. SEM: *Gypsum* in a pore of Clay Spur Bentonite, Crook County, Wyoming, USA

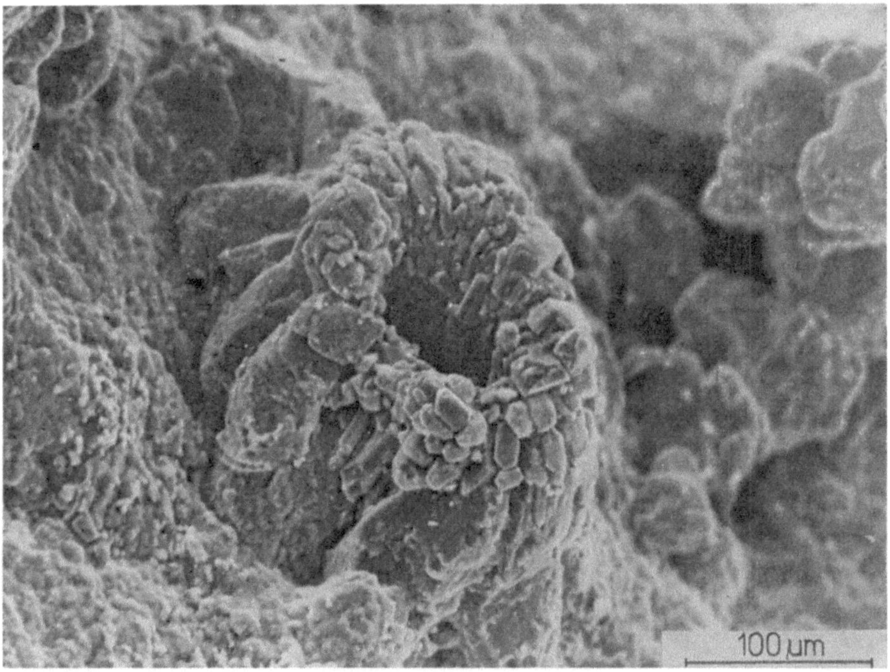

Fig. 473. SEM: *Gypsum* at the exit of a pore water channel from the surface of clay, Friedland, GDR

Fig. 474. SEM: *Gypsum* from kaolin, Fuchsberg near Halle, GDR

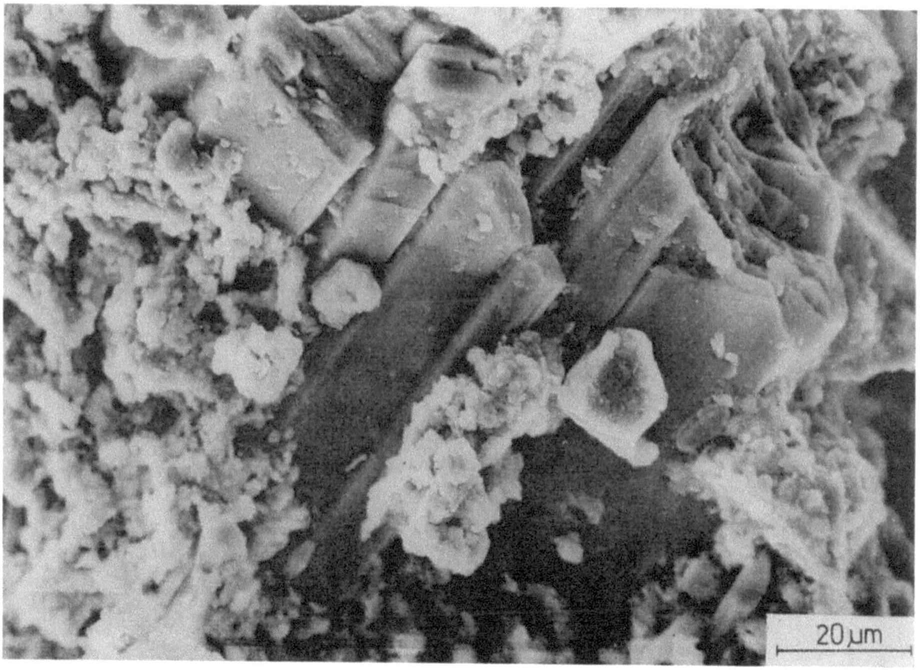

Fig. 475. SEM: *Mirabilite* from the surface of clay, Friedland, GDR

Fig. 476. SEM: *Parabutlerite* (grains) and *gypsum* from porphyry kaolin, Fuchsberg near Halle, GDR

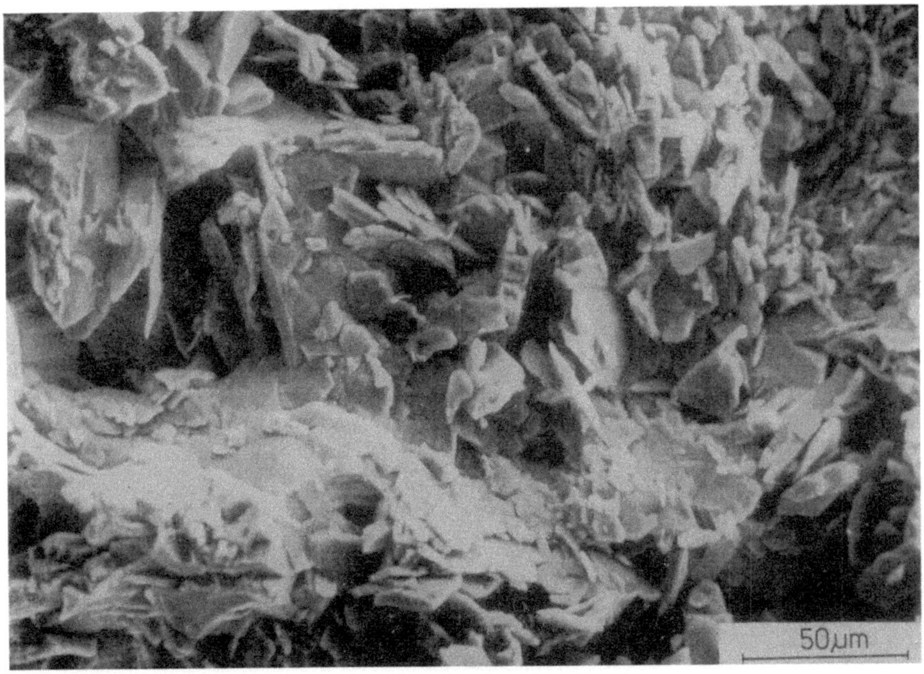

Figs. 477, 478. SEMs: *Aluminite* from a nodule of the hanging beds of porphyry kaolin, Fuchsberg near Halle, GDR

Fig. 478.

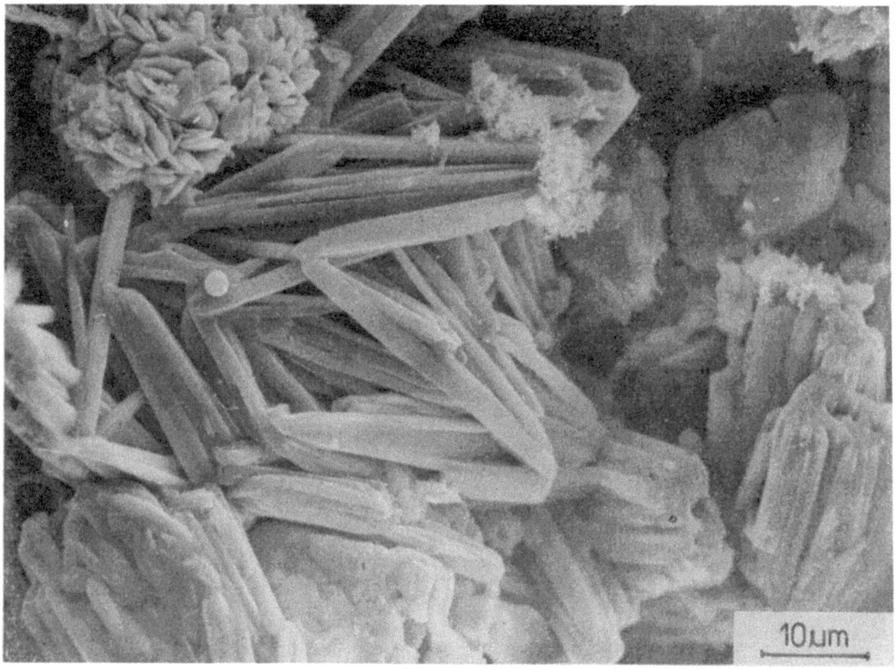

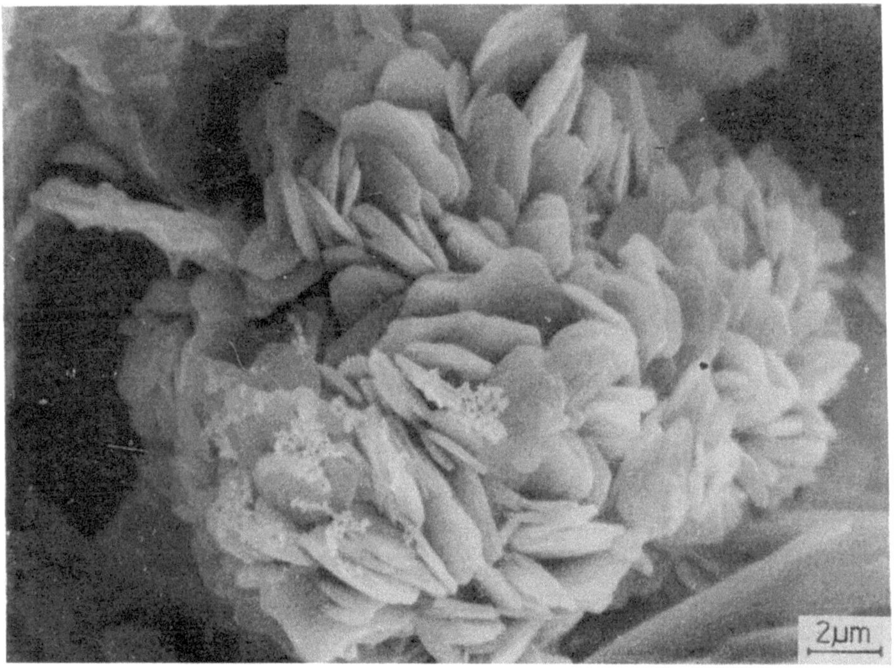

Fig. 479. SEM: *Roemerite* (needles) and greenish aggregates of *copiapite* from the surface of decomposed pyrite in clay, Friedland, GDR

Fig. 480. SEM: *Copiapite*, detail of Fig. 479

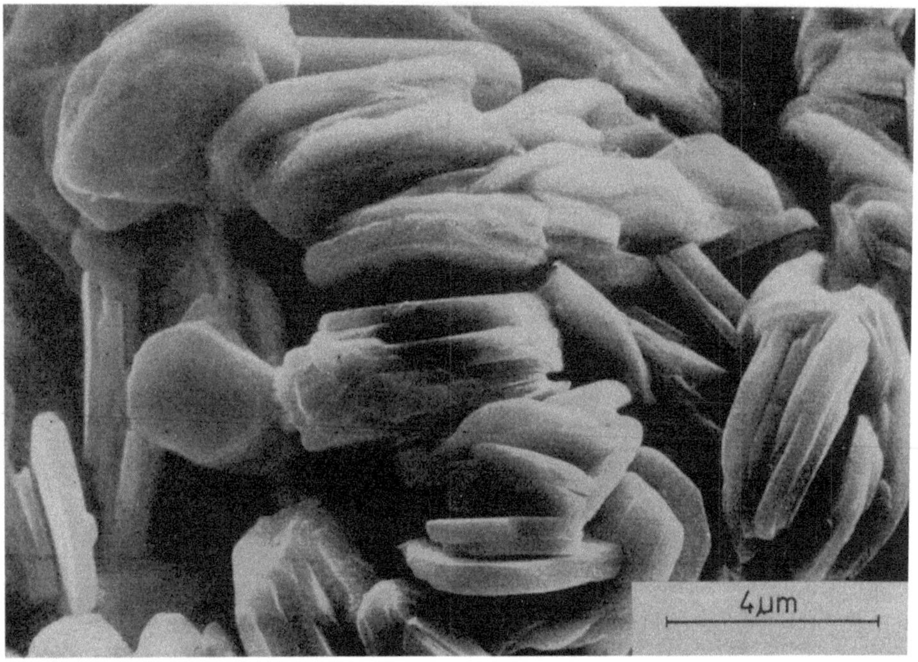

Fig. 481. SEM: *Copiapite* from a green crust on kaolin, Fuchsberg near Halle, GDR

Fig. 482. SEM: *Al-copiapite* on fissures in kaolin, Kemmlitz, GDR

Fig. 483. SEM: *Sideronatrite*, white efflorescence on the surface of kaolin, Fuchsberg near Halle, GDR

Fig. 484. SEM: *Metasideronatrite* on kaolin, Fuchsberg near Halle, GDR

Figs. 485, 486. SEMs: *Vivianite* from Tertiary clay, Profen near Halle, GDR

Fig. 486.

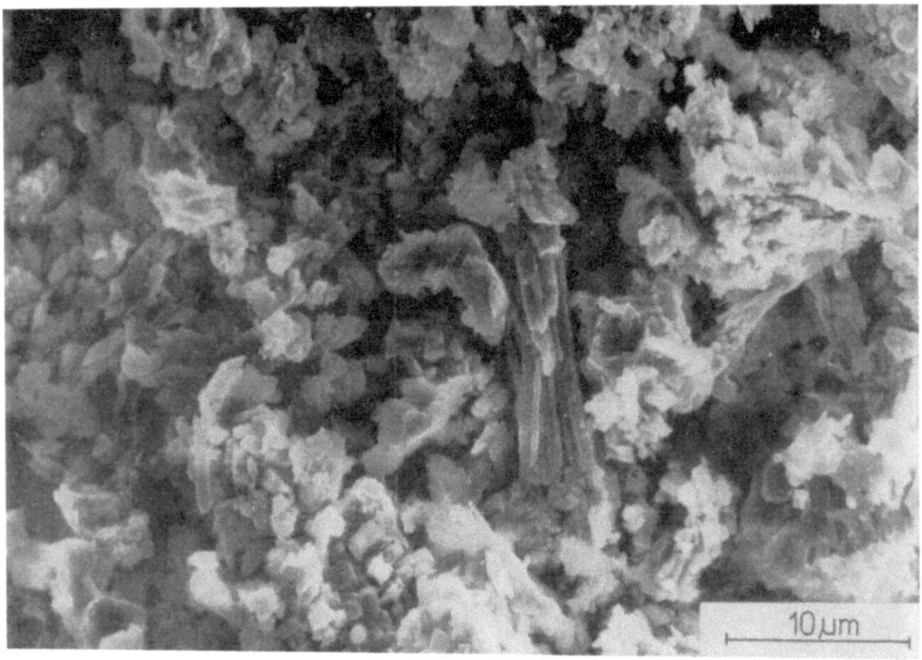

Fig. 487. SEM: *Vivianite* on wood from Quaternary clay, Wolin, Poland

Fig. 488. SEM: *Vivianite* from a joint in granodiorite kaolin, Caminau, GDR

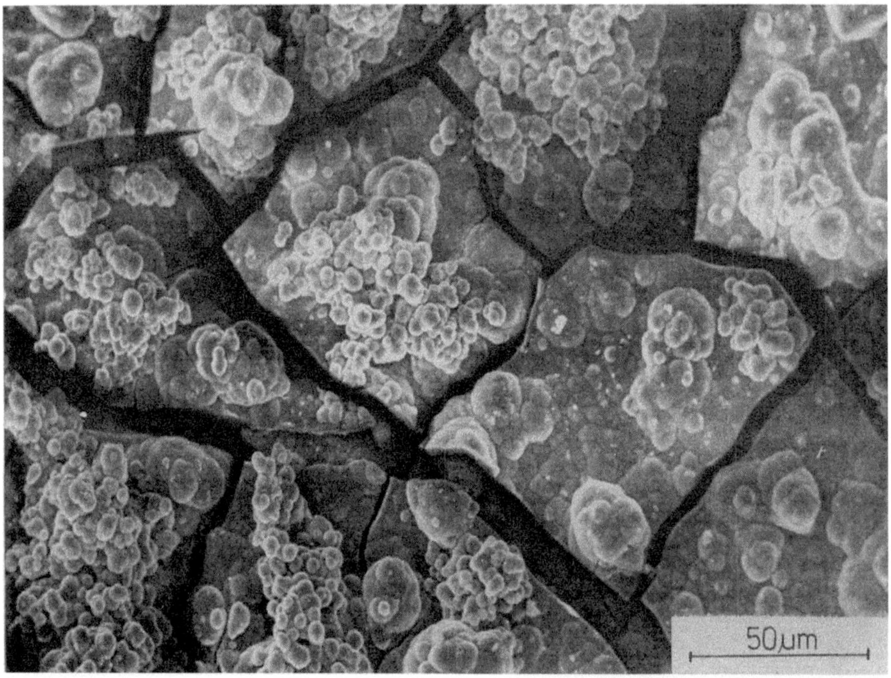

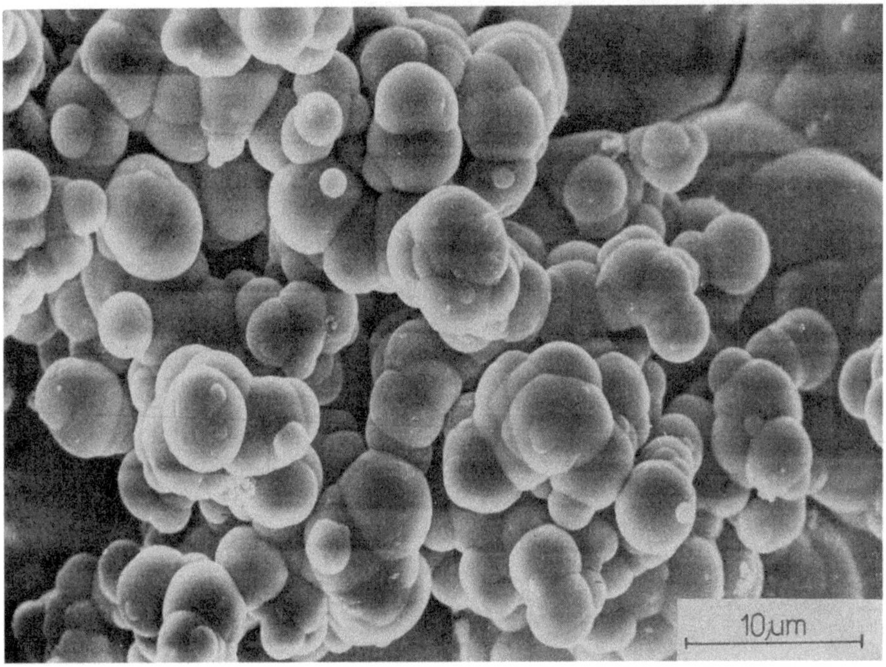

Figs. 489, 490. SEMs: Amorphous *iron phosphate* from alum schist, Gräfenthal, GDR

Fig. 490.

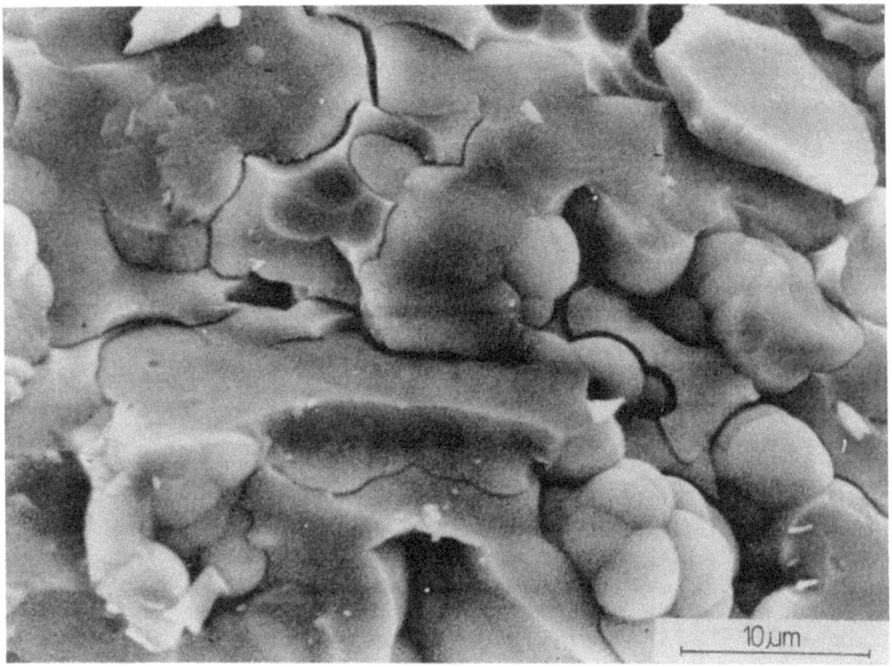

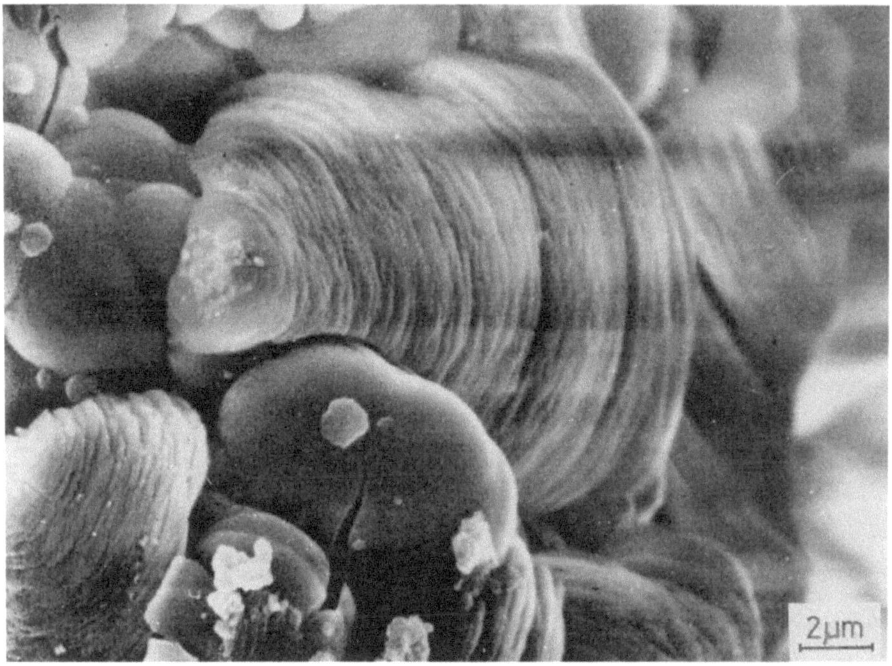

Fig. 491. SEM: *Allophane*, fabric of fracture, Leoben, Austria

Fig. 492. SEM: *Allophane* Leoben, Austria

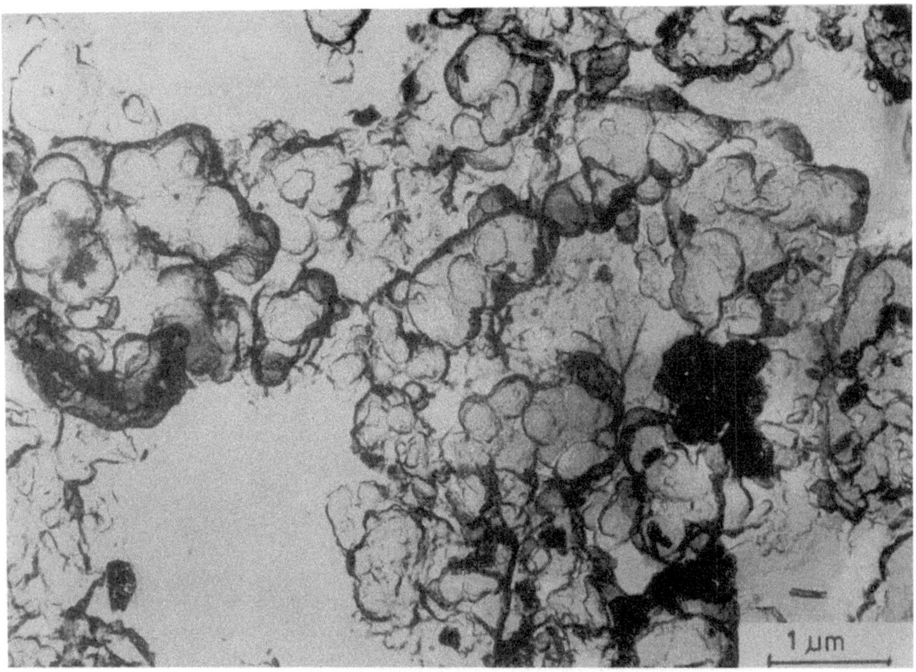

Fig. 493. Replica: *Allophane* from type locality, Gräfenthal, GDR

Fig. 494. SEM: *Allophane* from type locality, Gräfenthal, GDR

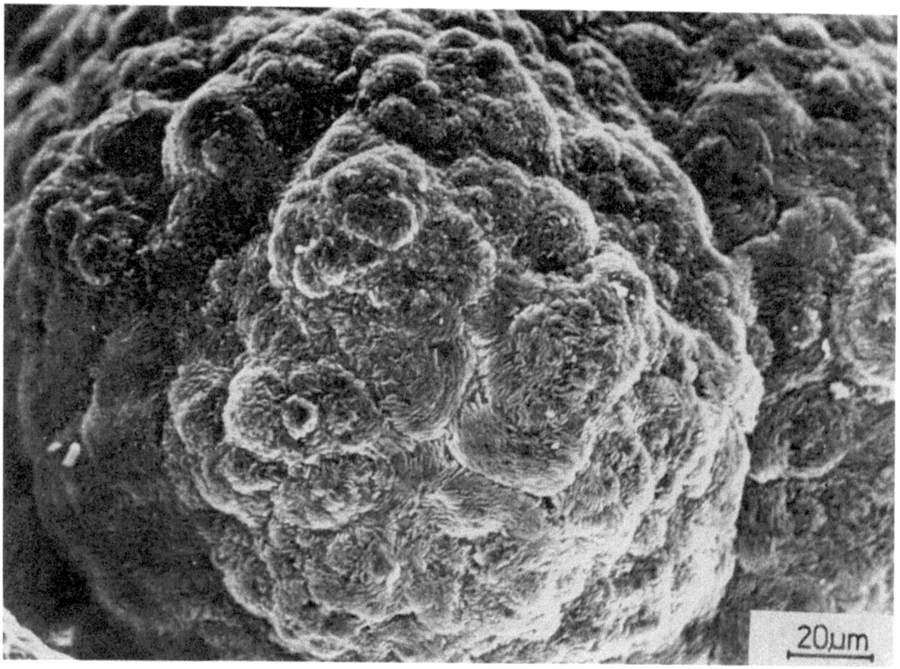

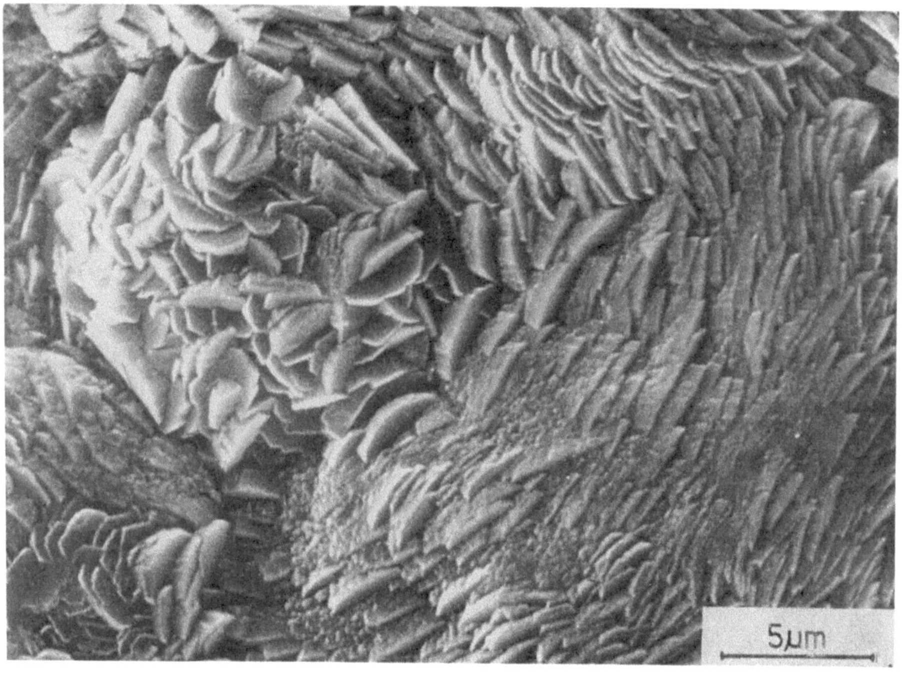

Figs. 495, 496, 497. SEMs: *Hydrargillite* from allophane, Gräfenthal, GDR

Fig. 496.

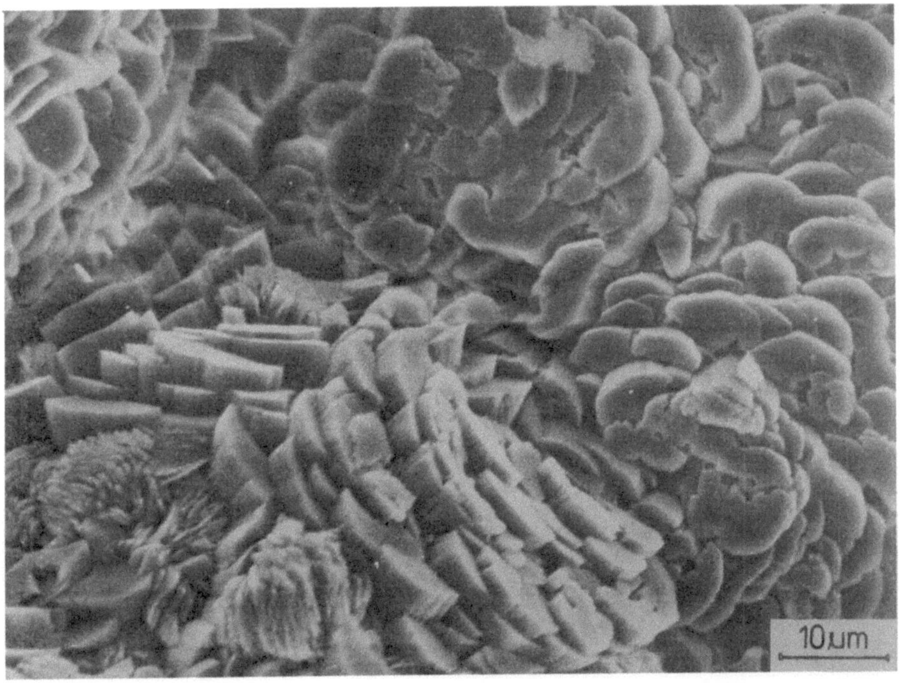

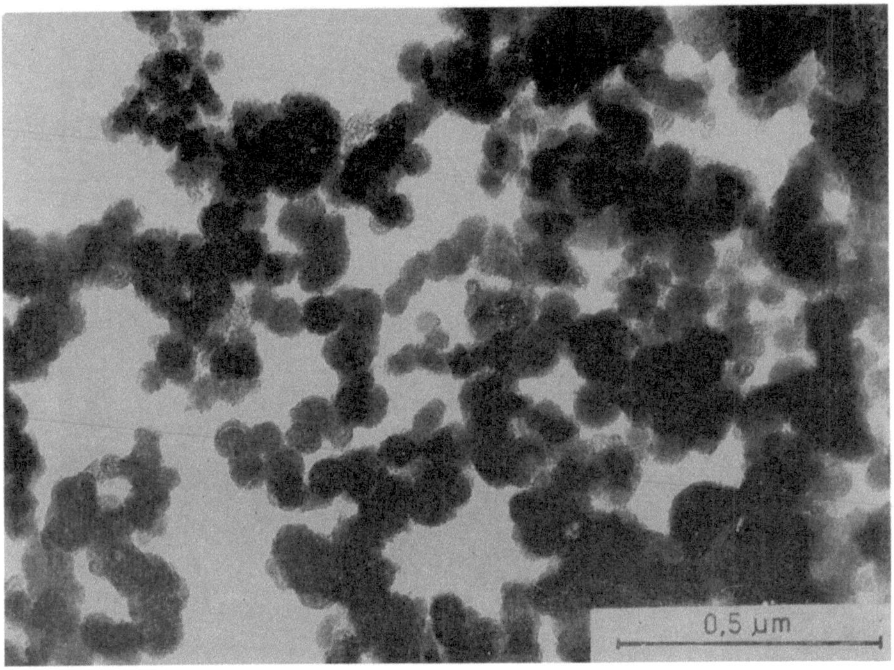

Fig. 497. (*explained at p. 289*)

Fig. 498. TEM: *Allophane*, Zlate Hory, Czechoslovakia

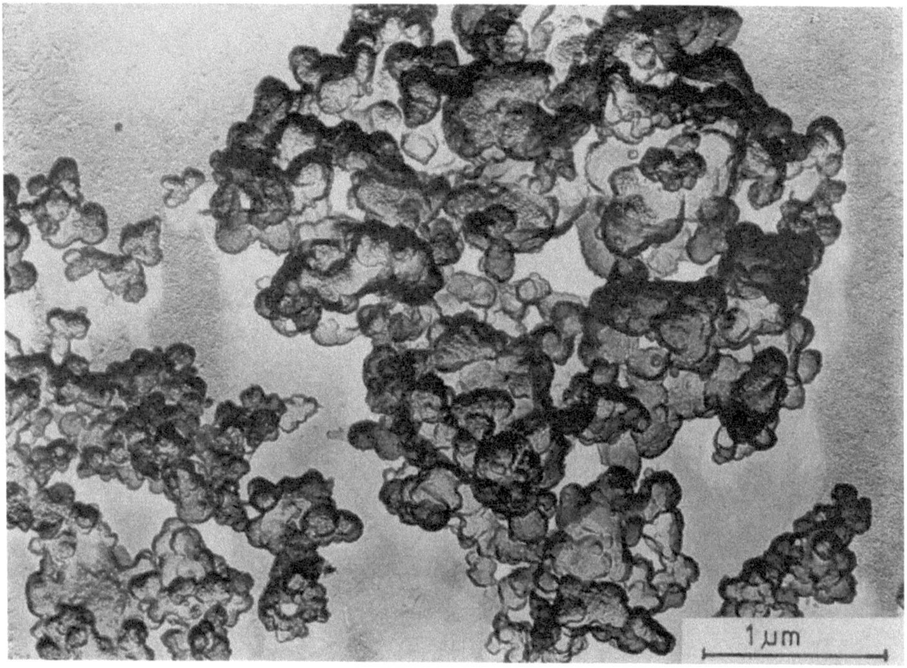

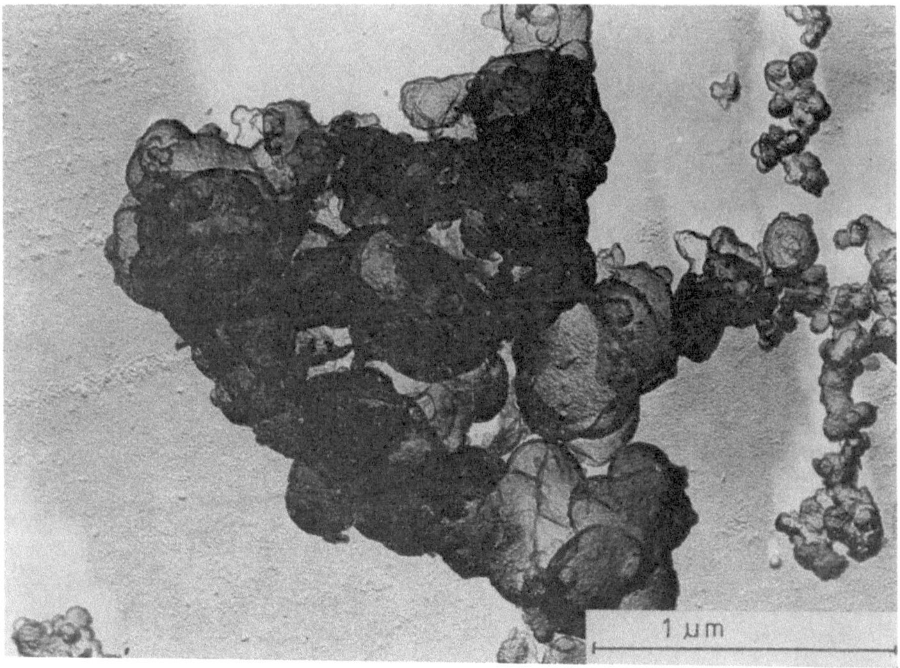

Figs. 499, 500. Replicas: *Allophane*, Zlate Hory, Czechoslovakia

Fig. 500.

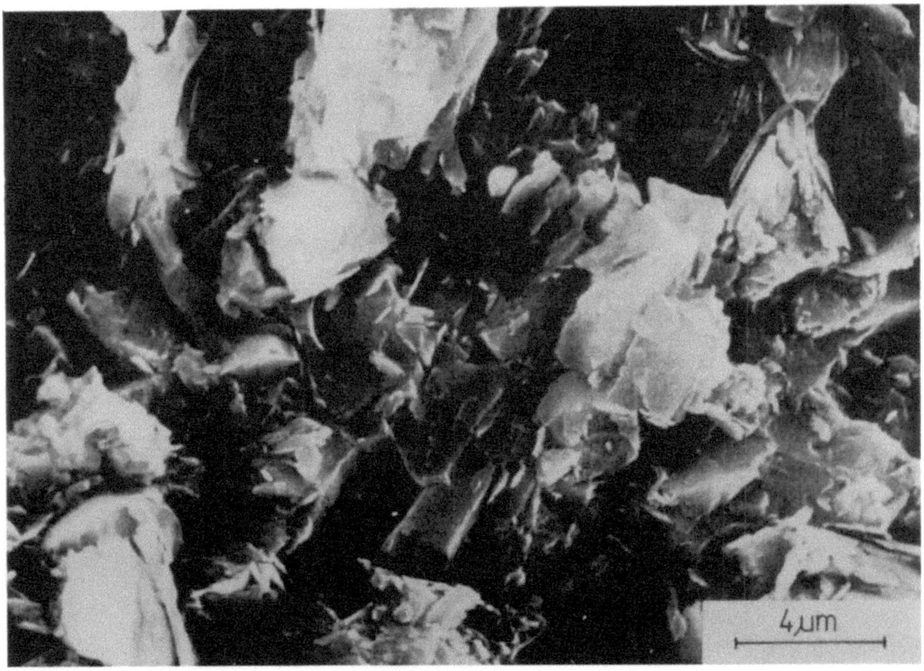

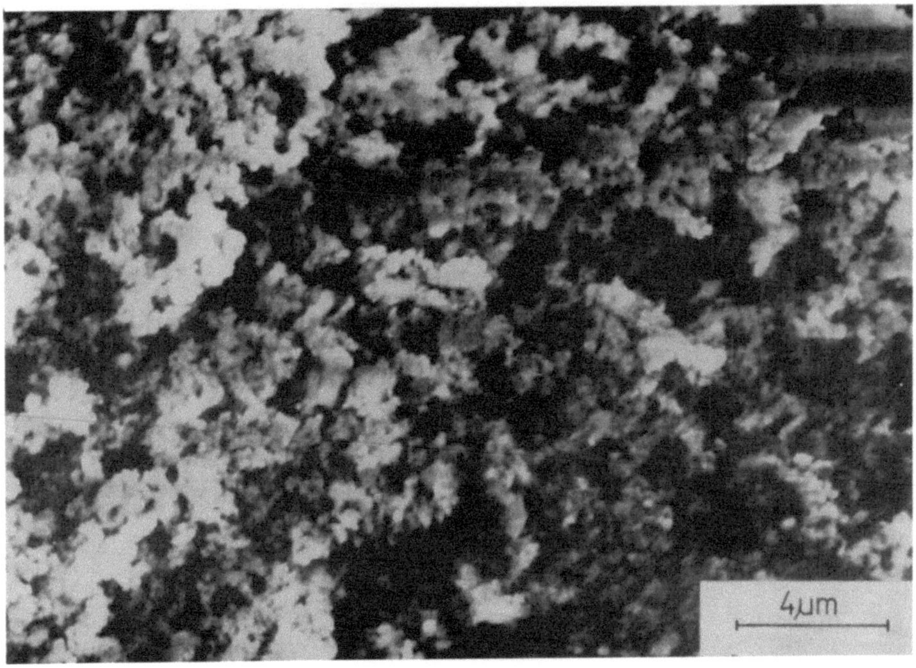

Fig. 501. SEM: *Diaspore*, Flint clay, Missouri, USA

Fig. 502. SEM: *Allophane*, Zlate Hory, Czechoslovakia

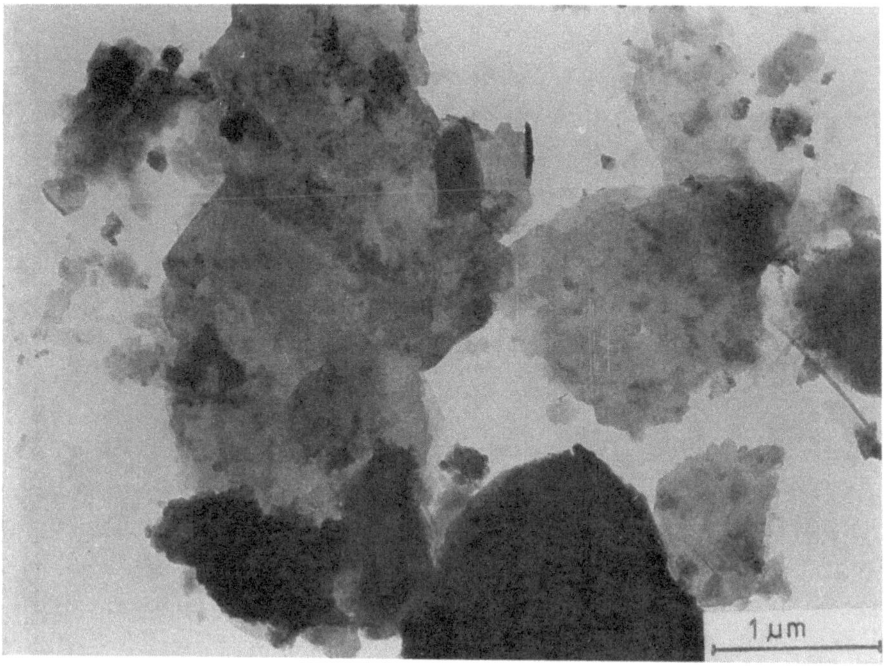

Fig. 503. TEM: *Amorphous silica* (small grains) and kaolinite (so-called "anauxite"), Bilina, Czechoslovakia

Fig. 504. Replica: *Amorphous silica* (small grains) on kaolinite (so-called "anauxite"), Bilina, Czechoslovakia

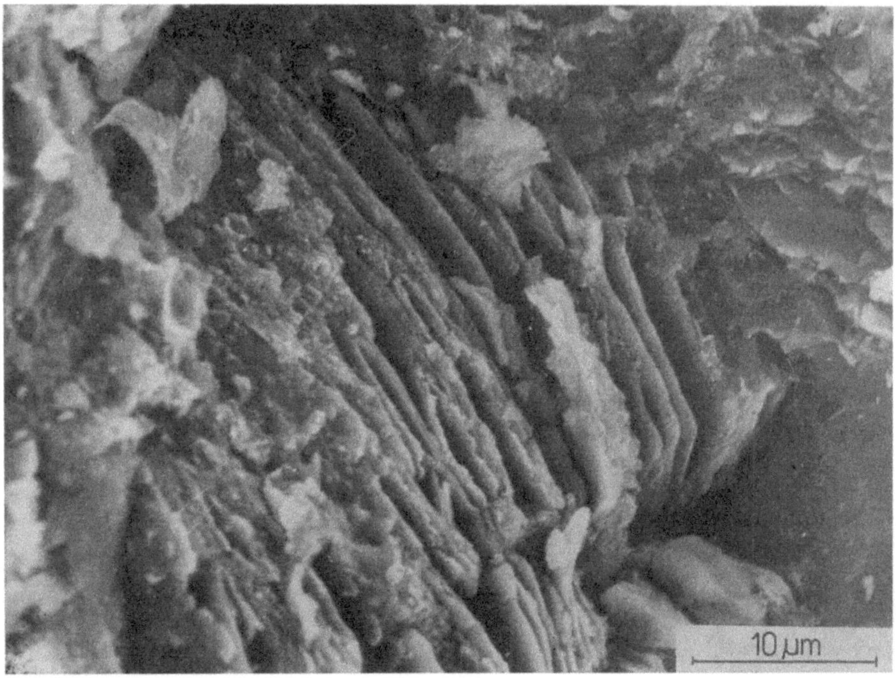

Fig. 505. SEM: *Amorphous silica* (small grains) on *kaolinite* (so-called "anauxite"), Bilina, Czechoslovakia

Fig. 506. SEM: *Silica lepispheres* from so-called "anauxite", Bilina, Czechoslovakia

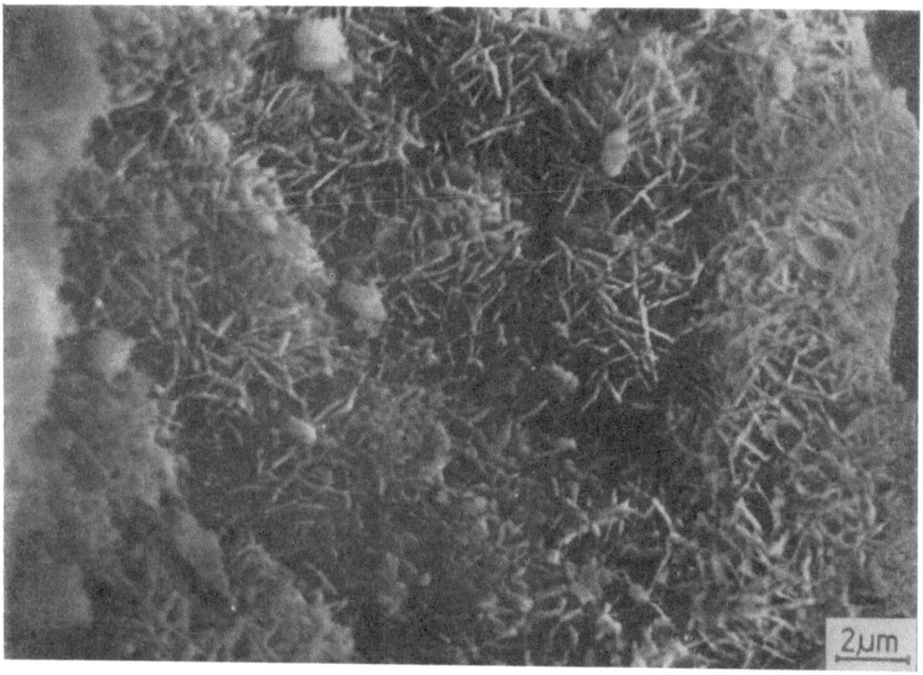

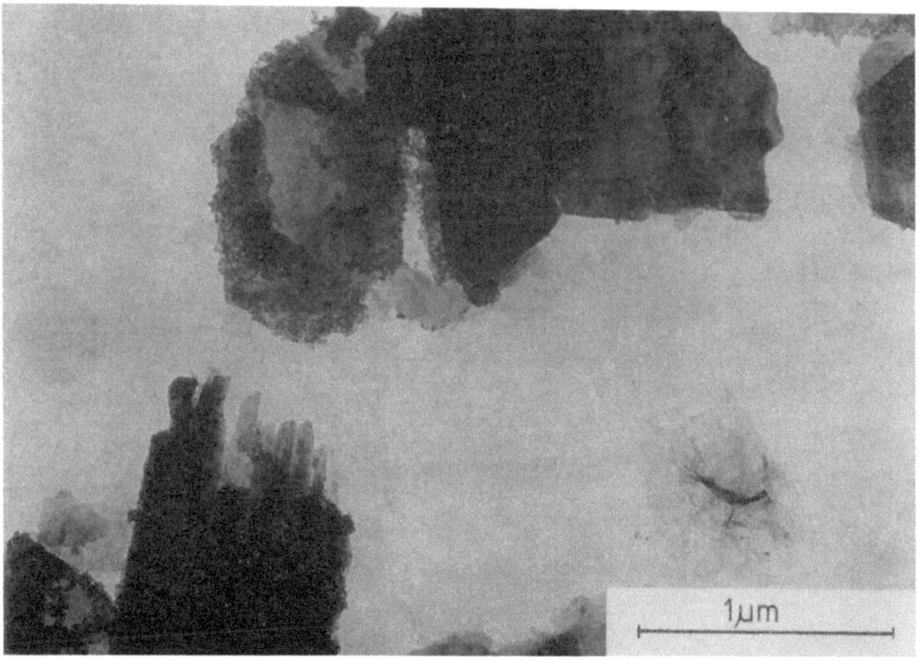

Fig. 507. SEM: *Silica lepispheres*, detail of Fig. 506

Fig. 508. TEM: *Amorphous silica* coating clay minerals from kaolin, Ledče, Czechoslovakia

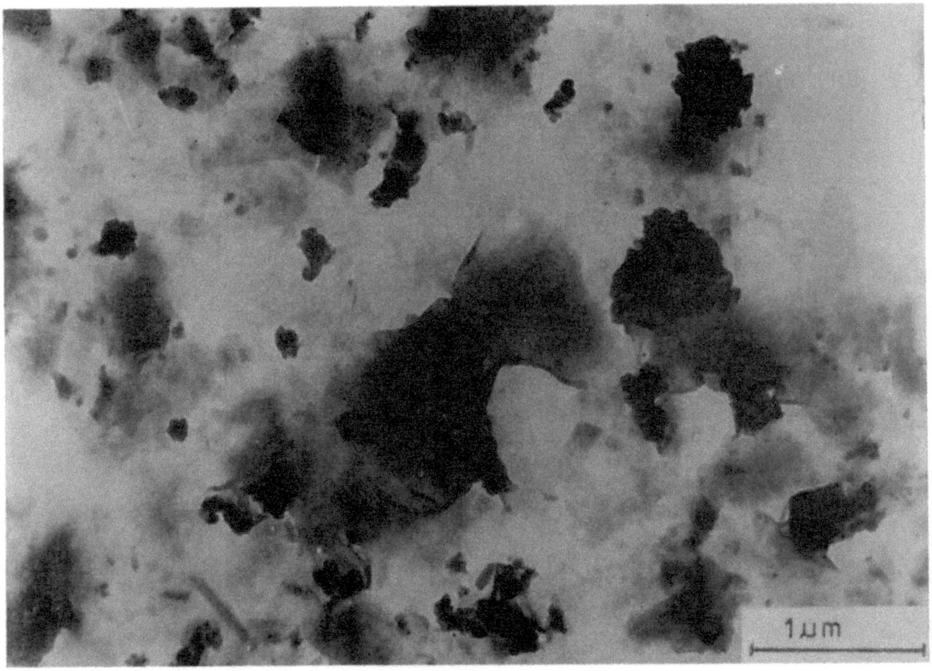

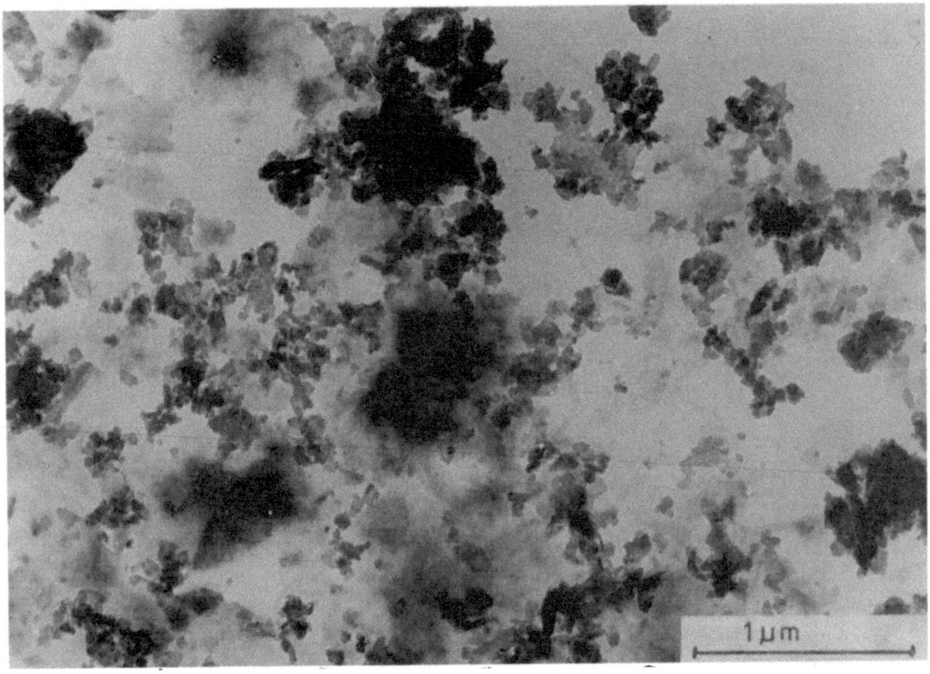

Fig. 509. TEM: *Opal-CT* and *montmorillonite* from Upper Cretaceons chalk, Rügen, GDR

Fig. 510. TEM: *Opal-CT* (small grains) and *muscovite-montmorillonite mixed-layer mineral* from Upper Cretaceons chalk, Rügen, GDR

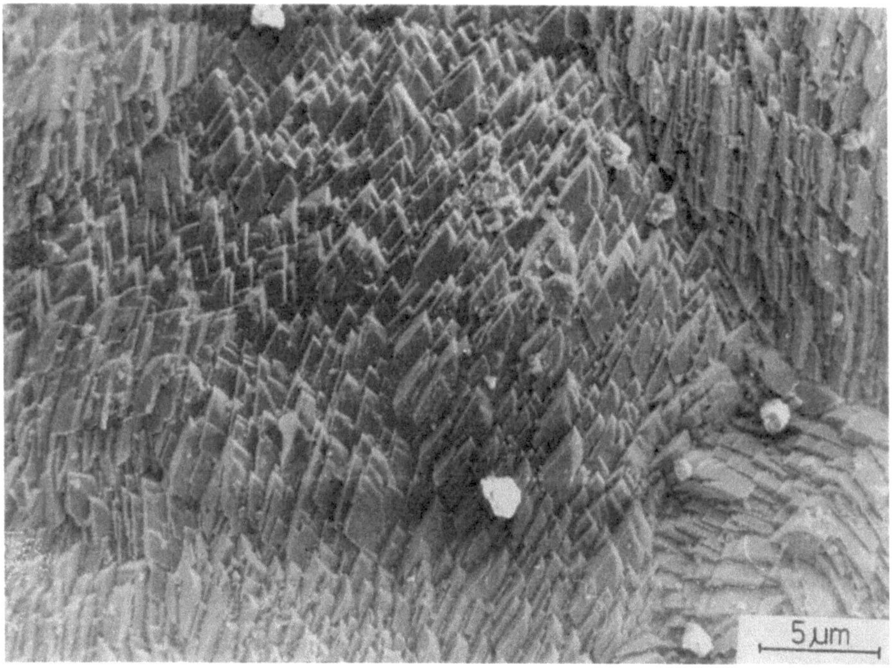

Fig. 511. Replica: *Calcite*, etched fracture from Zechstein, Thuringia, GDR

Fig. 512. SEM: *Calcite*, fissure mineral in a concretion from clay, Friedland, GDR

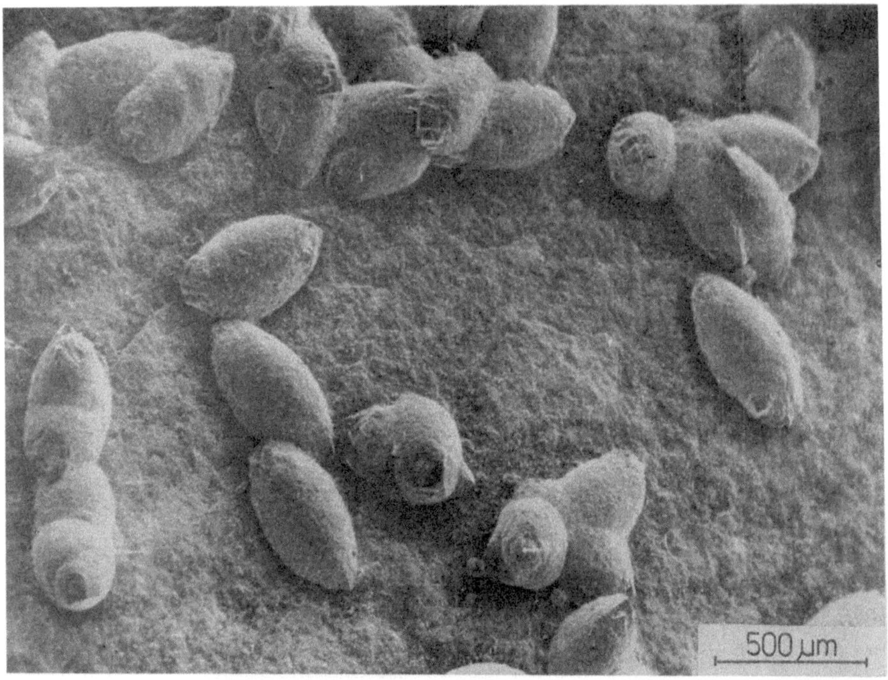

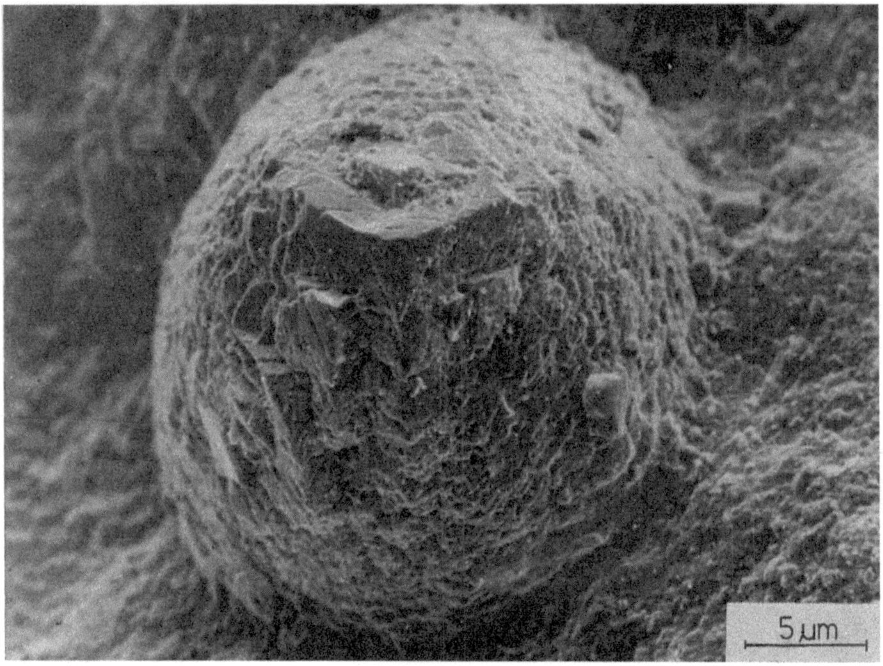

Fig. 513. SEM: *Calcite* aggregates on joints in a phosphate concretion from clay, Friedland, GDR

Fig. 514. SEM: *Calcite*, detail of Fig. 513

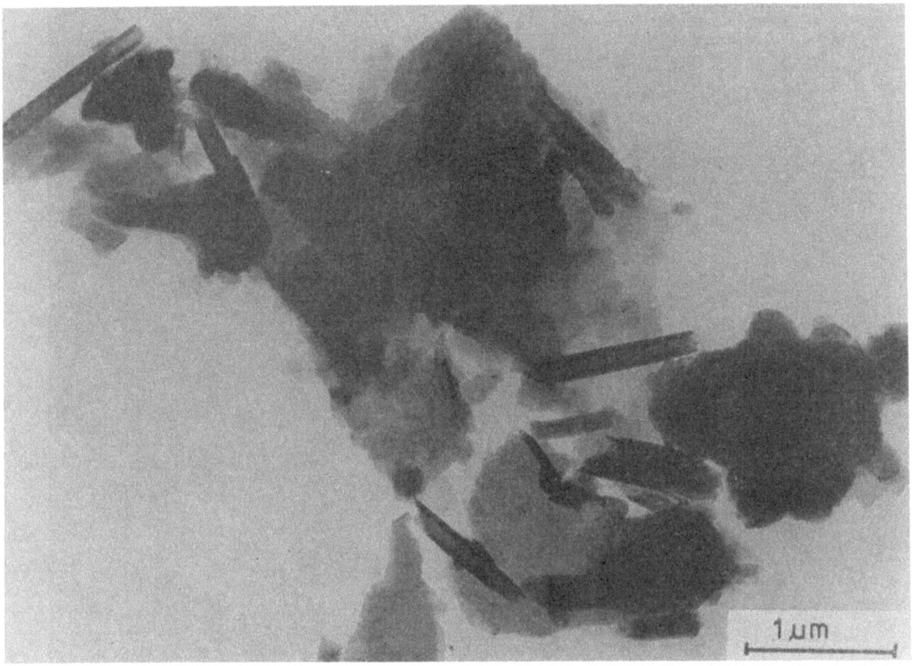

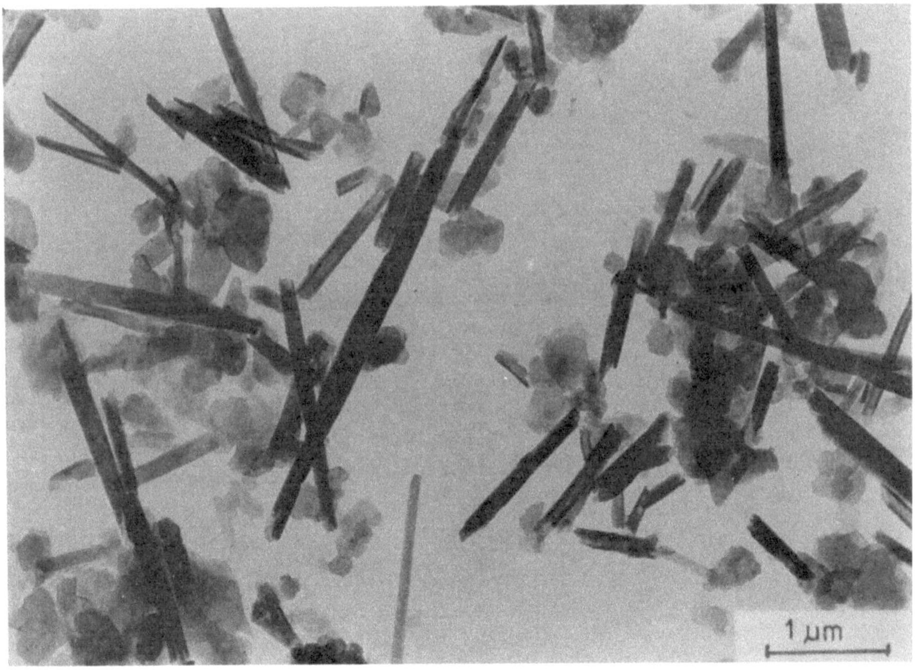

Figs. 515, 516. TEMs: *Kaolinite* and *halloysite* from porphyry kaolin, deposit "Frieden", Kemmlitz, GDR

Fig. 516.

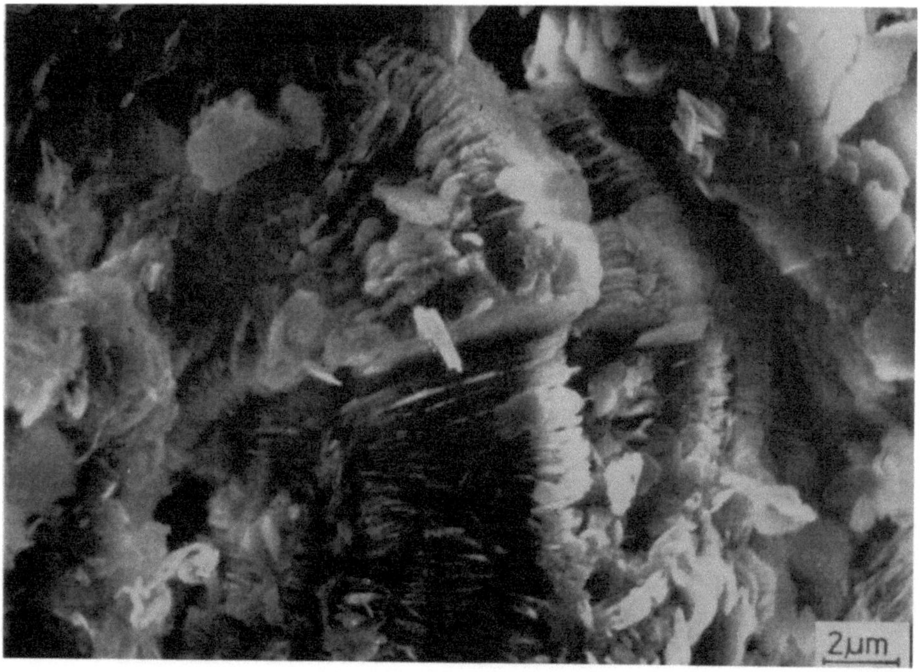

Figs. 517, 518. SEMs: *Porphyry kaolin*, deposit "Frieden", Kemmlitz, GDR

Fig. 518.

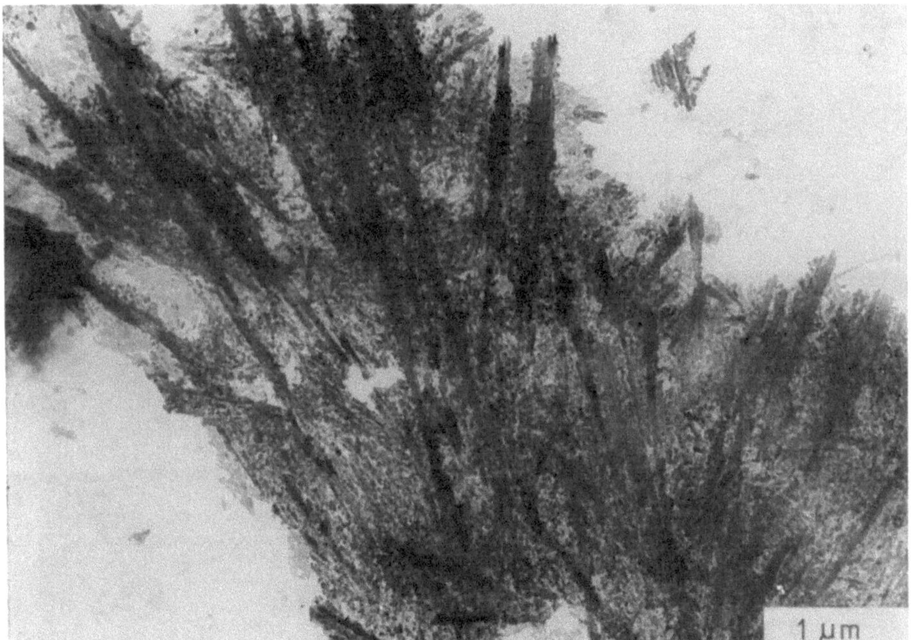

Fig. 519. TEM: *Quartze* and *kaolinite* from kaolin, Mandelowo, Mongolia

Fig. 520. TEM: *Crystobalite*, Callenberg, GDR

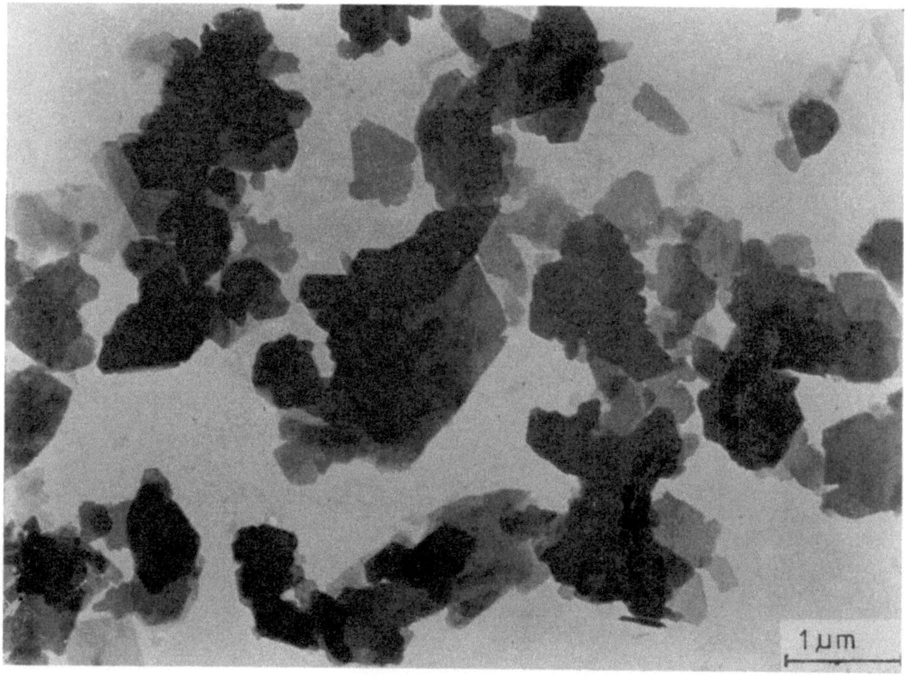

Fig. 521. TEM: Porphyry kaolin, *kaolinite* with some *mixed-layer slats*, Fuchsberg near Halle-GDR

Fig. 522. SEM: Porphyry kaolin, *kaolinite* with some *mixed-layer slats*, Fuchsberg near Halle-GDR

Fig. 523. SEM: Porphyry kaolin, *quartz* and *kaolinite*, Fuchsberg near Halle, GDR

Fig. 524. SEM: Sandstone kaolin, *kaolinite*, Spergau near Halle, GDR

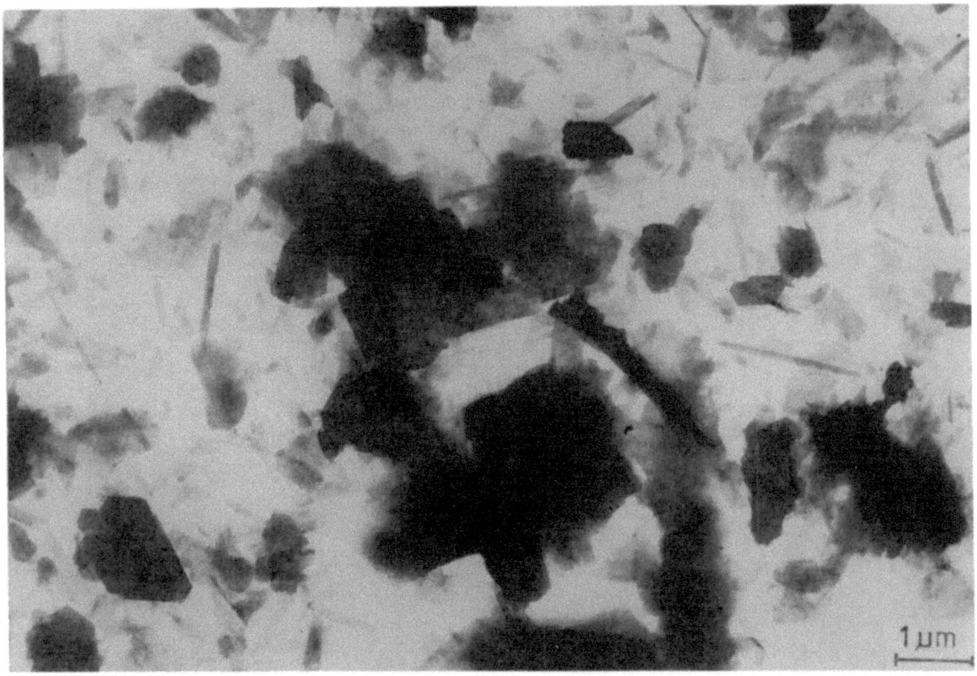

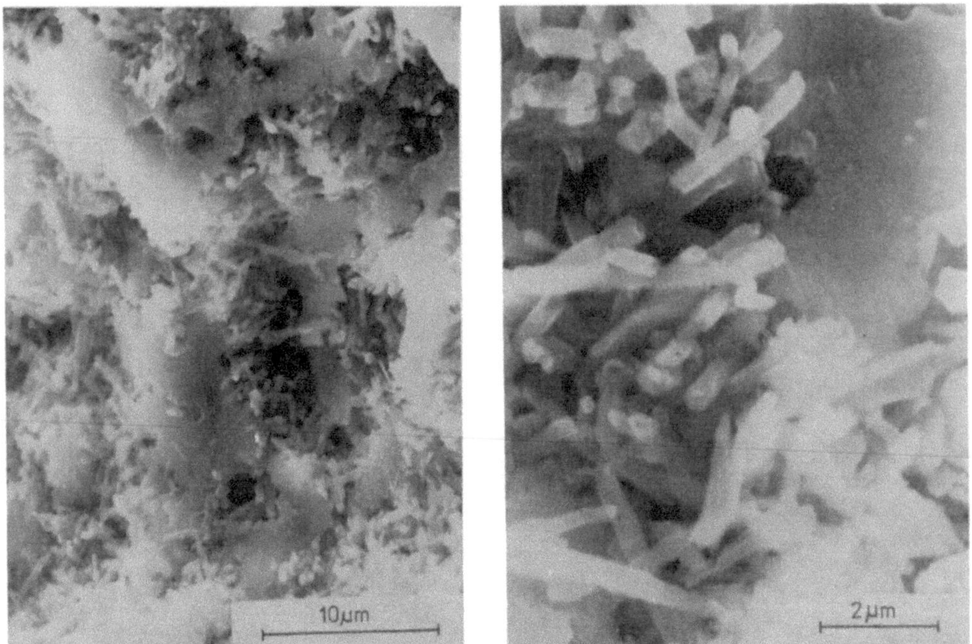

Fig. 525. TEM: *Muscovite-montmorillonite mixed-layer* and *kaolinite* from altered pitchstone, Garsebach near Meißen, GDR

Fig. 526. SEMs: (a) alteration of pitchstone to *mixed-layer*, (b) detail of (a), Garsebach near Meißen, GDR

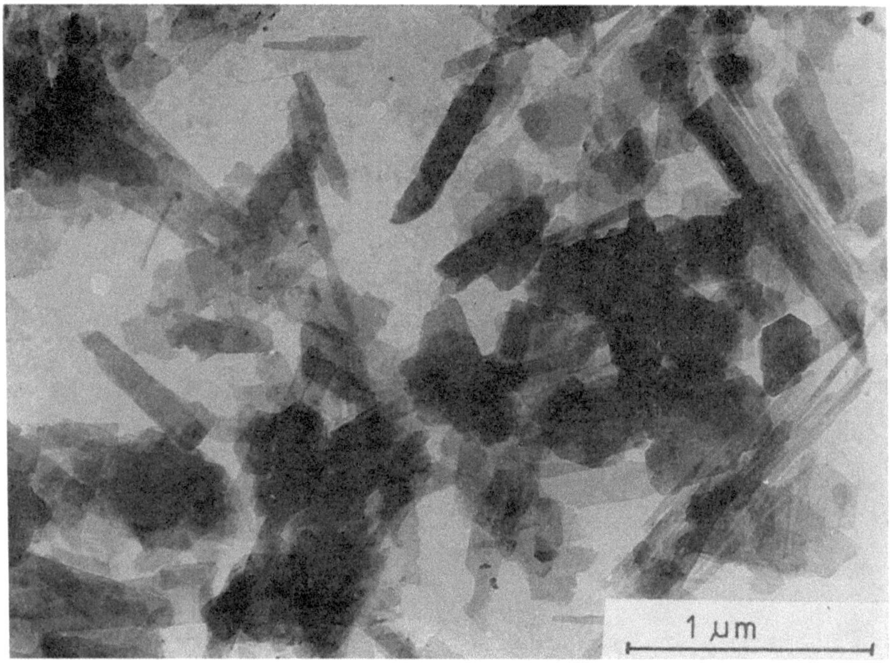

Fig. 527. TEM: Ignimbrite kaolin, *kaolinite* and *mixed-layer*, deposit "Glückauf", Kemmlitz GDR

Fig. 528. SEM: Ignimbrite kaolin, *mixed-layer* in a pore, deposit "Glückauf", Kemmlitz, GDR

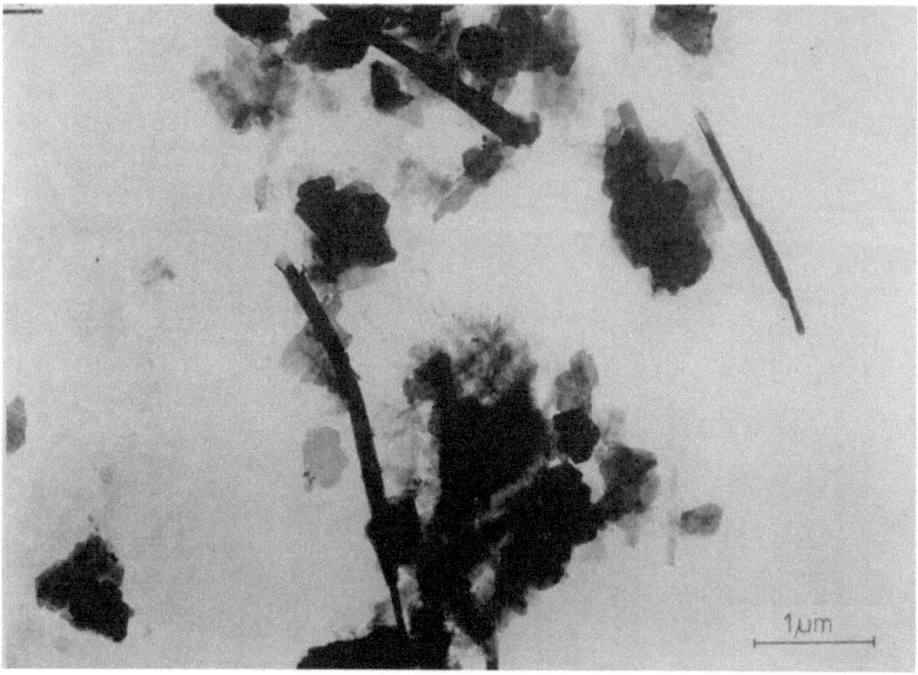

Fig. 529. TEM: Porphyry kaolin, *kaolinite, halloysite* and *mixed-layer*, deposit "Fortschritt", Kemmlitz, GDR

Fig. 530. SEM: Porphyry kaolin, *kaolinite*, deposit "Fortschritt", Kemmlitz, GDR

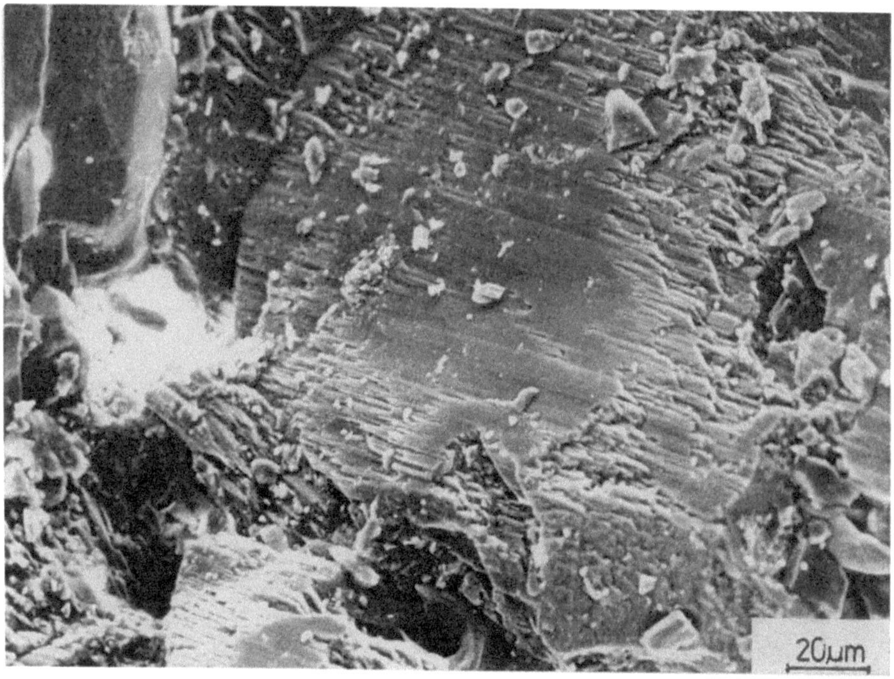

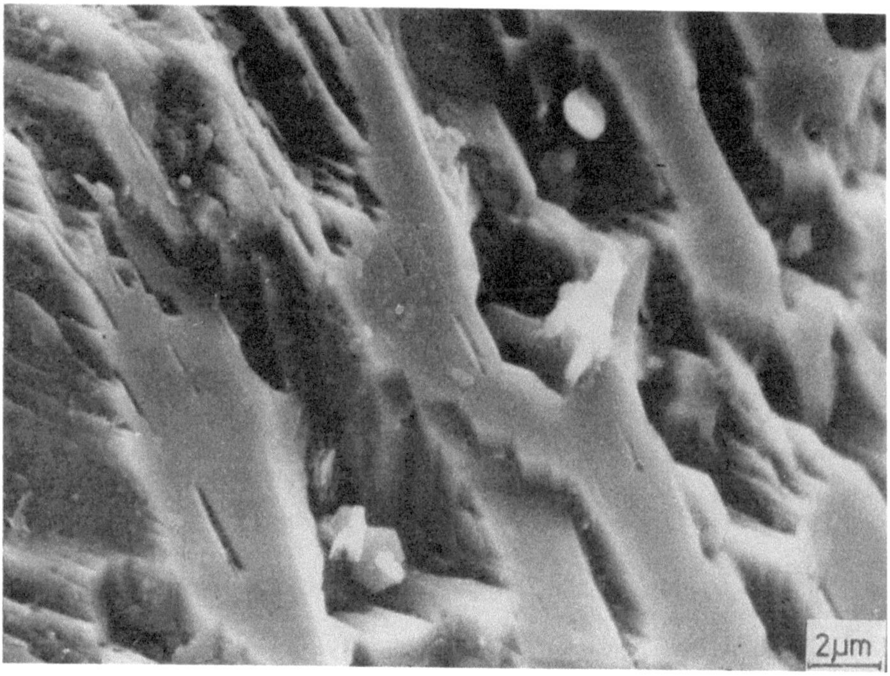

Figs. 531, 532. SEMs: *Feldspar* in granite kaolin, Bornholm, Denmark

Fig. 532.

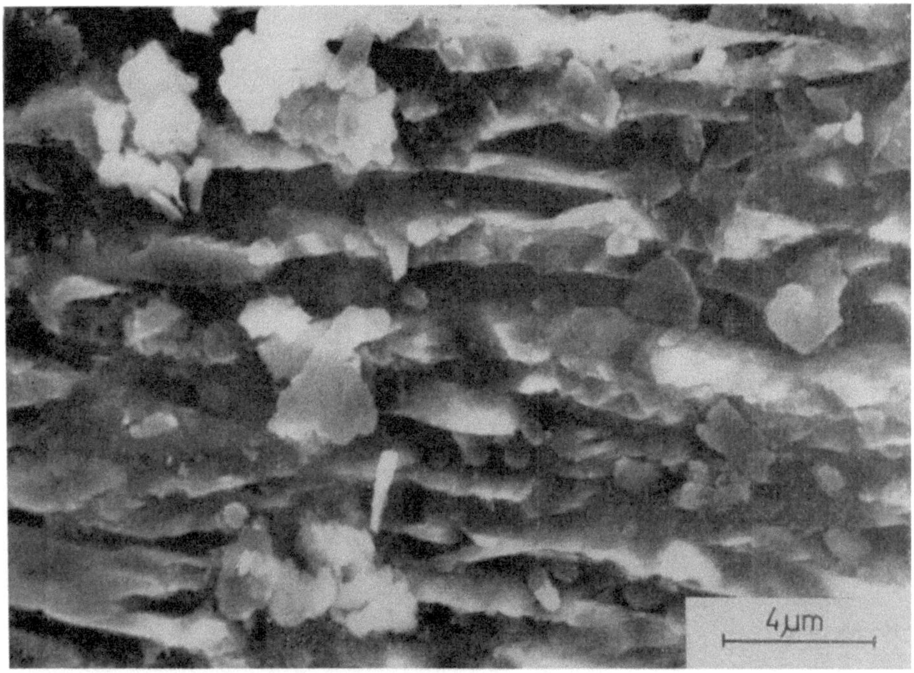

Fig. 533. SEM: Partly decomposed *feldspar* and *kaolinite* from granite kaolin, Bornholm, Denmark

Fig. 534. SEM: Partly decomposed *feldspar* and *halloysite* from granite kaolin, Bornholm, Denmark

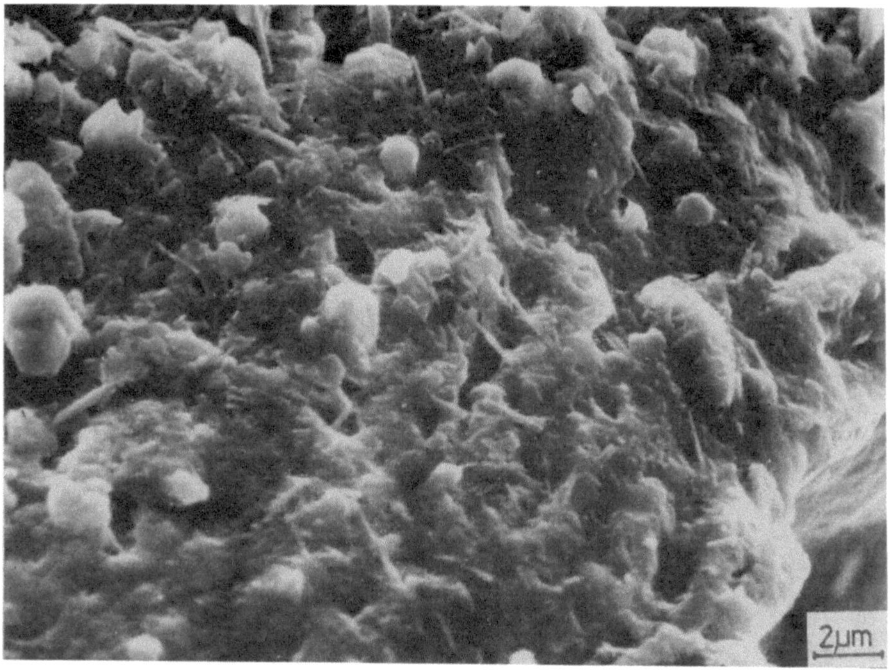

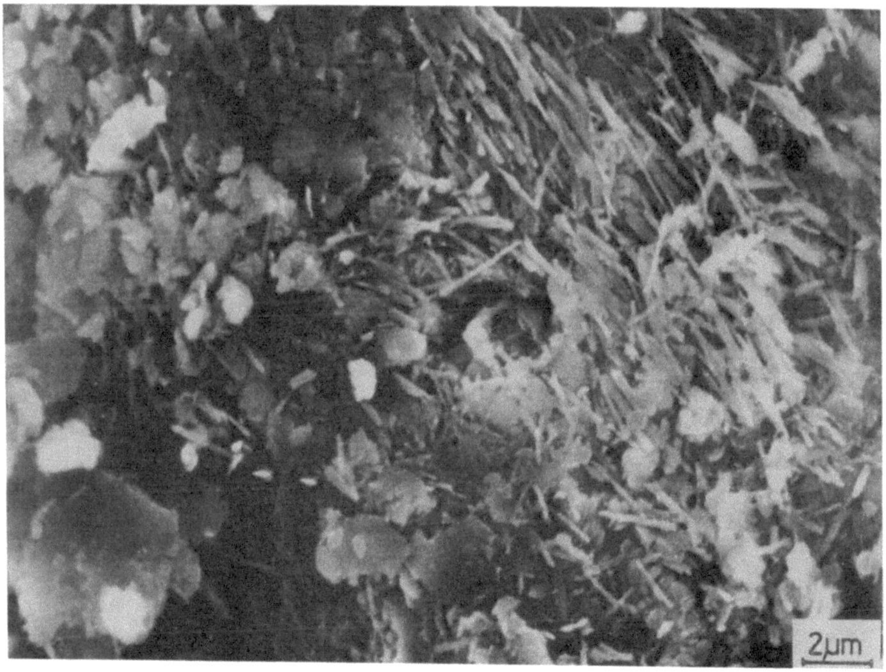

Fig. 535. SEM: Partly decomposed *feldspar* with *kaolinite* and *halloysite* from granite kaolin, Bornholm, Denmark

Fig. 536. SEM: Decomposed *feldspar*, *kaolinite* and *halloysite* from granite kaolin, Bornholm, Denmark

Fig. 537. SEM: *Halloysite* from granite kaolin, same specimen as Fig. 536, Bornholm, Denmark

Fig. 538. SEM: *Kaolinite* from granite kaolin, same specimen as Fig. 536, Bornholm, Denmark

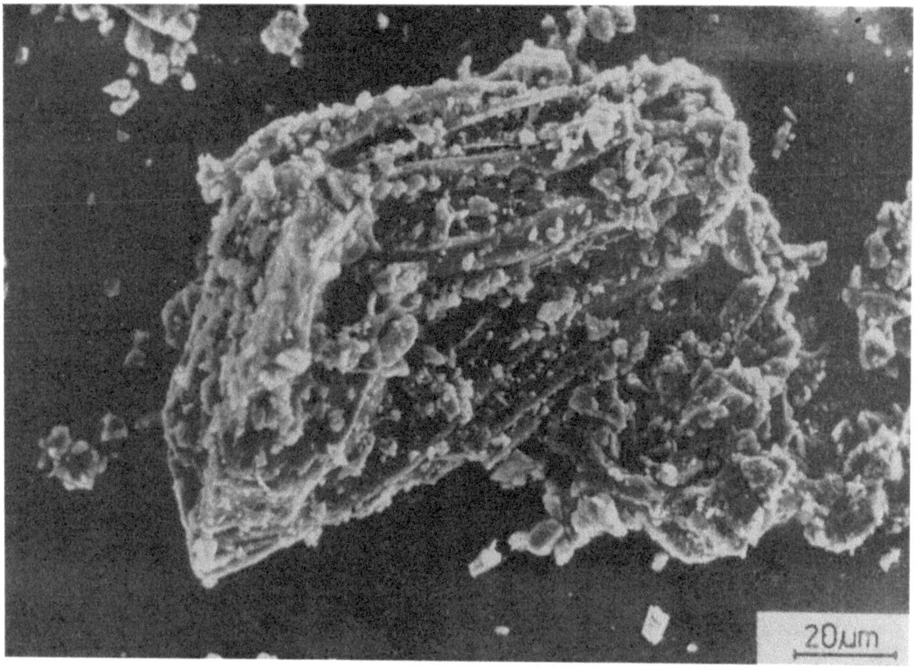

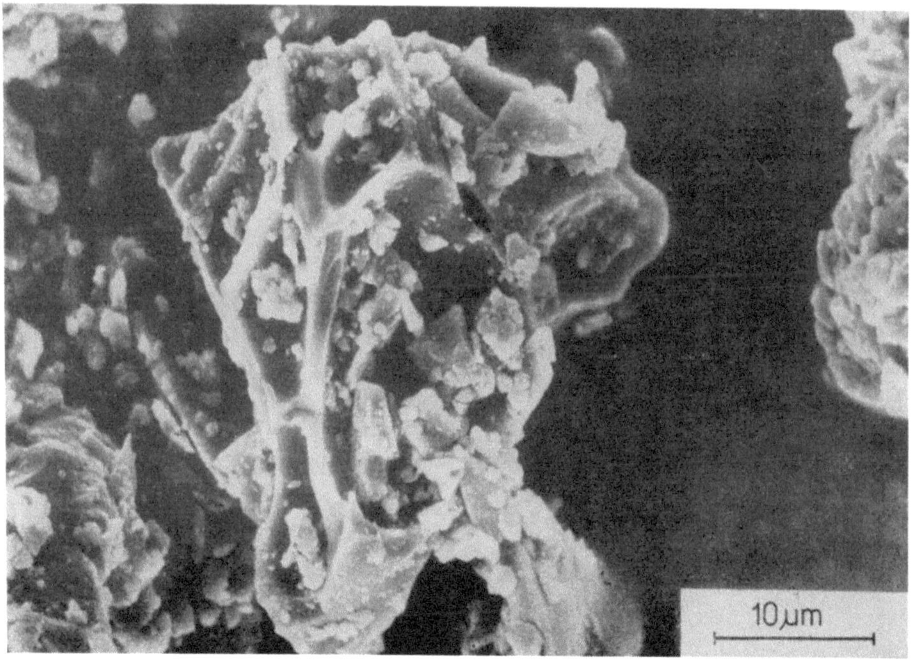

Figs. 539, 540. SEMs: *Volcanic ash* from Mt. St. Helen's eruption 1981 (sample from JOHN RUPP, Bloomington, USA)

Fig. 540.

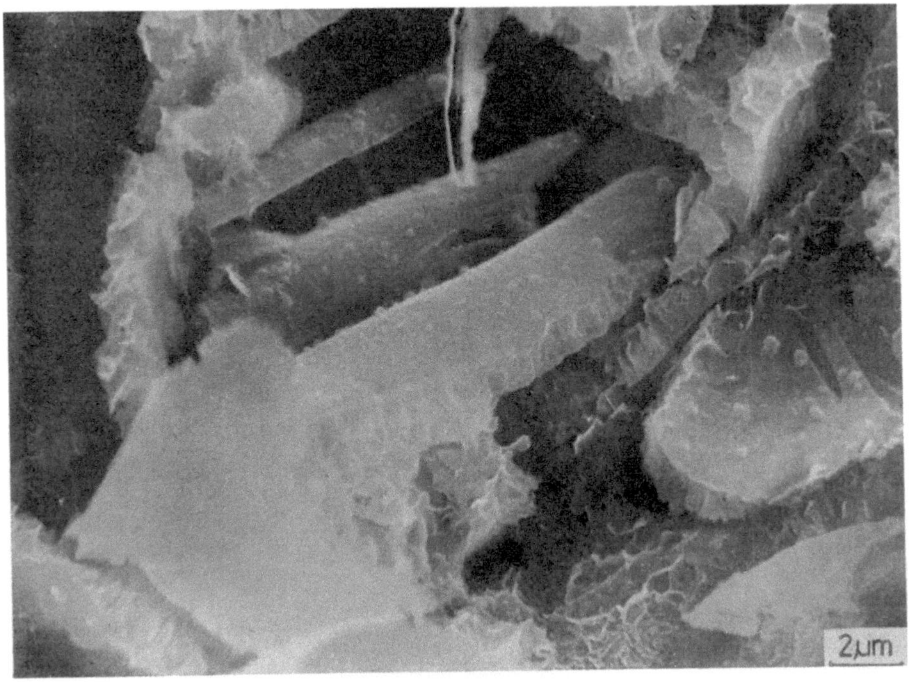

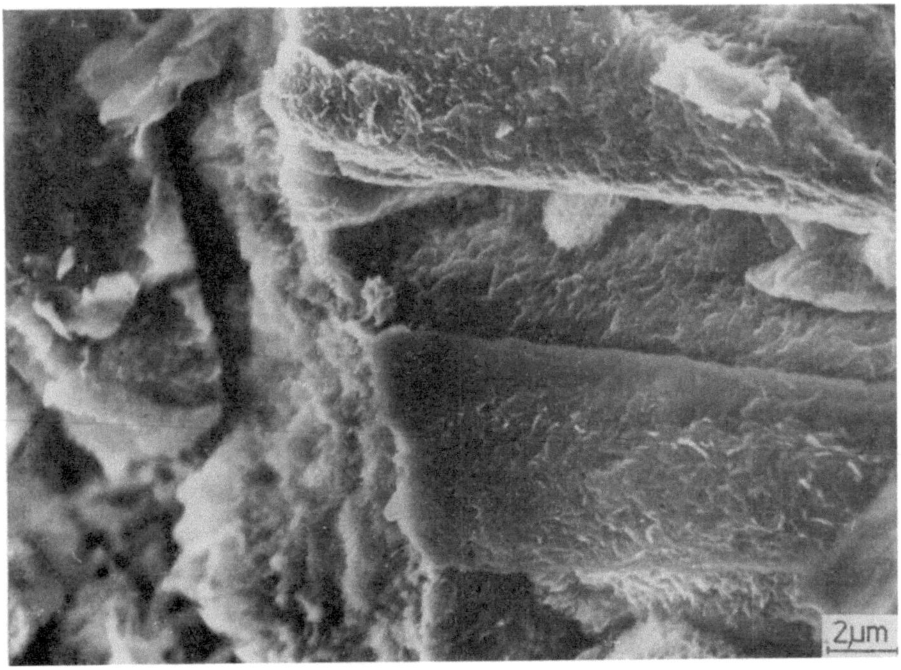

Fig. 541. SEM: *Bentonite*, so-called "hard plate" Moosburg, BRD

Figs. 542, 543, 544. SEMs: *Bentonite*, Mukalla, People's Dem. Rep. of Yemen

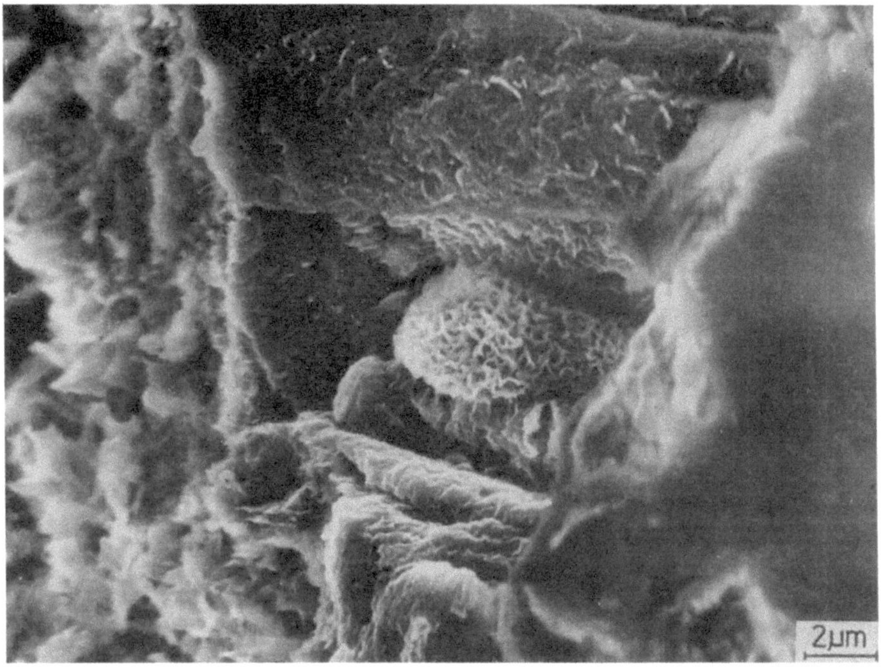

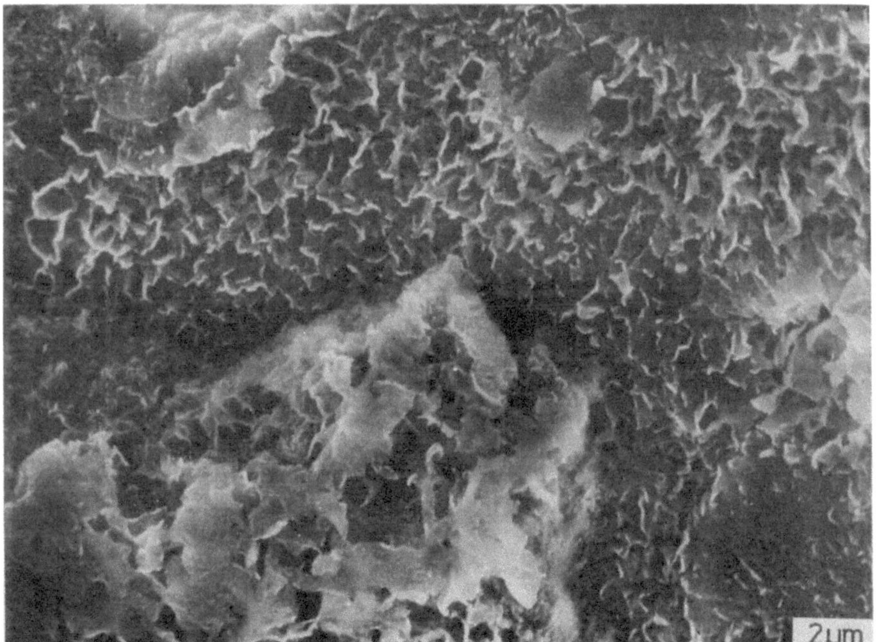

Fig. 543.

Fig. 544.

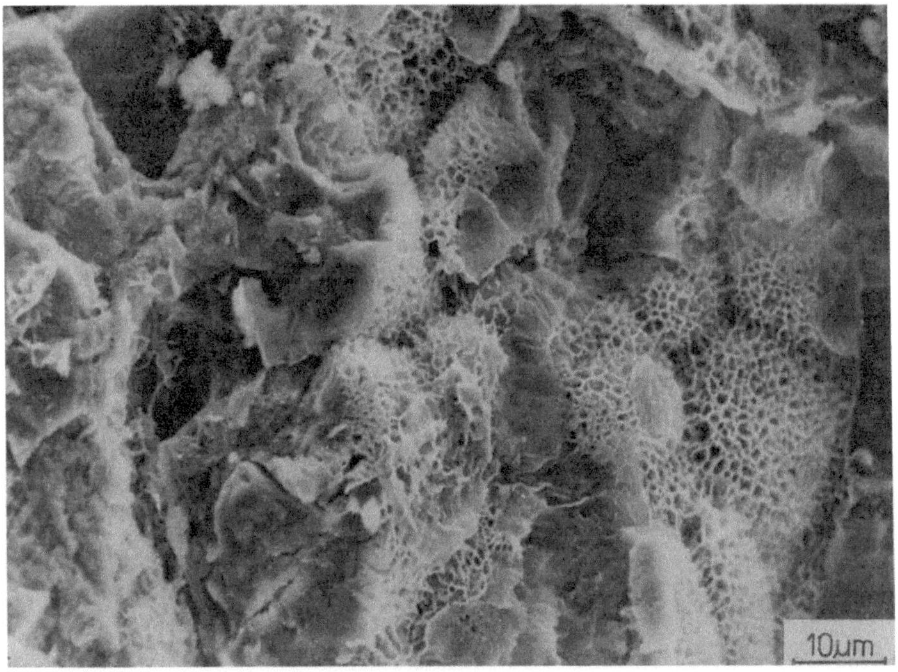

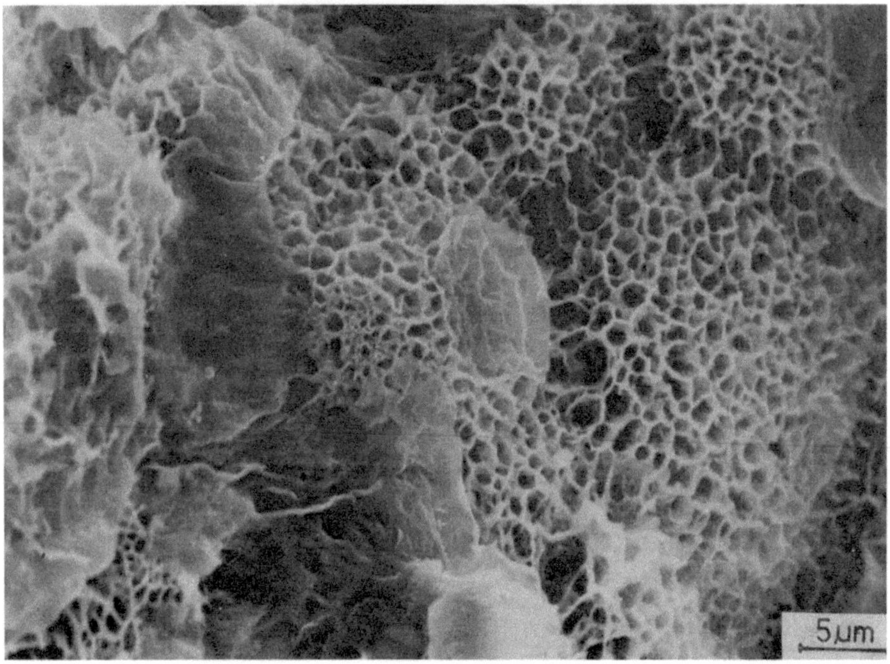

Figs. 545, 546, 547, 548. SEMs: *Bentonite* (montmorillonite and relics of volcanic glass) from Upper Pliocene Bakhtaire Conglomerate, Qhara-tapa, Iraq

Fig. 546.

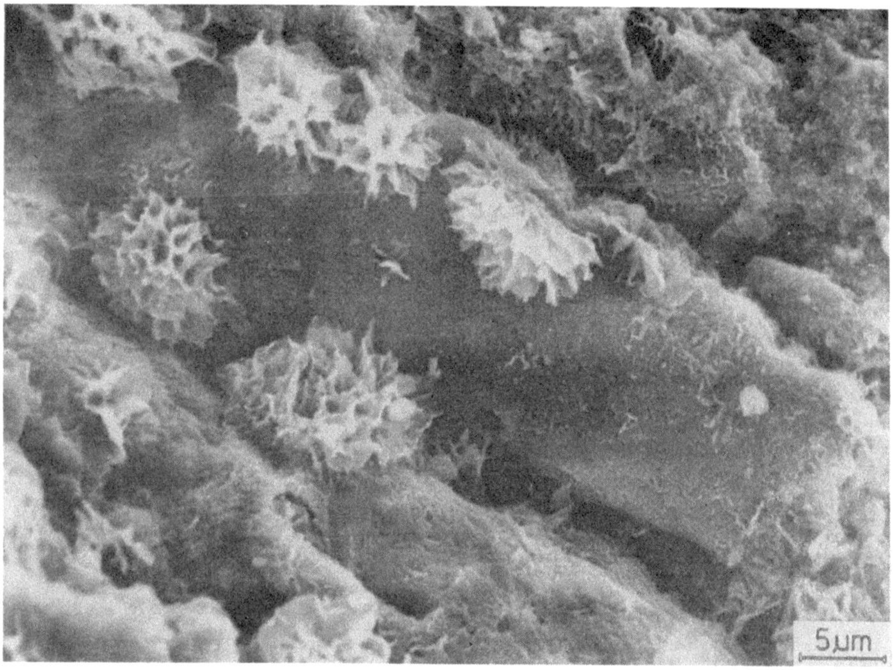

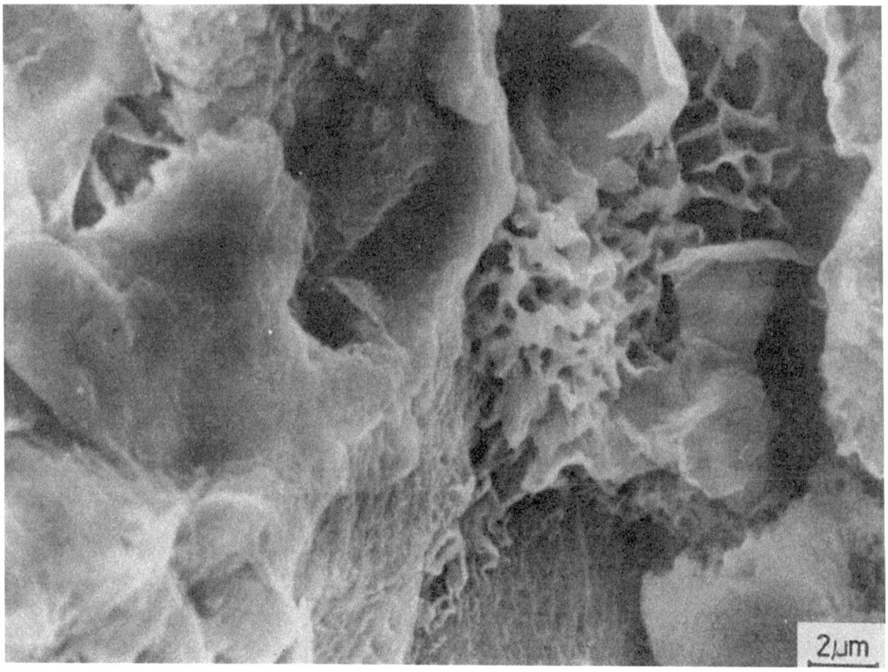

Fig. 547.

Fig. 548.

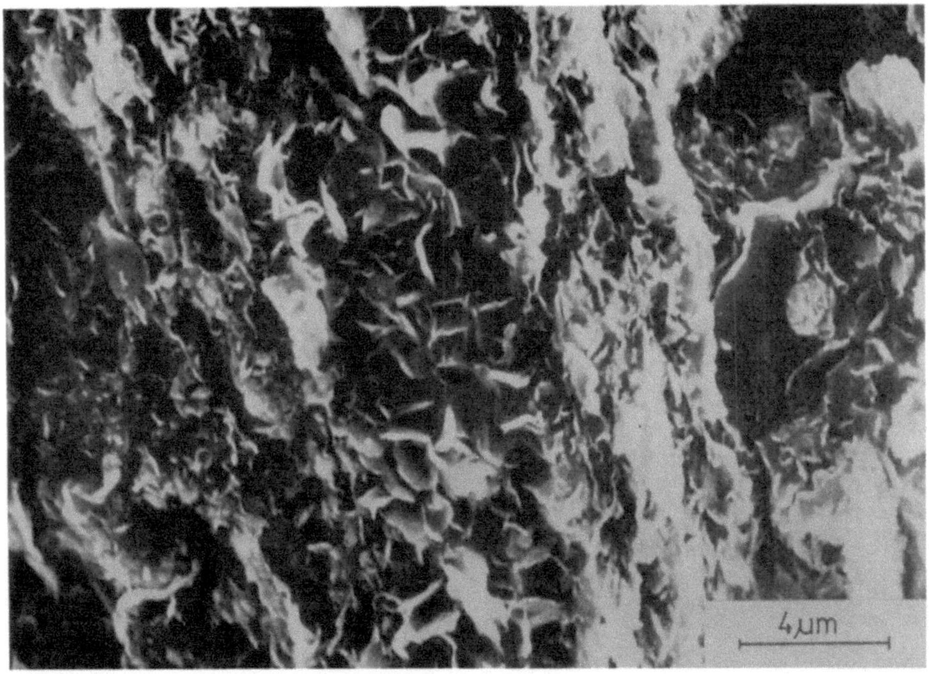

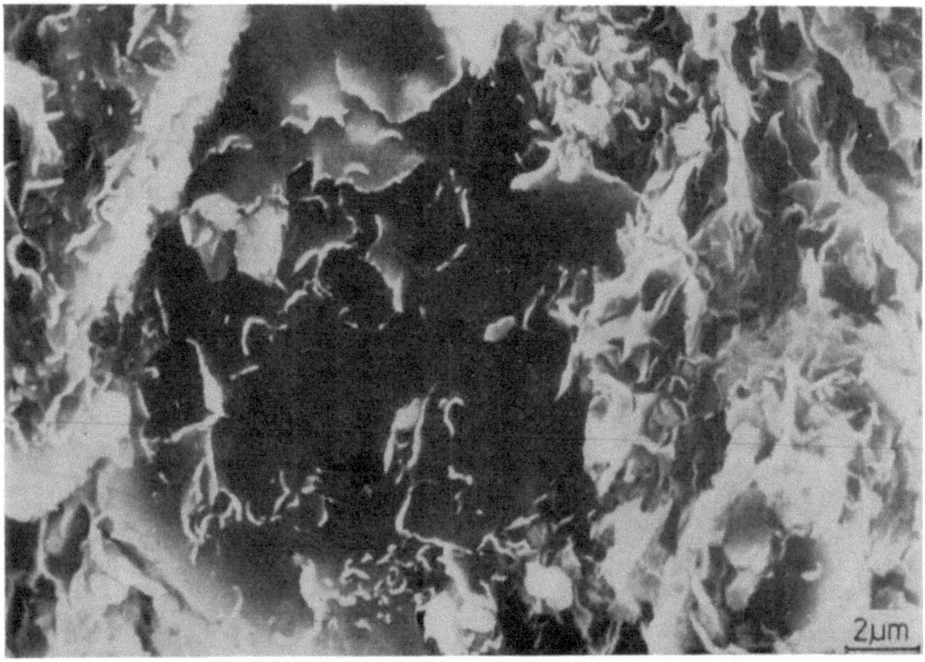

Fig. 549. SEM: *Bentonite*, bed "A", Royal Pit, Wyoming, USA

Fig. 550. SEM: *Bentonite*, bed "A", grain of slowly altered glas, Royal Pit, Wyoming, USA

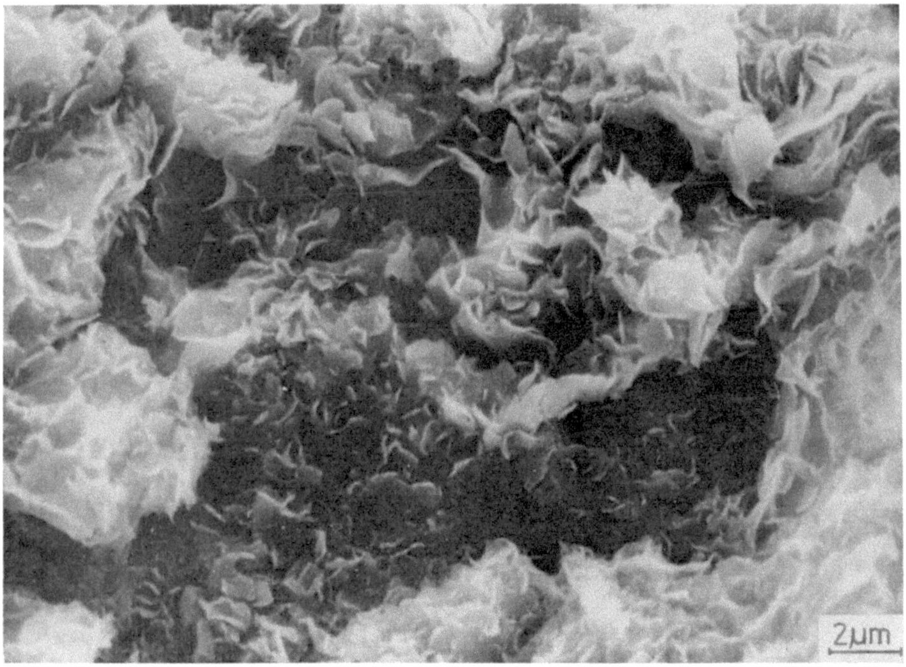

Figs. 551, 552. SEMs: Clay Spur *Bentonite*, fracture vertical to bedding, South State Lease Pit, Wyoming, USA

Fig. 552.

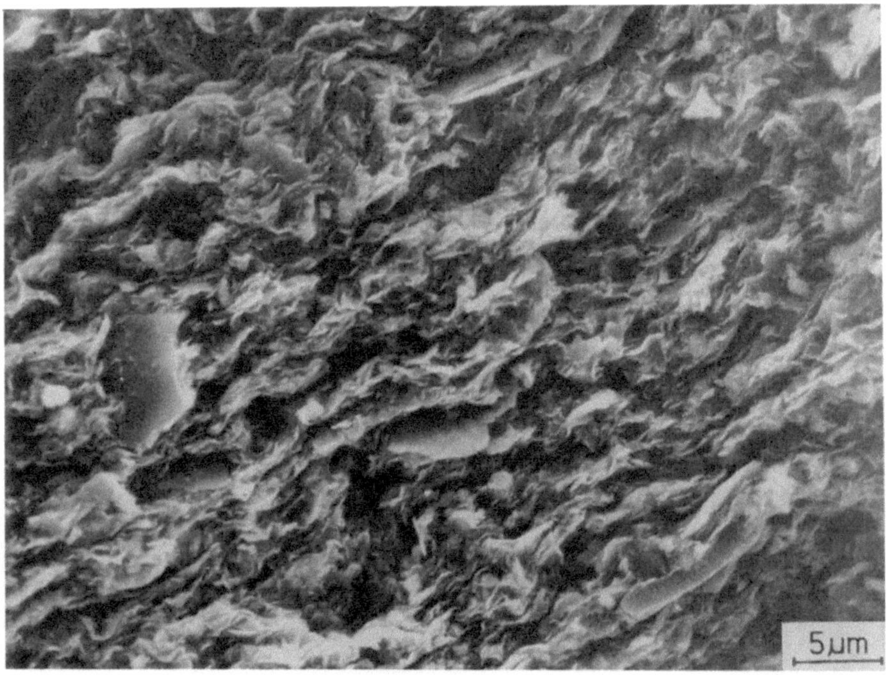

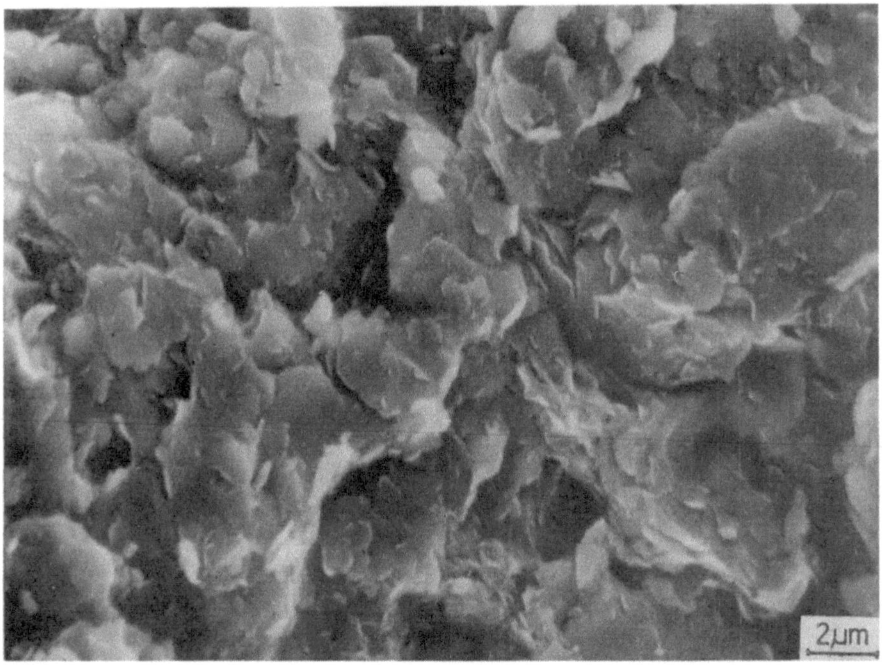

Fig. 553. SEM: *Montmorillonitic shale,* fracture vertical to bedding, Belle Fourche Shale, South State Lease Pit, Wyoming, USA

Fig. 554. SEM: *Montmorillonitic shale,* fracture parallel to bedding, Belle Fourche Shale, South State Lease Pit, Wyoming, USA

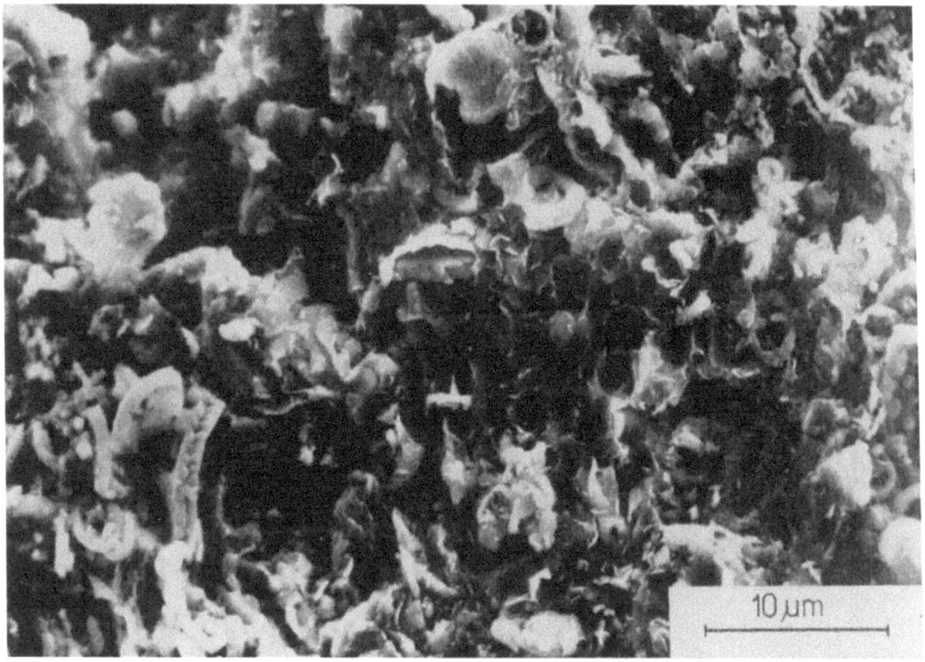

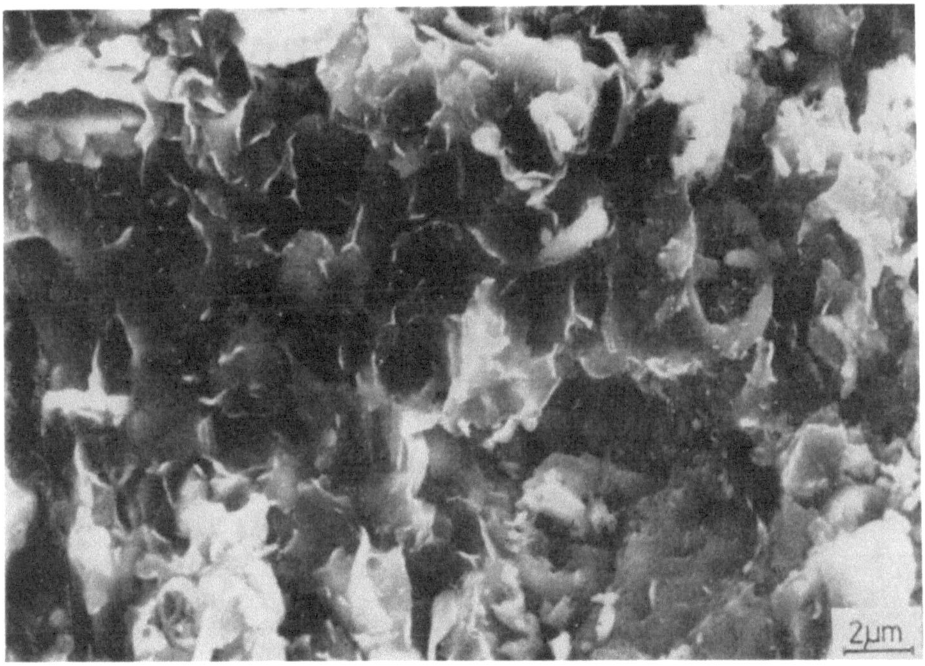

Fig. 555. SEM: Marl with small streaks of *montmorillonite* from the Greenhorn Fm., Upper Cenomanian, Russel County, Kansas, USA

Fig. 556. SEM: Detail of Fig. 555

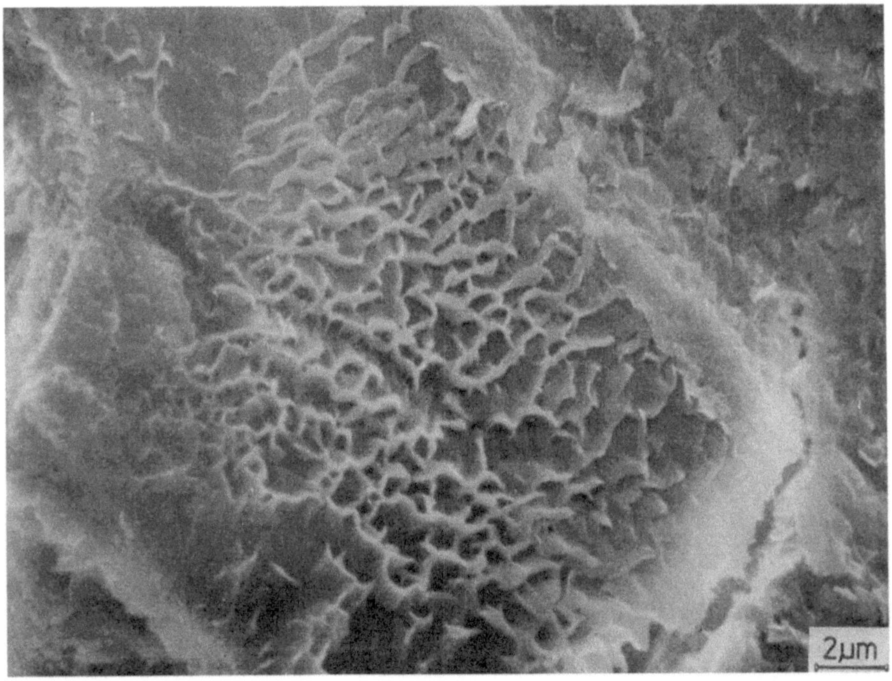

Fig. 557. SEM: *Smectite* on crystal surface, weathered basalt, Berzdorf, Lusatia GDR

Fig. 558. SEM: *Smectite* from weathered basalt, Berzdorf, Lusatia, GDR

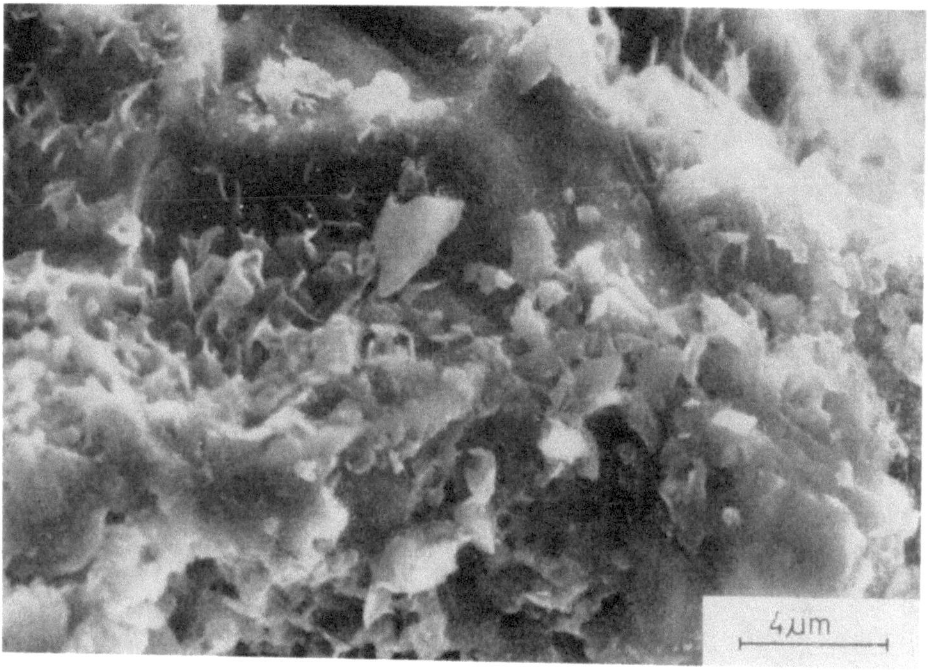

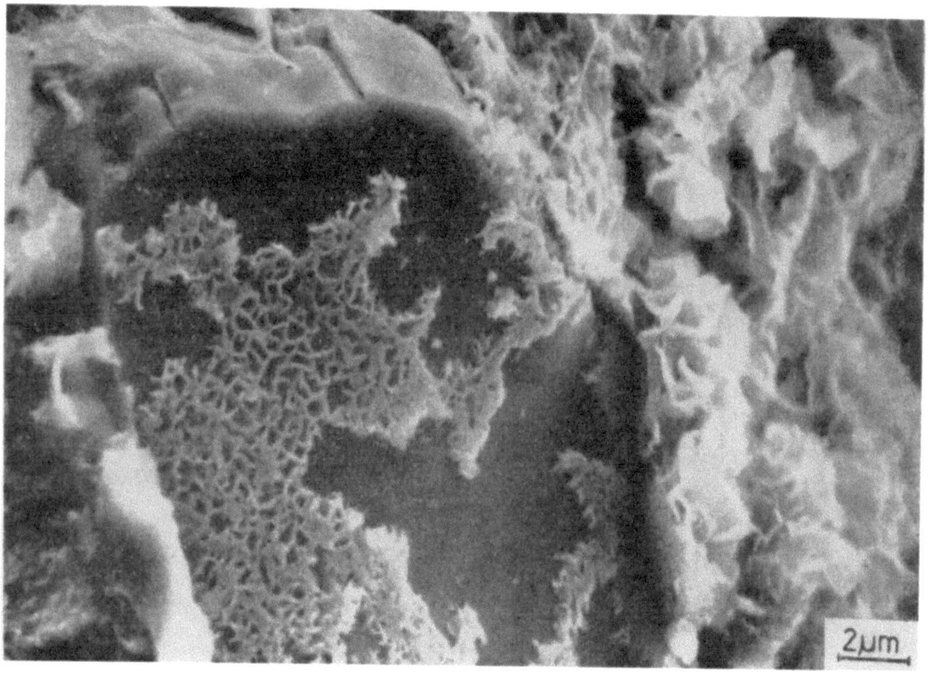

Fig. 559. SEM: *Smectite* on *plagioclase* from weathered granodiorite, Berzdorf, Lusatia, GDR

Fig. 560. SEM: *Smectite* on crystal surface from weathered basalt, Berzdorf, Lusatia, GDR

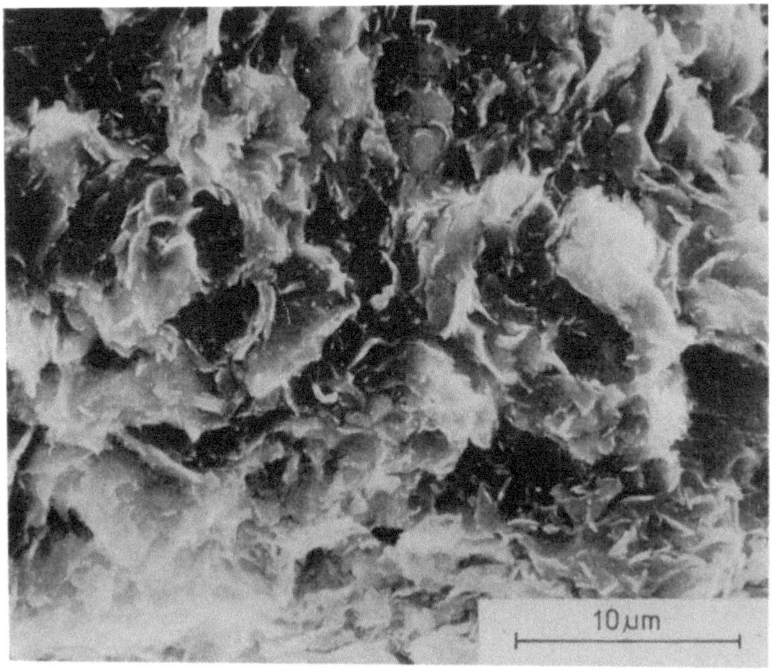

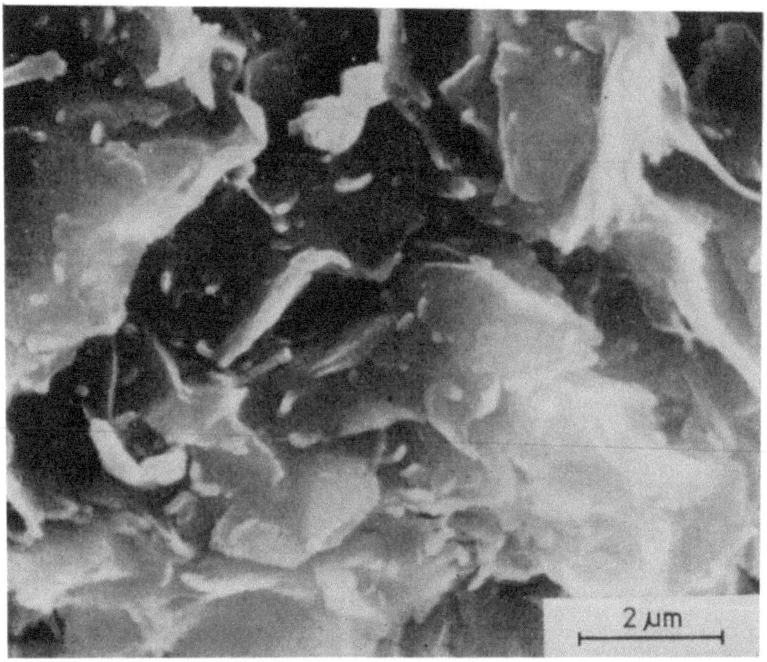

Fig. 561. SEM: *Montmorillonized pitchstone*, Garsebach near Meißen, GDR

Fig. 562. SEM: Detail of Fig. 561

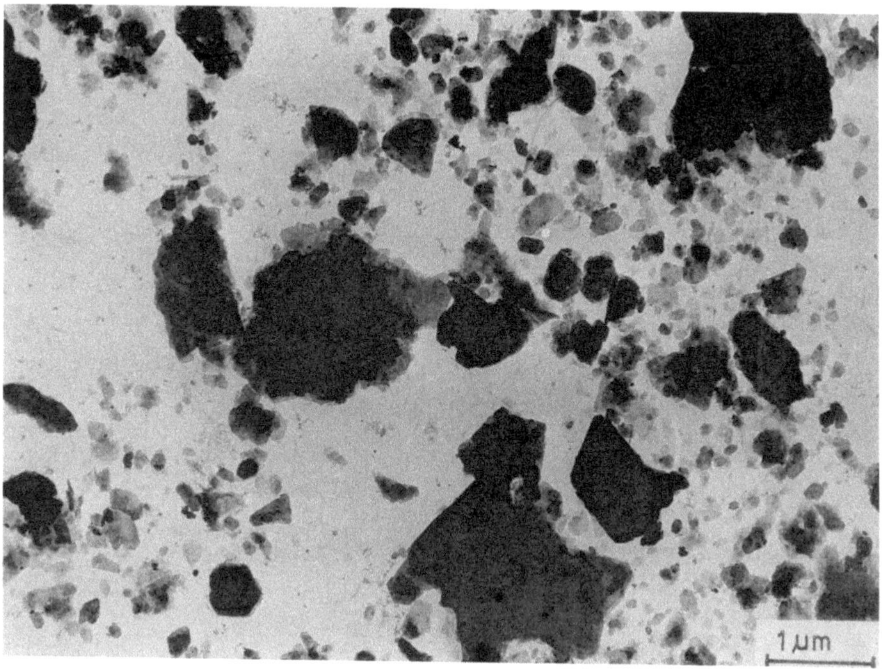

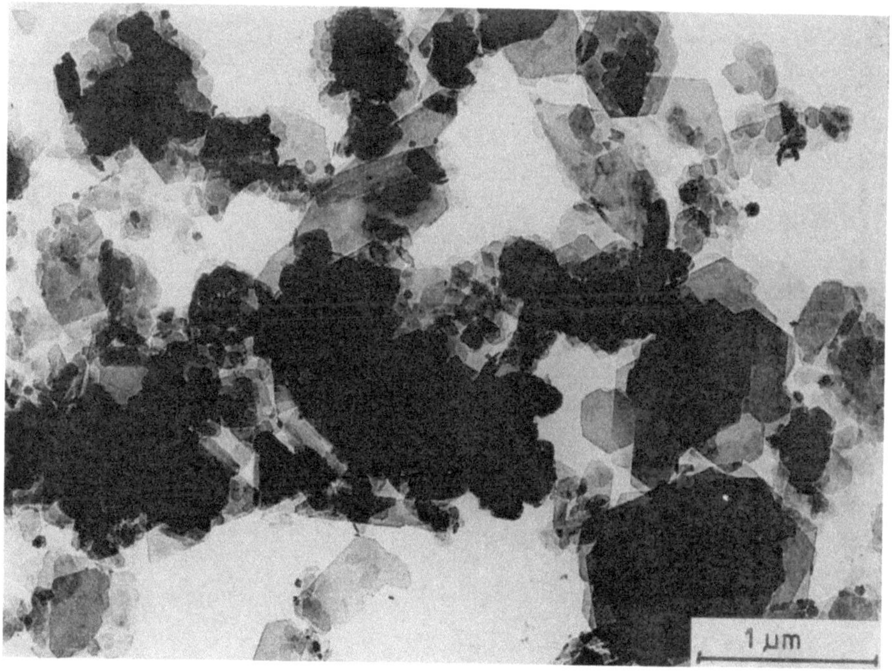

Fig. 563. TEM: *Kaolinitic Tertiary clay*, Brandis, GDR

Fig. 564. TEM: *Kaolinitic Tertiary clay*, Haselbach, GDR

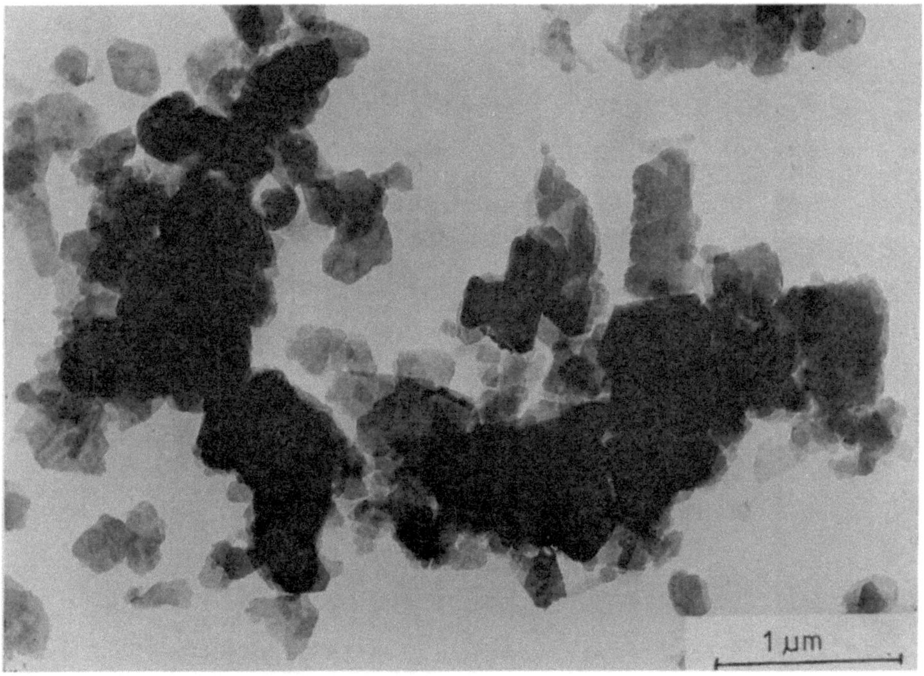

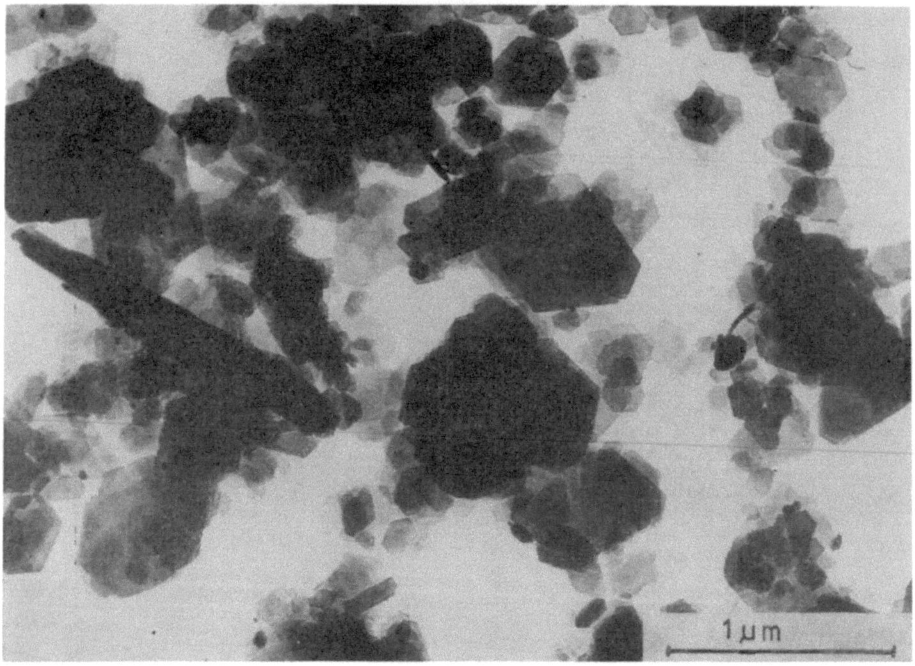

Fig. 565. TEM: *Kaolinitic Tertiary clay*, Bennstedt near Halle, GDR

Fig. 566. TEM: *Kaolinitic Tertiary clay*, Löthain near Meißen, GDR

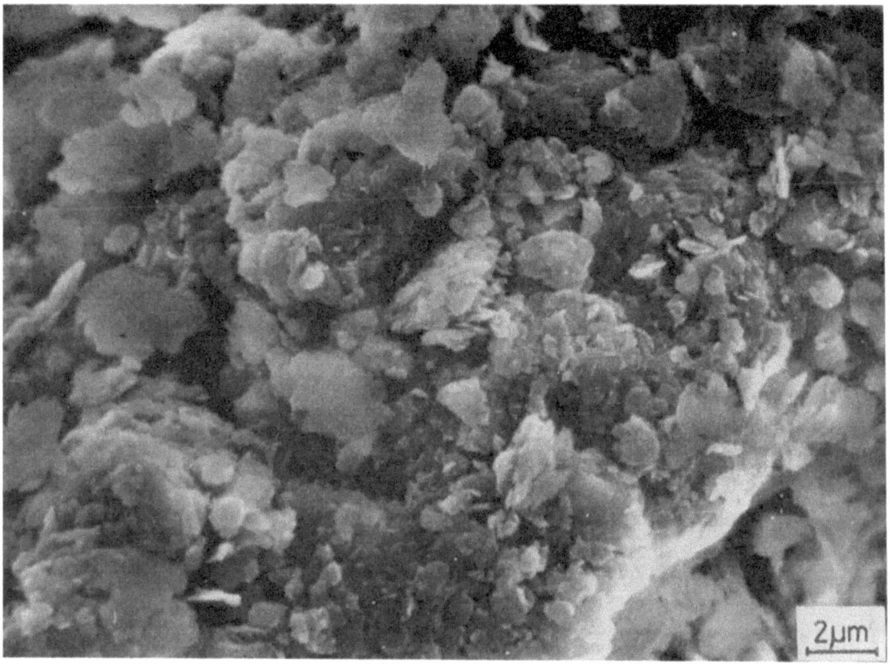

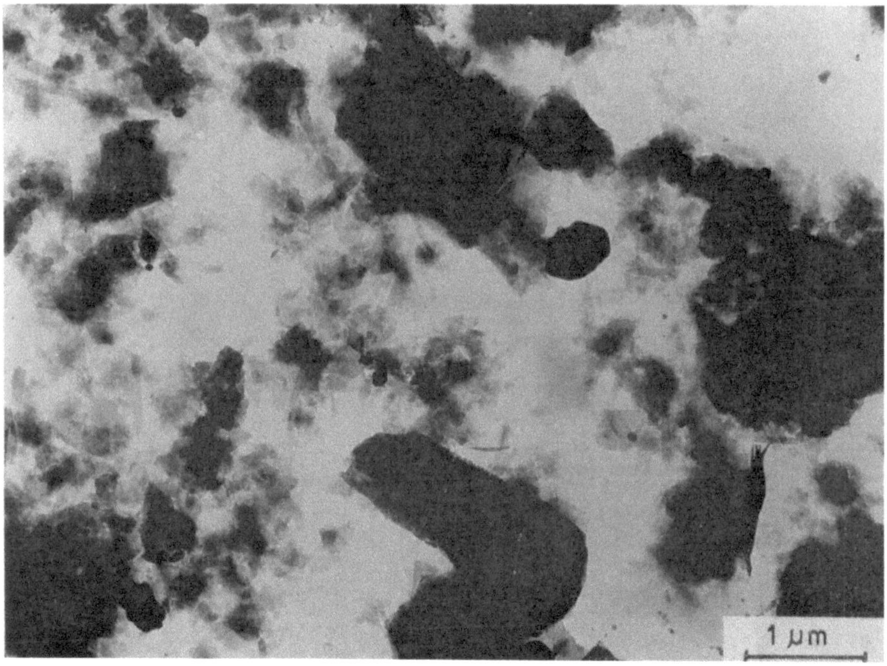

Fig. 567. SEM: *Kaolinitic Tertiary clay*, Löthain near Meißen, GDR

Fig. 568. TEM: *Blue Cambrian clay*, rich in *muscovite-montmorillonite mixed-layer* (slats), near Leningrad, USSR

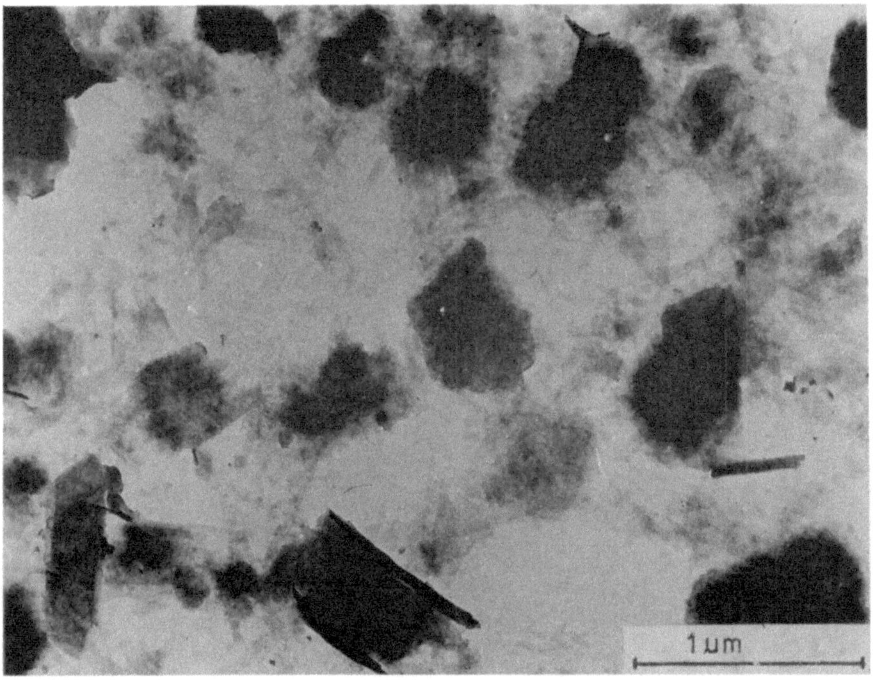

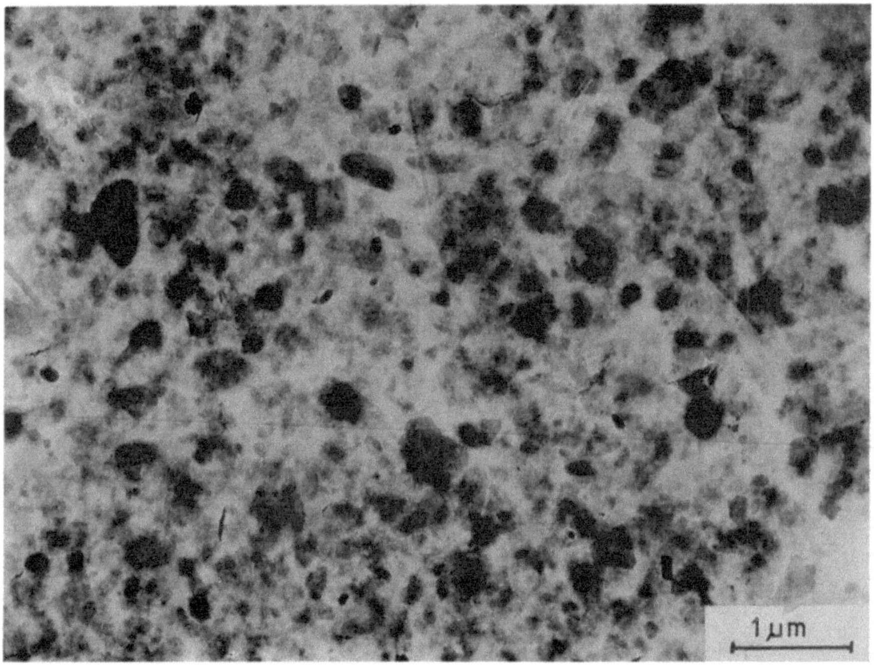

Fig. 569. TEM: *Tertiary clay*, rich in *muscovite-montmorillonite mixed-layer* (slats), some *mica*, Friedland, GDR

Fig. 570. TEM: Grain fraction < 0,2 µm of *Tertiary clay*, Friedland, GDR

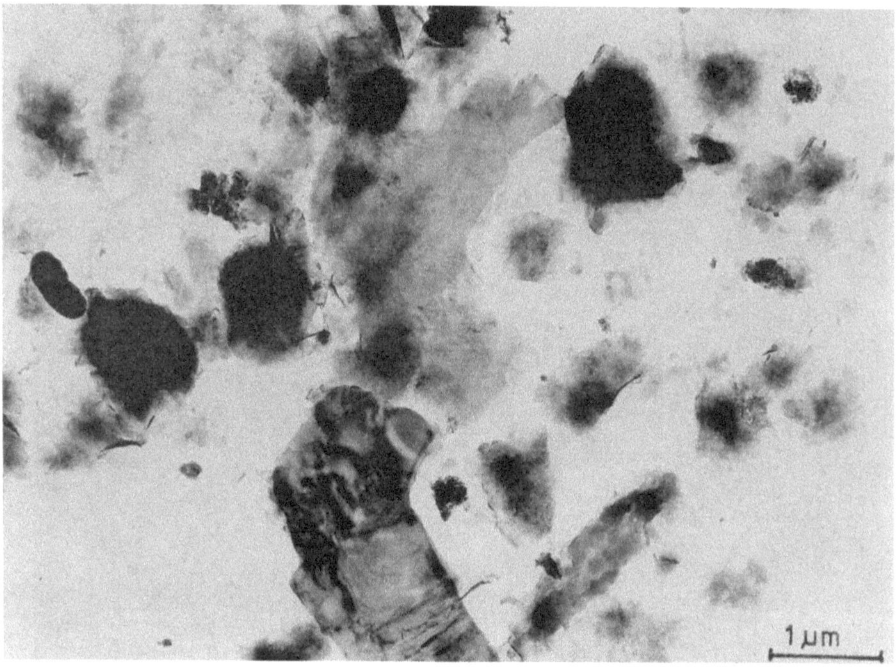

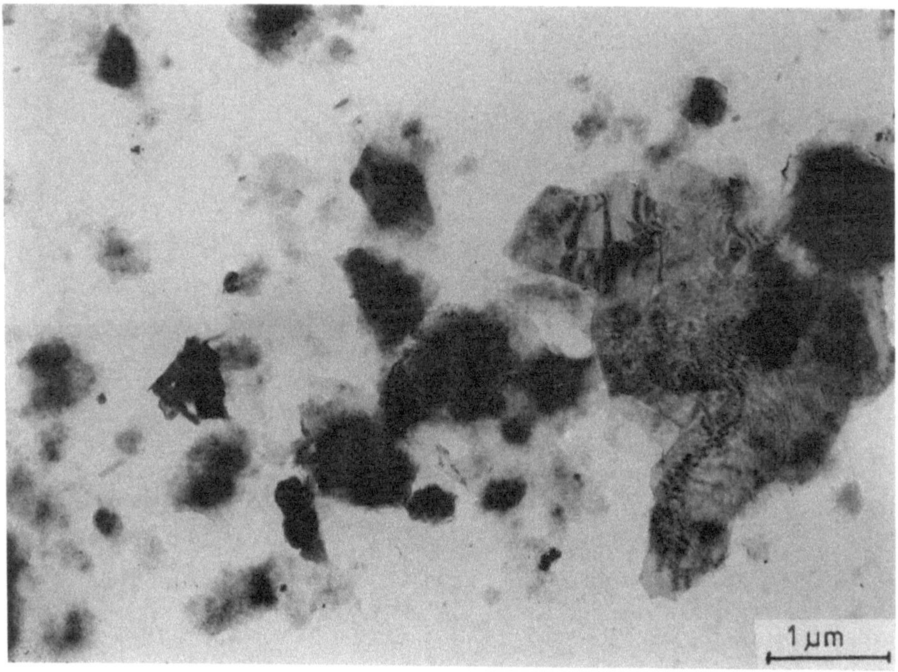

Figs. 571, 572. TEMs: *Tertiary clay*, rich in *muscovite-montmorillonite mixed-layer*, some *mica*
Dolgen near Rostock, GDR

Fig. 572

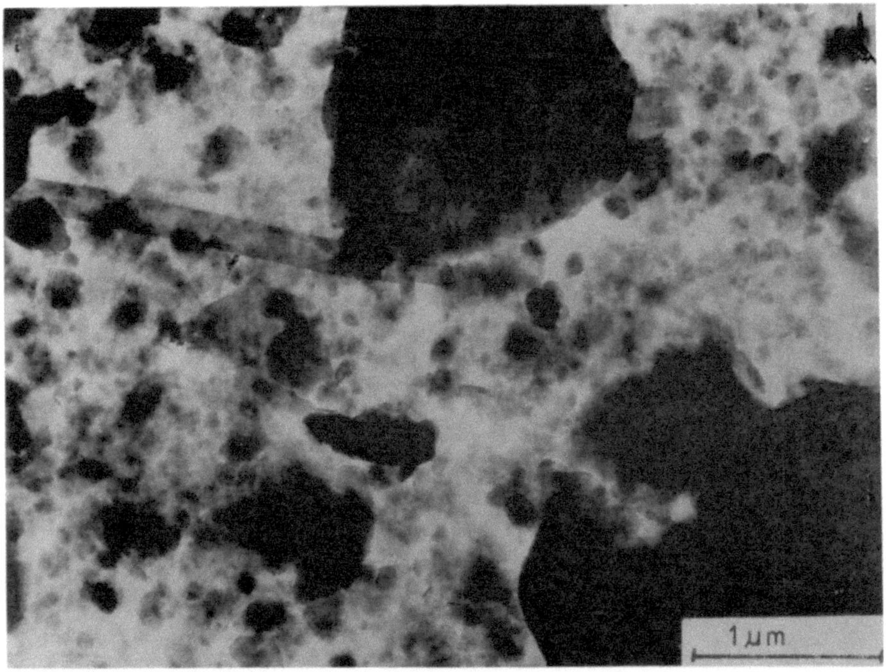

Fig. 573. TEM: *Liassic clay,* rich in *muscovite-montmorillonite mixed-layer,* some *kaolinite* and *mica,* Grimmen, GDR

Fig. 574. Replica: *Liassic clay,* same specimen as Fig. 573, Grimmen, GDR

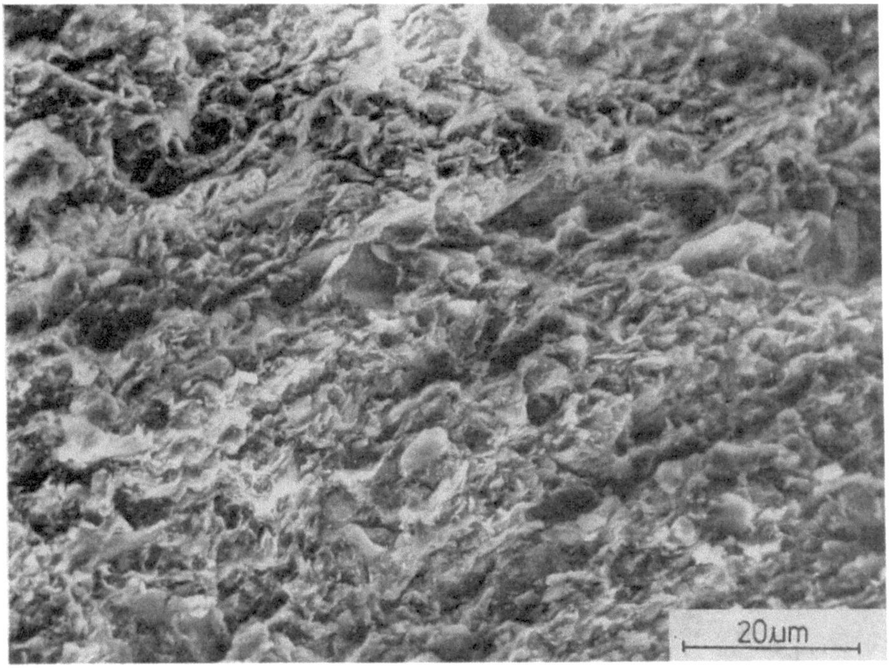

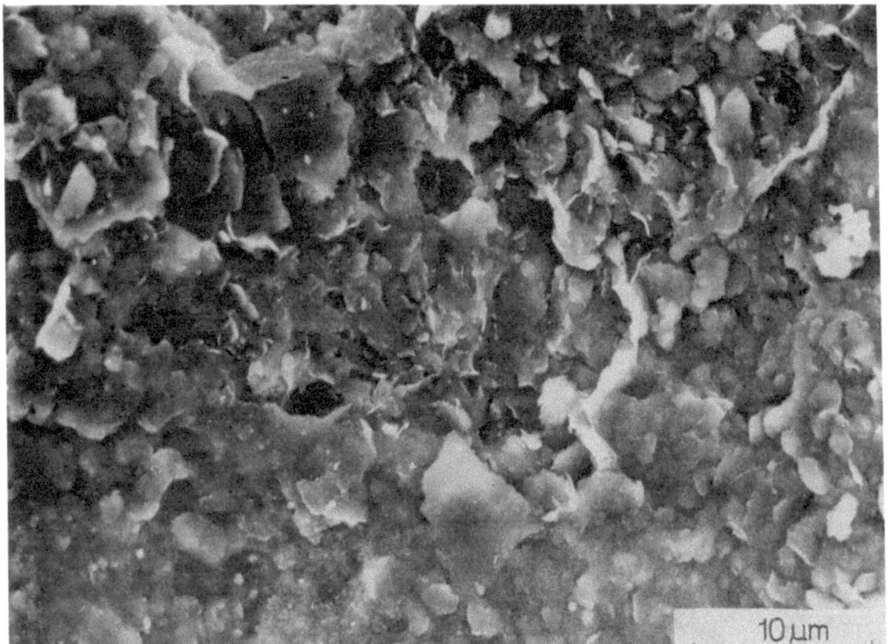

Fig. 575. SEM: *Liassic clay*, fracture vertical to bedding, Grimmen, GDR

Fig. 576. SEM: *Liassic clay*, fracture parallel to bedding, Grimmen, GDR

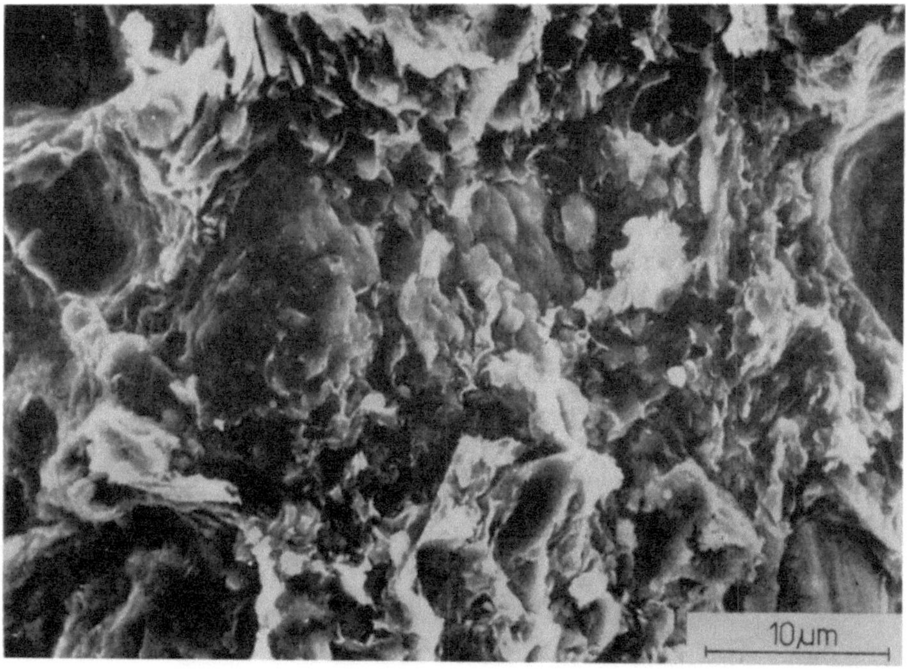

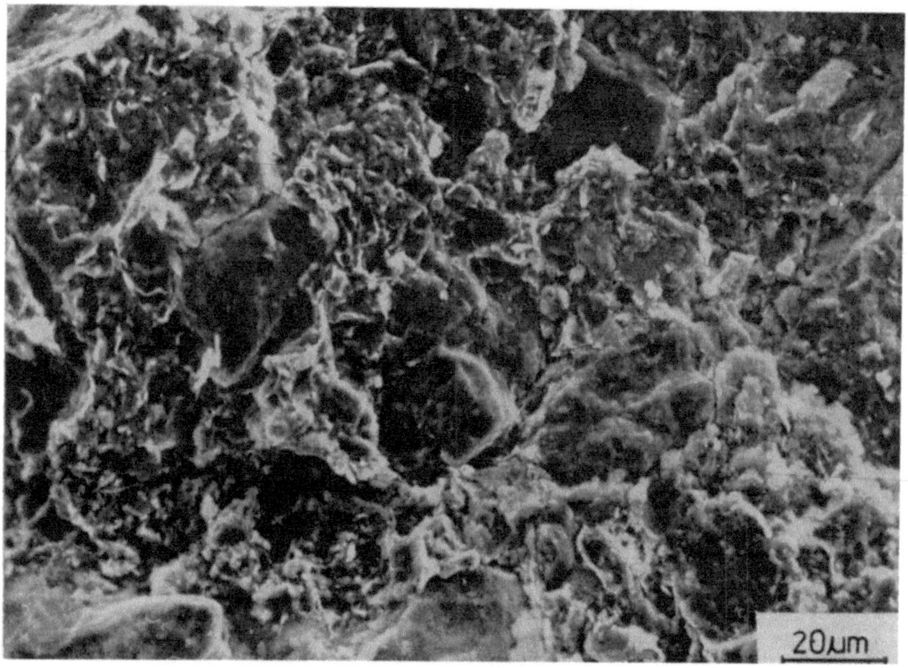

Fig. 577. SEM: *Liassic clay*, fracture vertical to bedding with clay coated quartz grains, Grimmen, GDR

Fig. 578. SEM: *Liassic clay*, fracture parallel to bedding with clay coated quartz grains, Grimmen, GDR

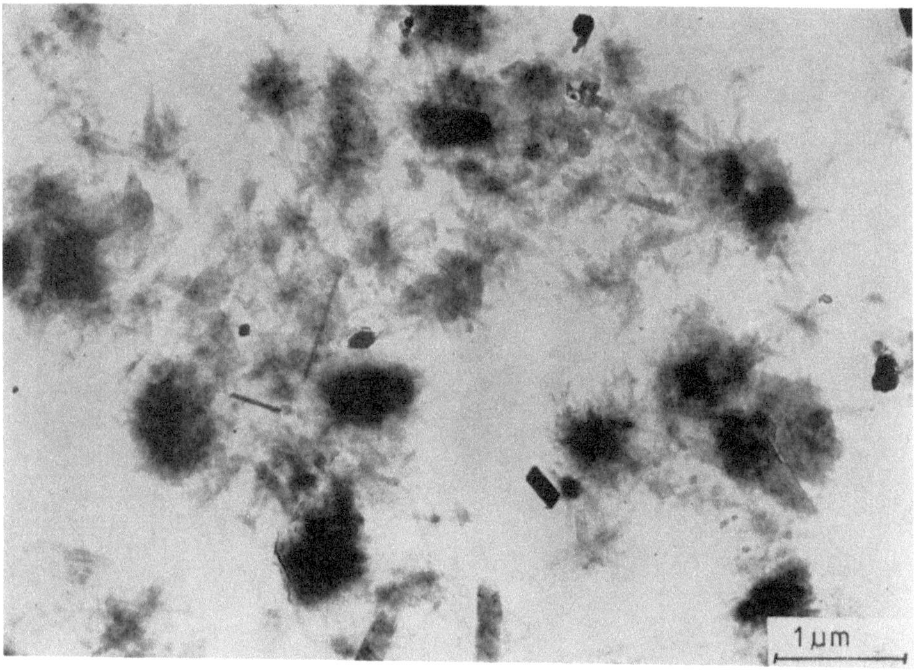

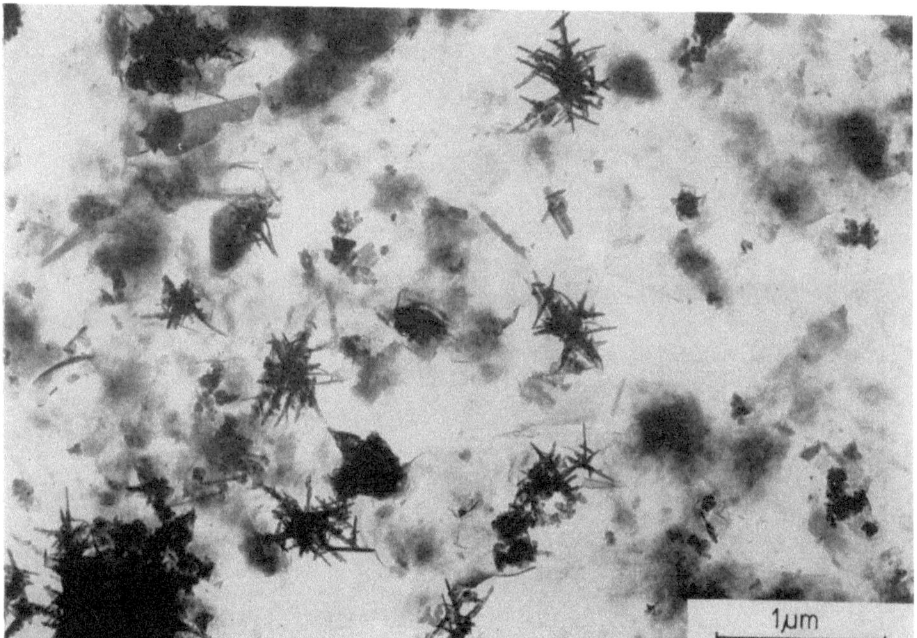

Fig. 579. TEM: Noncarbonates from Upper Cretaceous chalk, *muscovite-montmorillonite mixed-layer* (slats), Rügen, GDR

Fig. 580. TEM: Noncarbonates from Upper Cretaceous chalk, *muscovite-montmorillonite mixed-layer*, *mica* and *iron minerals* (star-shaped), Rügen, GDR

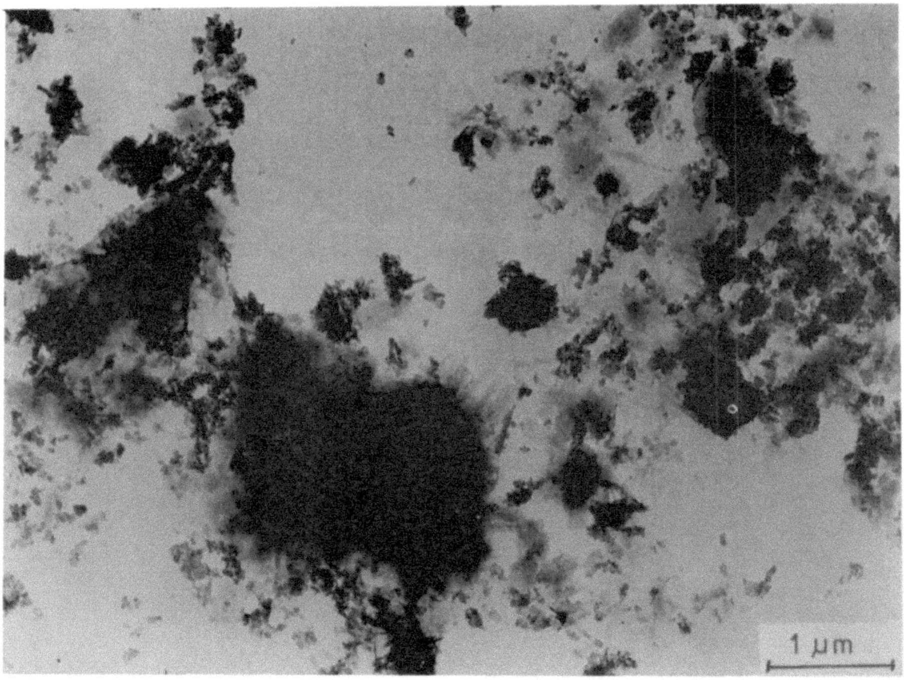

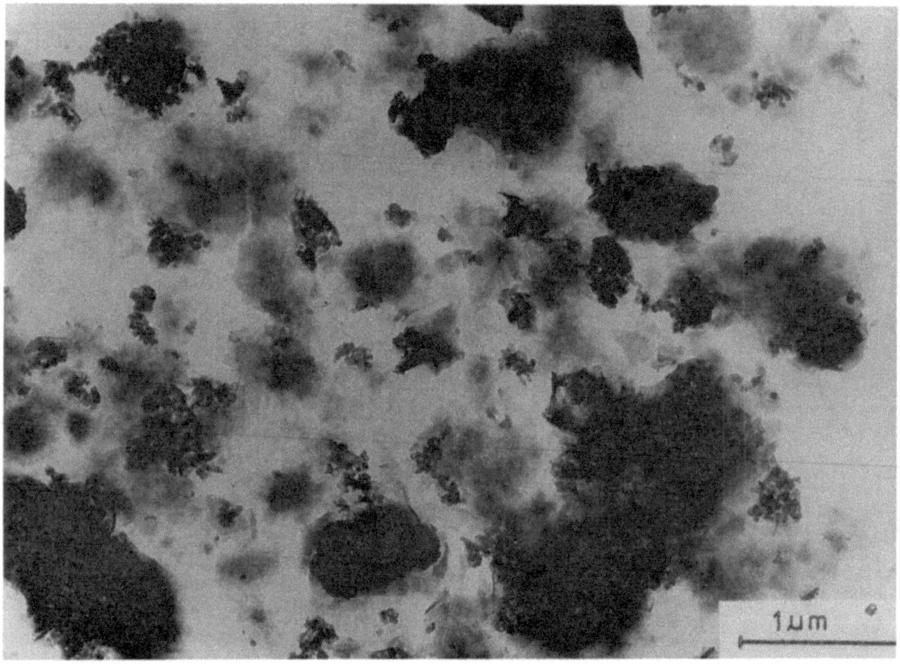

Fig. 581. TEM: Noncarbonates from Upper Cretaceous chalk, *muscovite-montmorillonite mixed-layer* and *opal-CT* (very small grains), Rügen, GDR

Fig. 582. TEM: Noncarbonates from Upper Cretaceous chalk, *muscovite-montmorillonite mixed-layer, montmorillonite, opal-CT*, Rügen, GDR

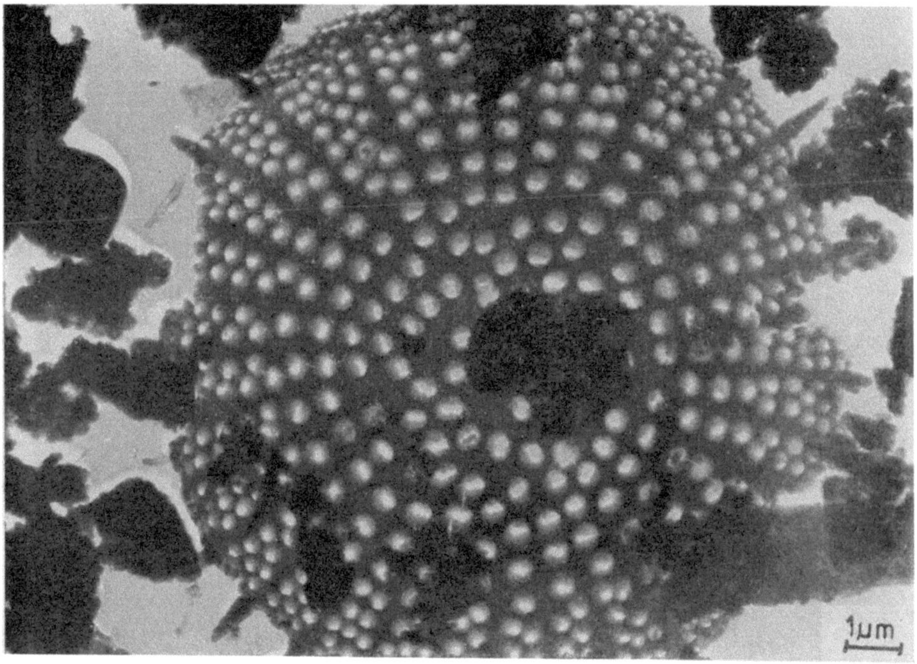

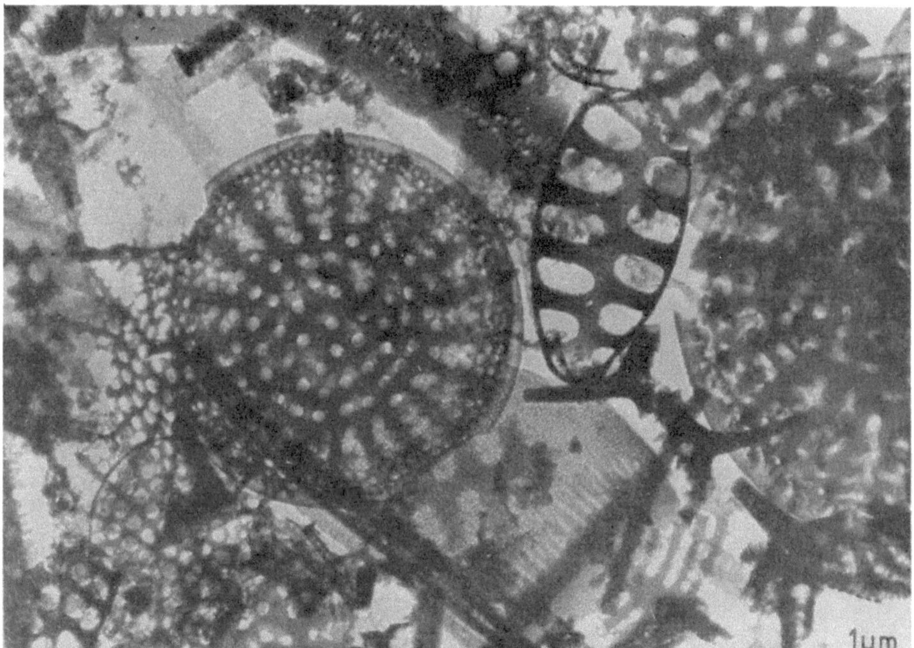

Figs. 583, 584. TEMs: *Diatom shells*, Klieken, GDR

Fig. 584.

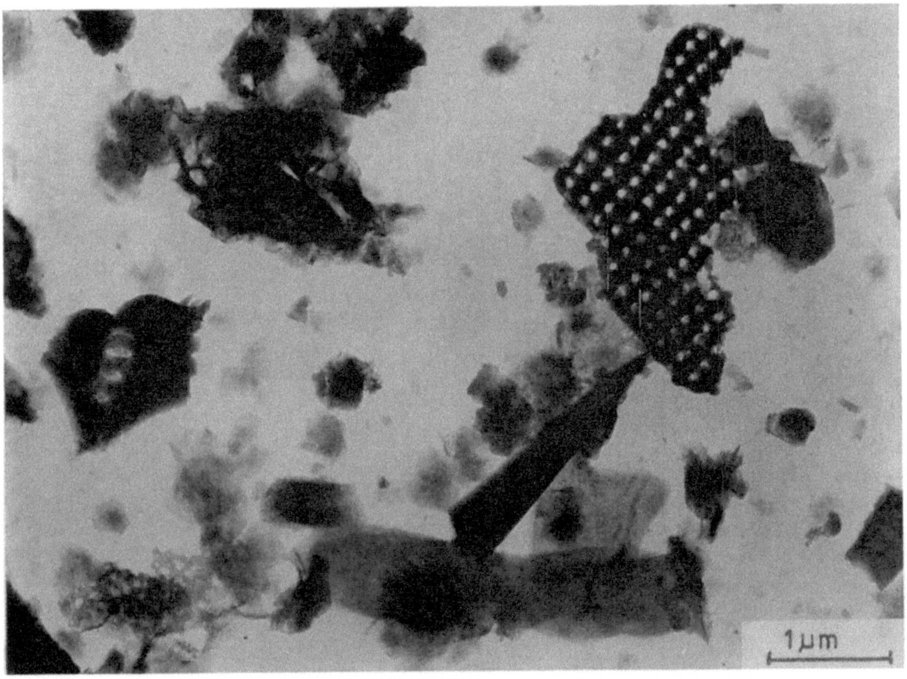

Fig. 585. TEM: Tripoli with *diatom shells, quartz* and *mica,* Seifhennersdorf, GDR

Figs. 586, 587, 588. Replicas: *Diatom shells,* Klieken, GDR

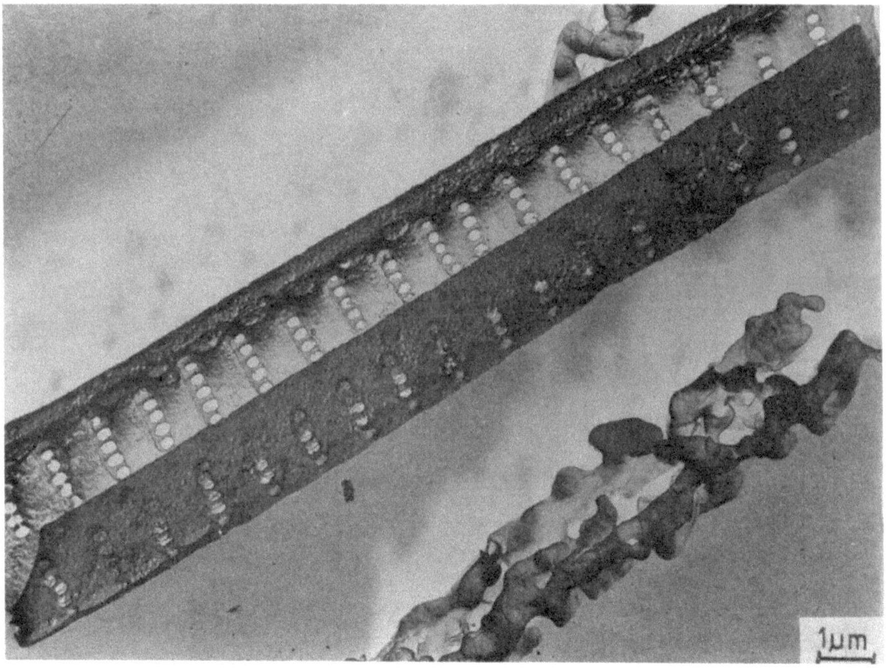

Fig. 587.

Fig. 588.

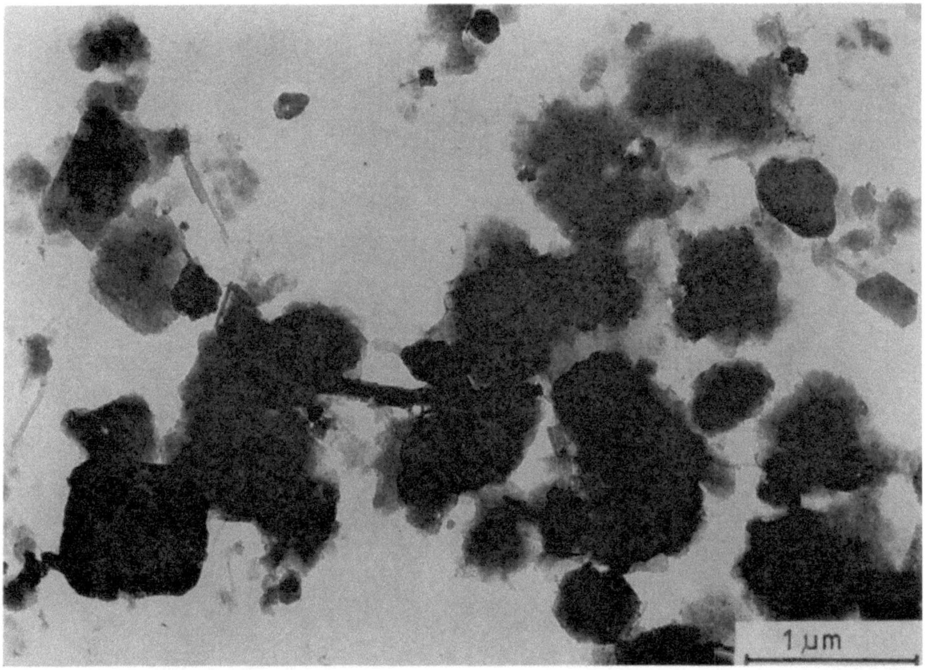

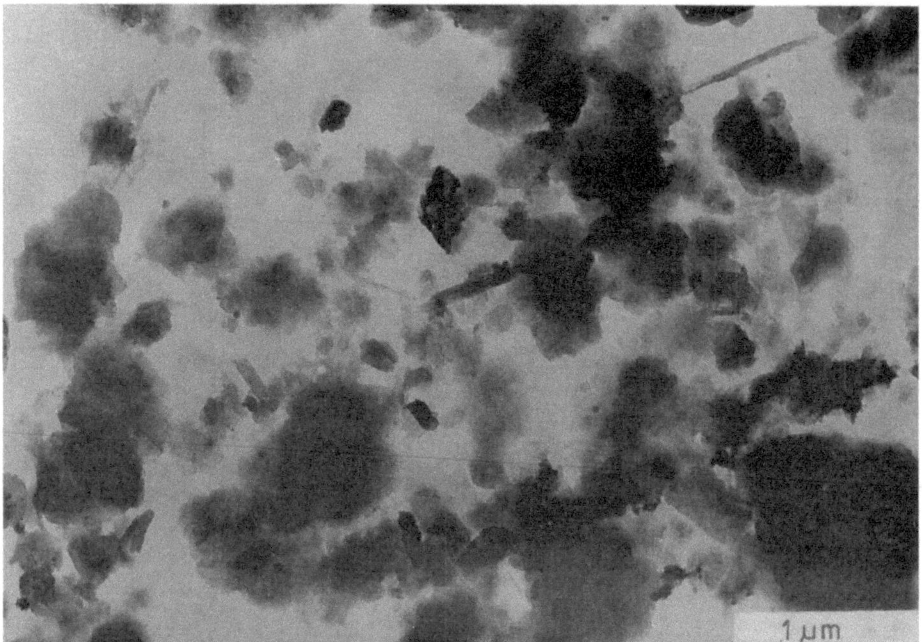

Fig. 589. TEM: Deep-sea clay, *muscovite-montmorillonite mixed-layer, kaolinite,* Atlantic Ocean, Canary Basin

Fig. 590. TEM: Deep-sea clay, *muscovite-montmorillonite mixed-layer, kaolinite,* Atlantic Ocean, North American Basin

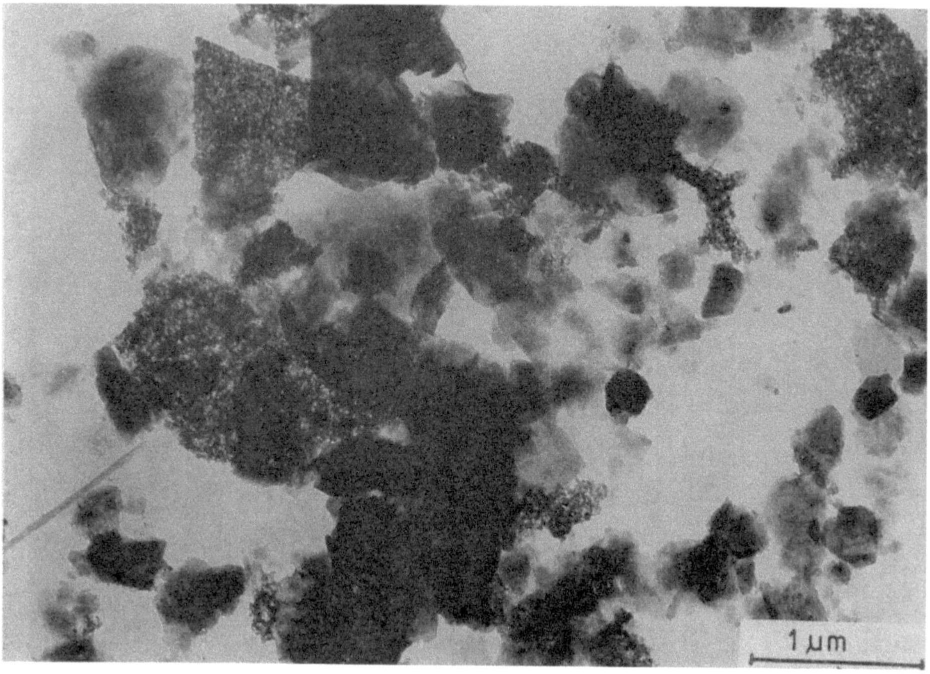

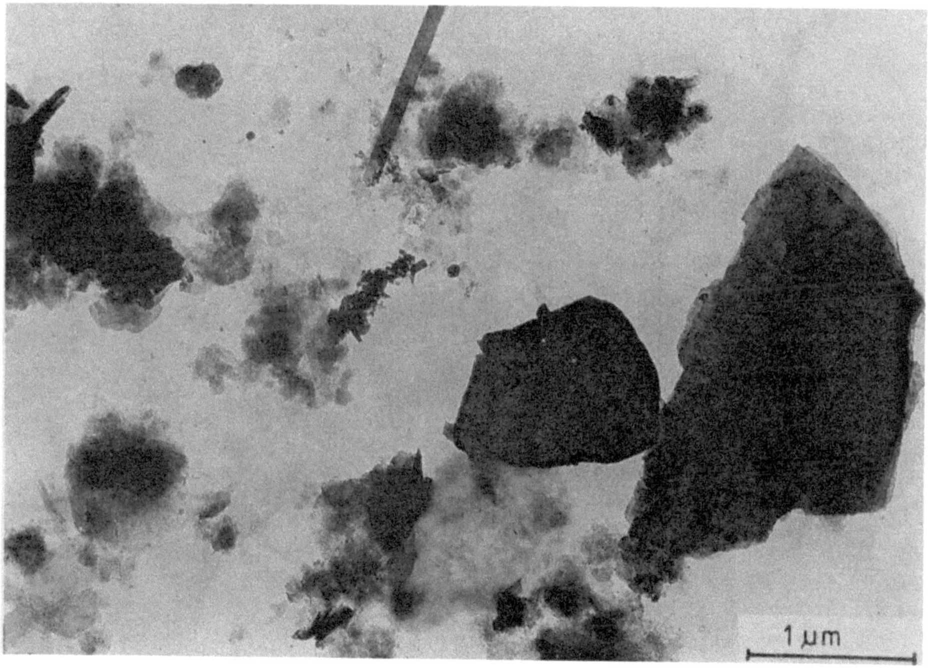

Fig. 591. TEM: Deep-sea clay, *smectite*, *illite*, *muscovite-montmorillonite mixed-layer*, relics of diatom shells, Pacific Ocean, North Equatorial

Fig. 592. TEM: Clay from the Baltic Sea, *muscovite-montmorillonite mixed-layer*, *mica*, *kaolinite*, Bay of Mecklenburg

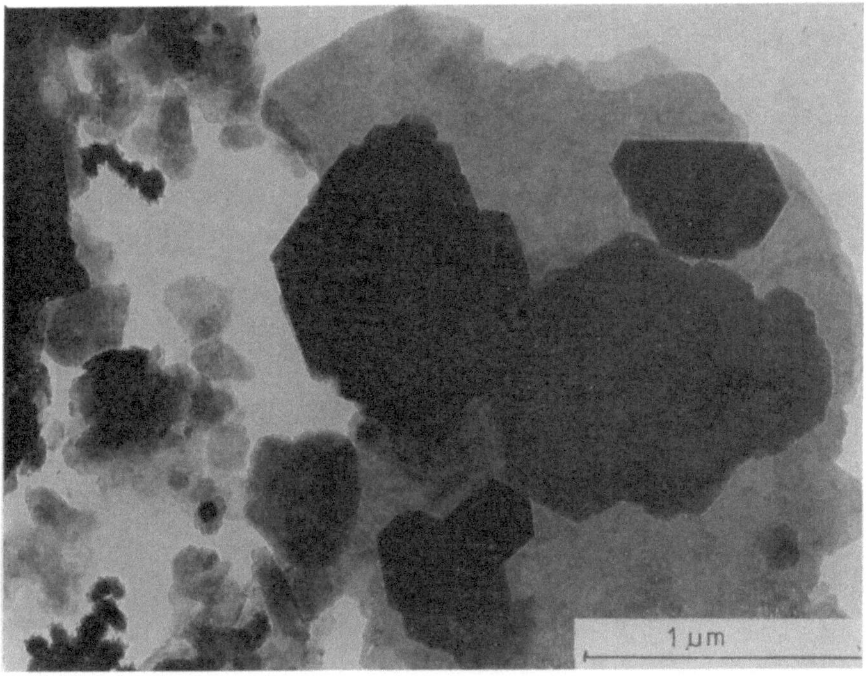

Fig. 593. TEM: *Kaolinite* from Tertiary brown coal (low temperature ash), Geiseltal near Halle, GDR

Fig. 594. Replica: *Kaolinite* from Tertiary brown coal (low temperature ash), Geiseltal near Halle, GDR

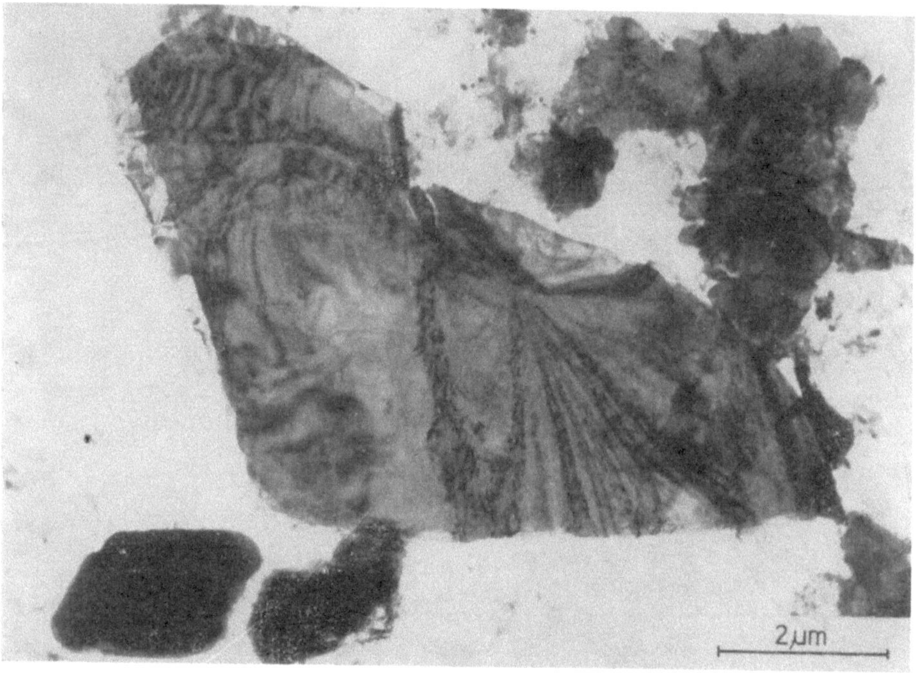

Fig. 595. TEM: *Mica* from Tertiary brown coal (low temperature ash), Geiseltal near Halle, GDR

Fig. 596. Replica: *Anhydrite* (idiomorphous crystals), *kaolinite* and very fine-grained *bassanite* from brown coal (low temperature ash), Geiseltal near Halle, GDR

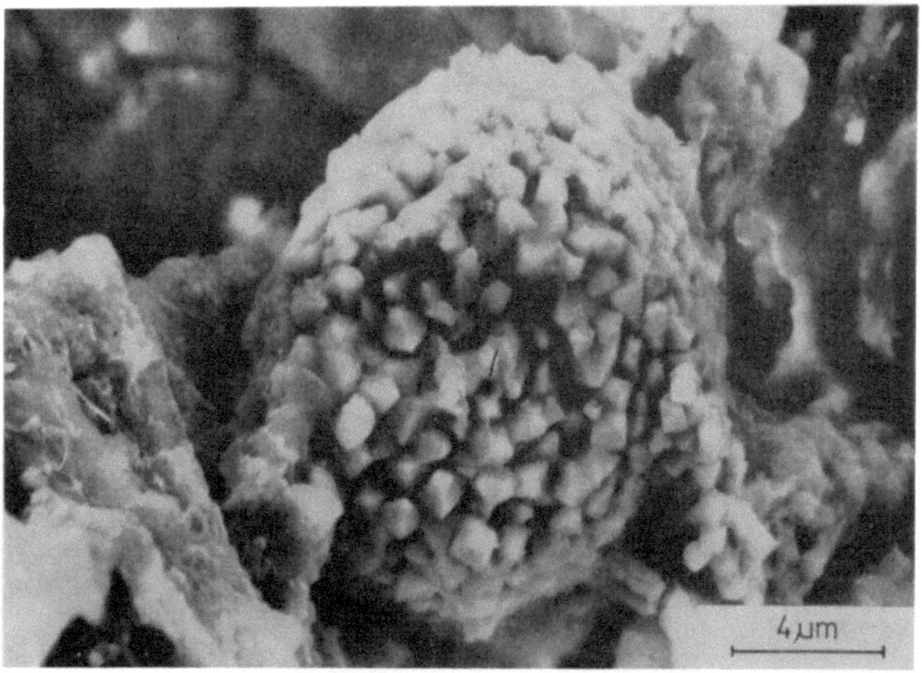

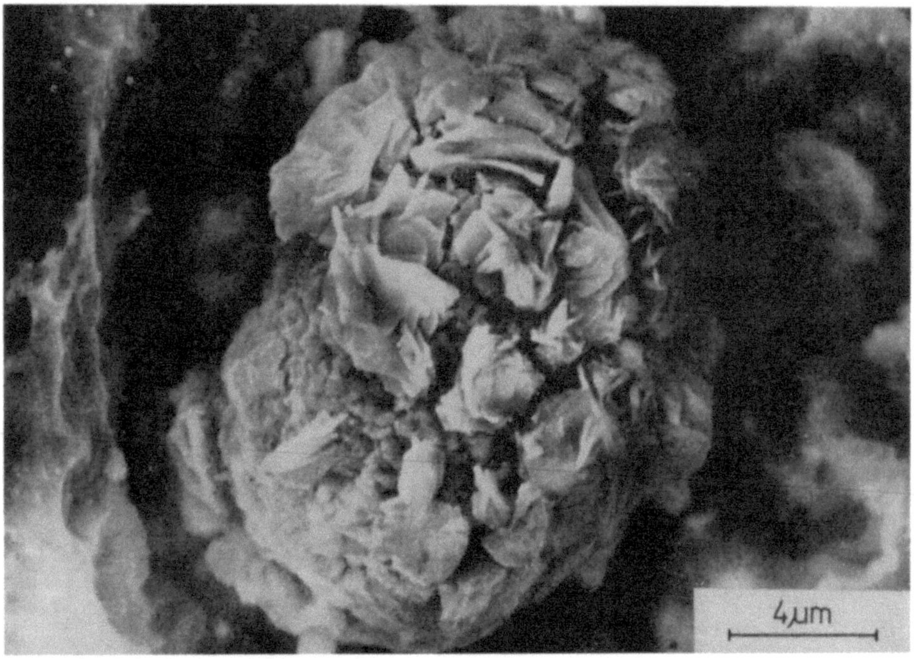

Fig. 597. SEM: Framboidal *pyrite* from Tertiary brown coal (low temperature ash), Geiseltal near Halle, GDR

Fig. 598. SEM: *Jarosite*-encrusted *pyrite* from Tertiary brown coal (low temperature ash), Geiseltal near Halle, GDR

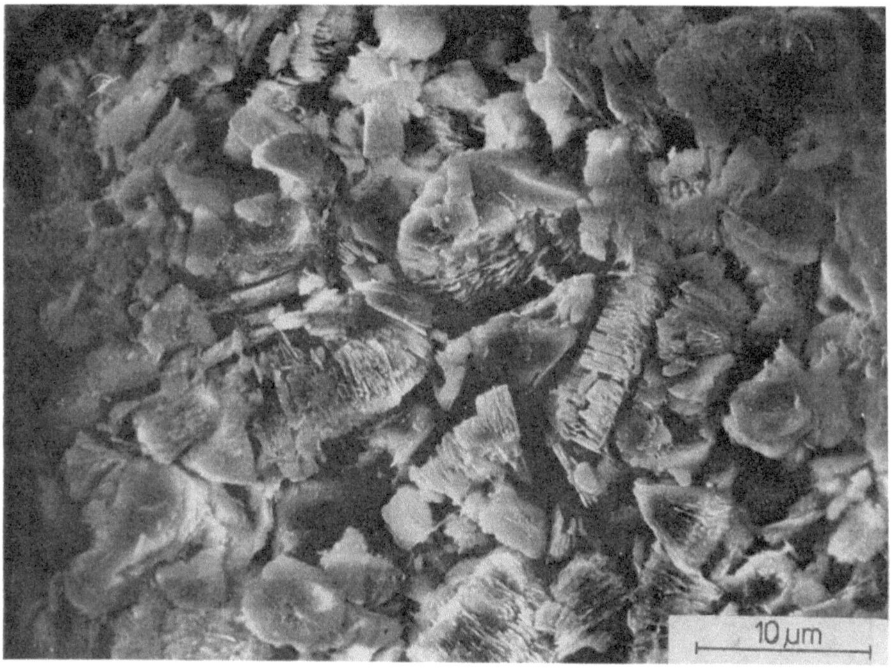

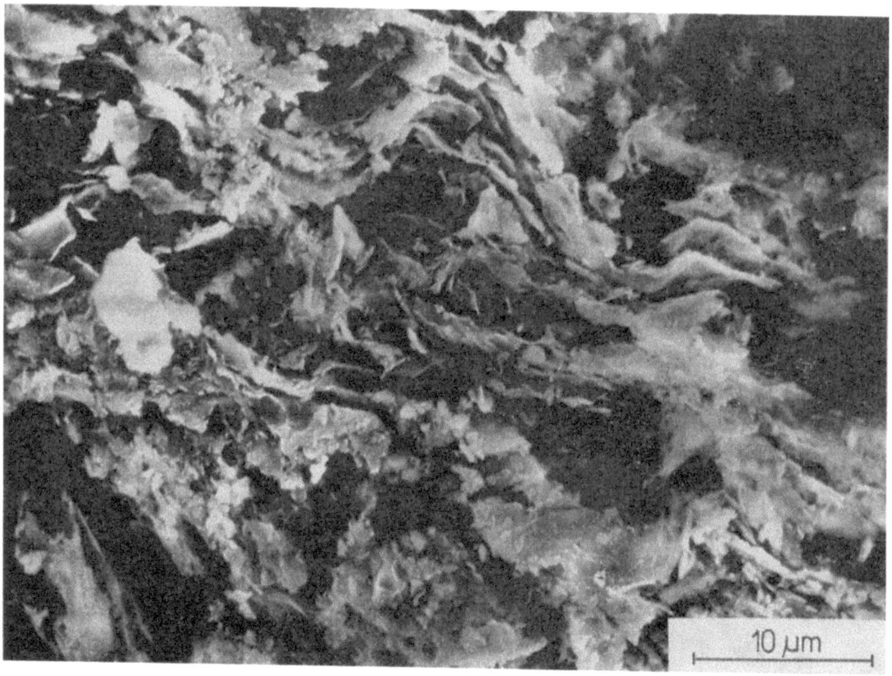

Fig. 599. SEM: *Vermicular kaolinite* in hard coal, specimen etched with active oxygen in the low temperature ashing apparatus, Ostrava, Czechoslovakia

Fig. 600. SEM: *Kaolinite* fabric of hard coal, specimen etched with active oxygen in the low temperature ashing apparatus, Recklinghausen, FRG

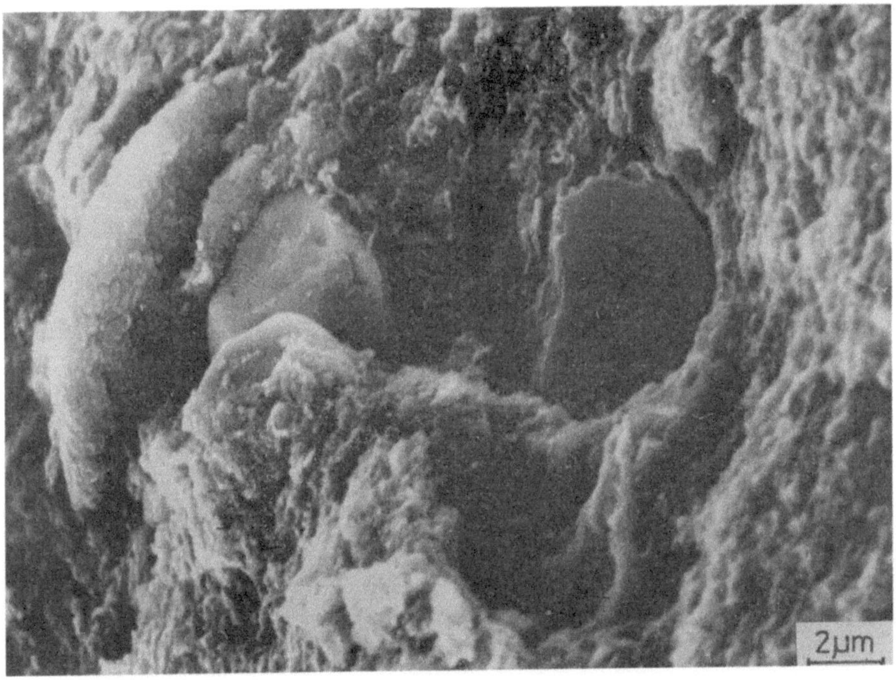

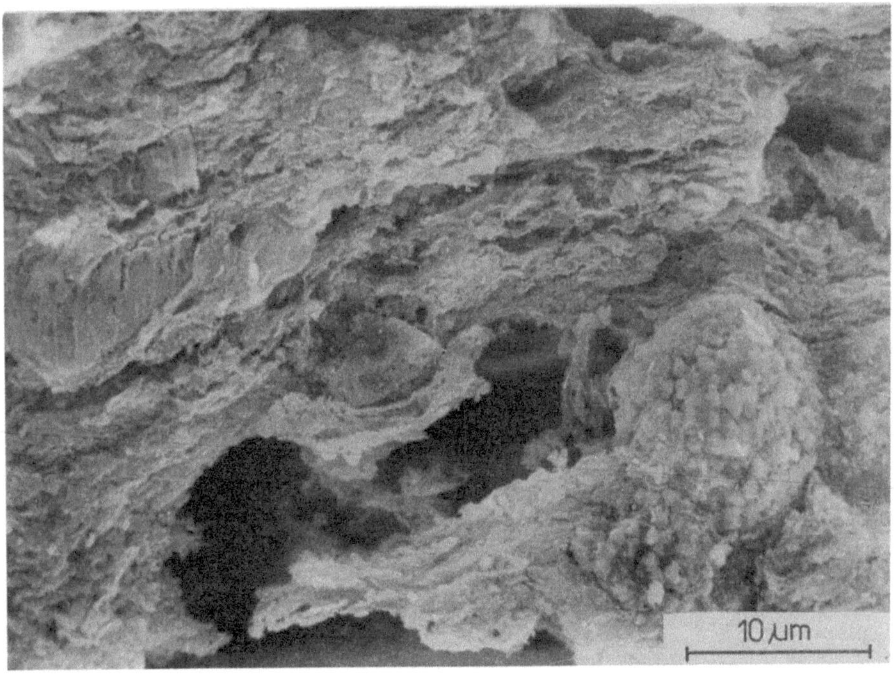

Fig. 601. SEM: Fabric of oil shale, *montmorillonite* and grains of ***bitumen***, Messel, FRG

Fig. 602. SEM: Fabric of oil shale after etching with active oxygen in the low temperature ashing apparatus, *montmorillonite* and *framboidal pyrite*, Messel, FRG

Subject index

In dieser Reihe erschienen / Published in this series

Heft 1*

Eberhard Grumbt
Sedimentgefüge im Buntsandstein Südwest- und Südthüringens. Ein Beitrag zur Untersuchung von Rotsedimenten

1974. 205 Seiten — 80 Abbildungen, davon 11 Faltabbildungen — 15 Falttafeln — 18 Tabellen — 16,7 × 24 cm — DDR 36,— M

Heft 2*

Lothar Eissmann
Das Quartär der Leipziger Tieflandsbucht und angrenzender Gebiete um Saale und Elbe. Modell einer Landschaftsentwicklung am Rand der europäischen Kontinentalvereisung

1975. 228 Seiten — 58 Abbildungen, davon 4 Falttafeln im Text und 10 Falttafeln separat 17 Tafeln — 23 Tabellen — 16,7 × 24 cm — DDR 48,— M

Heft 3*

Autorenkollektiv
Zur biostratigraphischen Untergliederung des Rotliegenden

1975. 138 Seiten — 12 Abbildungen, davon 2 Faltabbildungen — 18 Tafeln — 16 Tabellen 16,7 × 24 cm — DDR 27,— M

Heft 4*

Hans-Jürgen Teschke
Entwicklung und tektonischer Bau des südwestlichen Randbereichs der Osteuropäischen Tafel

1975. 151 Seiten — 44 Abbildungen — 4 Tabellen — 16,7 × 24 cm — DDR 30,— M

Heft 5*

Autorenkollektiv
Tonminerale. Genese, Lagerstätten, industrielle Bedeutung und Nutzung. 3. Tonmineraltagung, Greifswald, 16.—18. 7. 1973

1976. 291 Seiten — 119 Abbildungen — 1 Tafel 46 Tabellen — 16,7 × 24 cm — DDR 57,— M

Heft 6*

Walter Hoppe
Die paläogeographisch-fazielle Entwicklung im Südteil des Germanischen Buntsandsteinbeckens

58 Seiten — 9 Abbildungen — 8 Tabellen

Hartmut Haubold, Peter Puff
Zur Genese der Solling-Folge (Untere Trias, Buntsandstein) in Thüringen

18 Seiten — 6 Abbildungen — 1 Tabelle

Peter Puff
Gliederung der Solling-Folge (Trias, Buntsandstein) im südlichen Thüringen mittels Violetter Horizonte

16 Seiten — 3 Abbildungen — 1 Tabelle
1976. 16,7 × 24 cm — DDR 18,— M

Heft 7*

Yanaki Tenčov
Flora und Biostratigraphie des Oberkarbons im Svoge-Becken (VR Bulgarien)

1977. 168 Seiten — 2 Abbildungen — 23 Tafeln — 3 Anlagen — 16,7 × 24 cm — DDR 30,— M

Heft 8*

Karin Mayer
Entstehung, Eigenschaften und Verteilung der Edelgase und ihrer Isotope. — Ein Beitrag zur Geochemie der Edelgase

73 Seiten — 19 Abbildungen — 22 Tabellen — 3 Anlagen

Peter Beuge; Wolgang Kramer
Lamprophyre Ostthüringens und ihre anomalen Quecksilbergehalte im Ergebnis endogener und exogener Anreicherungsprozesse

20 Seiten — 12 Abbildungen — 5 Tabellen
Buchbesprechungen
1977. 16,7 × 24 cm — DDR 21,— M

Heft 9*

Autorenkollektiv
100 Jahre Glazialtheorie im Gebiet der skandinavischen Vereisungen

1978. 368 Seiten — 77 Abbildungen — 30 Tabellen — 8 Tafeln — 16,7 cm × 24 cm — DDR 72,— M

Heft 10*

Autorenkollektiv
Beiträge zur Paläontologie

1978. 122 Seiten — 22 Abbildungen — 6 Tabellen — 3 Tafeln — 16,7 × 24 cm — DDR 24,— M

Heft 11*

Manfred Störr (Hrsg.)
Genese des Kaolins
26 Beiträge von 41 Autoren

1978. 350 Seiten — 164 Abbildungen — 58 Tabellen — 16,7 × 24 cm — DDR 68,— M

Heft 12*

Autorenkollektiv
21. Geophysikalisches Symposium Leipzig, 14. bis 17. September 1976
15 Beiträge von 36 Autoren

1978. 180 Seiten – 96 Abbildungen – 2 Tabellen – 16,7 × 24 cm – DDR 36,– M

Heft 13*

HANNELORE PETER
Die Florenentwicklung – pflanzengeographische Differenzierung – im Zeitraum des Jura

1979. 84 Seiten – 22 Abbildungen – 3 Tabellen – 16,7 × 24 cm – DDR 18,– M

Heft 14*

MARTIN GUNTAU (Hrsg.)
Biographien hervorragender russischer und sowjetischer Geowissenschaftler
19 Beiträge von 16 Autoren

1979. 199 Seiten – 26 Abbildungen – 16,7 × 24 cm – DDR 38,– M

Heft 15*

HERMANN BRAUSE
Probleme des Krustenbaus und der geotektonischen Entwicklung auf der Geotraverse Baltikum – DDR – Česky massif

32 Seiten – 14 Abbildungen

HELMUT LÜTZENS
Zur geotektonischen Entwicklung des Harzvariszikums mit besonderer Berücksichtigung synparoxysmaler Resedimentationsprozesse im Mittelharz

66 Seiten – 15 Abbildungen – 3 Tafeln
1979. 16,7 × 24 cm – DDR 22,– M

Heft 16*

RUDOLF DABER (Hrsg.)
Evolution/Naturgeschichte höherer Pflanzen. Zum 100. Geburtstag von WALTHER GOTHAN
13 Beiträge von 15 Autoren

1980. 439 Seiten – 65 Abbildungen – 7 Tabellen – 16,7 × 24 cm – DDR 84,– M

Heft 17*

HARALD LÜTZNER
Sedimentation der variszischen Molasse im Thüringer Wald

1981. 217 Seiten – 46 Abbildungen – 8 Tabellen – 21 Tafeln – 16,7 × 24 cm – DDR 45,– M

Heft 18

MANFRED STÖRR
Die Kaolinlagerstätten der Deutschen Demokratischen Republik

1983. 226 Seiten – 108 Abbildungen, davon 1 separat – 75 Tabellen – 16,7 × 24 cm – DDR 45,– M

Heft 19/20*

WOLF-DIETER HEINRICH (Hrsg.)
Wirbeltier-Evolution und Faunenwandel im Känozoikum. Gedenkschrift zum 100. Geburtstag von WILHELM OTTO DIETRICH (1881–1964)
15 Beiträge von 19 Autoren

1983. 394 Seiten – 73 Abbildungen – 18 Tafeln – 21 Tabellen – 16,7 × 24 cm – DDR 75,– M

Heft 21

JOACHIM SCHOMBURG, MANFRED STÖRR
Dilatometerkurvenatlas der Tonmineralrohstoffe

1984. 88 Seiten – 119 Abbildungen – 4 Tabellen – 17 × 24 cm – DDR 18,– M

Heft 22

MARTIN GUNTAU
Die Genesis der Geologie als Wissenschaft. Studie zu den kognitiven Prozessen und gesellschaftlichen Bedingungen bei der Herausbildung der Geologie als naturwissenschaftliche Disziplin an der Wende vom 18. zum 19. Jahrhundert

1984. 132 Seiten – 17 × 24 cm – DDR 27,– M

Heft 23

HARTMUT HAUBOLD
Stratigraphische Grundlagen des Stefan C und Rotliegenden im Thüringer Wald

1985. 111 Seiten – 15 Abbildungen – 7 Tabellen – 16,7 × 24 cm – DDR 22,– M

Heft 24*

EGINHARD FABIAN / MARTIN GUNTAU / MANFRED STÖRR (Hrsg.)
Philosophie und Geologie. Beiträge zum III. DDR-UdSSR-Symposium zur Geschichte der Geologischen Wissenschaften „Entwicklungsgeschichte von philosophisch-methodologischen Ideen in den Geowissenschaften" 25.–27. 10. 1983 in Greifswald.
24 Beiträge von 34 Autoren

1985. 176 Seiten – 11 Abbildungen – 2 Tabellen – 16,7 × 24 cm – DDR 33,– M

---

*) beim Verlag vergriffen/out of print

Schriftenreihe
für Geologische Wissenschaften
Series in Geological Sciences

Karl-Heinz Henning (1939)
Mineralogiestudium in Halle, Freiberg und Greifswald; Promotion A 1968, Promotion B 1980 in Greifswald, Dozent für Lagerstättenlehre seit 1980; über 30 Publikationen; Arbeiten zur Elektronenmikroskopie, Phasenanalyse und Tonmineralforschung

Studied mineralogy in Halle, Freiberg and Greifswald.
Doctorate A in 1968, doctorate B in 1980, both at Greifswald University.
Senior Lecturer of economic geology at this university since 1980. Research work on geology, mineralogy and phase analysis of clay minerals and other minerals.
More than 30 publications.

Manfred Störr (1934)
Geologiestudium in Greifswald; Promotion A 1967, Promotion B 1976 in Greifswald; o. Professor für Lagerstättenlehre seit 1977; über 80 Publikationen; Arbeiten zur Geologie, Mineralogie und Technologie von Ton- und Industriemineralen

Studied geology in Greifswald.
Doctorate A in 1967, doctorate B in 1976, both at Greifswald University.
Professor of economic geology at this university since 1977. Research work on geology, mineralogy and technology of clay and other industrial minerals.
More than 80 publications.

Die Elektronenmikroskopie ist für die submikroskopischen Tonminerale eine fundamentale Identifizierungsmethode. Die Autoren behandeln die morphologische Vielfalt der Tonminerale sowohl aus der Sicht der Genese und Mineralogie als auch ihrer Verwendung in der Industrie. Nach einer Einführung in die Präparationstechnik und der Beschreibung der Morphologie werden anhand von 600 transmissions- und rasterelektronenmikroskopischen Bildern die Erscheinungsformen der Tonminerale und in den Tonen vorkommender Nebenbestandteile (Fe-, Ti-Minerale, Sulfate, Phosphate u. a.) dargestellt. Das Buch wendet sich an Geologen, Mineralogen, Bodenkundler, Keramiker, Chemiker und Techniker, die sich mit Tonmineralen befassen.

Electron microscopy is an essential identification method for submicroscopic clay minerals. The authors consider the morphological variety of clay minerals with regard to their uses in industry and under genetic and mineralogical aspects. After an introductory chapter on preparation techniques and general outline of morphology the clay minerals and their admixtures (Fe-, Ti-minerals, sulfates, phosphates etc.) are depicted in no less than 600 transmission and scanning electron micrographs. This book is a source of valuable information for all who are concerned with clay minerals, e.g. geologists, mineralogists, soil scientists, ceramics engineers and technicians.

ISBN 3-05-500168-0
ISSN 0323-8946

Akademie-Verlag Berlin

Bei Fragen zur Produktsicherheit wenden Sie sich bitte an:
If you have any questions regarding product safety,
please contact:

Walter de Gruyter GmbH
Genthiner Straße 13
10785 Berlin
productsafety@degruyterbrill.com